T0262580

IET HISTORY OF TECHNOLOGY SERIES 24

Series Editor: Dr B. Bowers

The Life and Times of A.D. Blumlein

Other volumes in this series:

The Life and Times of A.D. Blumlein

Russell Burns

The Institution of Engineering and Technology in association with The Science Museum

Published by The Institution of Engineering and Technology, London, United Kingdom

First edition © 2000 The Institution of Electrical Engineers
Reprint with new cover © 2006 The Institution of Engineering and Technology

First published 2000
Reprinted 2006

The Institution of Engineering and Technology
Michael Faraday House
Six Hills Way, Stevenage
Herts, SG1 2AY, United Kingdom

www.theiet.org

British Library Cataloguing in Publication Data
A catalogue record for this product is available from the British Library

ISBN (10 digit) 0 85296 773 X
ISBN (13 digit) 978-0-85296-773-7

Printed in the UK by Bookcraft Ltd, Bath
Reprinted in the UK by Lightning Source UK Ltd, Milton Keynes

Alan Dower Blumlein (1903–1942)

Contents

List of illustrations

Foreword

I was six when my father was killed in the Halifax crash on the 7th June 1942. Because of the war I saw very little of him. My recollections of my father come mainly from the few weeks before his death.

He had no obituary due to a security blackout and it was not until the end of the war that short pieces about him appeared in print, and then normally in specialist publications. Even when stereo recording was first marketed in 1958, the daily papers were not really interested in his pioneer work in this field.

In the early 1970s my mother and I co-operated with the late Francis Thompson who wished to write a biography. This biography never appeared, and Thompson discouraged anyone else who wished to publish something about my father. It is not clear what material he collected. I have never had access to it, even though much was requested in our joint names, and it has remained unavailable despite numerous appeals to Thompson by various professional and learned bodies.

I regret that my mother did not live long enough to see the culmination of this long saga, but had she lived, she would certainly have joined me in thanking Russell Burns for his perseverance in piecing together the story and writing this biography.

Simon J L Blumlein
Petersfield, October 1999

Preface

When A D Blumlein and two of his colleagues were killed in an accident, in June 1942, the *Daily Telegraph* referred to Blumlein's death as a 'National loss'; Air Chief Marshal Sir Philip Joubert described the accident as 'a catastrophe'; and the Secretary of State for Air, Sir Archibald Sinclair, wrote that 'it would be impossible to over-rate the importance of the work on which they were engaged', their loss was a 'National disaster'.

Blumlein's contributions to radar were of crucial importance to the war effort and undoubtedly saved thousands of lives. The abandonment, in the spring of 1941, of the German 'Blitz' attacks on British cities and towns; the AA gunnery defence of the United Kingdom; the achievement of victory in the Battle of the Atlantic; the transformation of the bombing offensive over Germany; and the air offensive against Axis shipping in the Mediterranean Sea in the autumn of 1942, all owed much to the endeavours of Blumlein.

By any definition he was a genius. During his short working life from 1924 to 1942 Blumlein's work ranged over the fields of telephony, electrical measurements, monophonic sound recording and reproduction, stereophonic sound recording and reproduction, high-definition television, electronics generally, antennas, cables and radar. After his death Mr I Shoenberg, who was the Director of Research at Electric and Musical Industries Ltd, where Blumlein had been employed from 1931, said: 'There was not a single subject to which he turned his mind that he did not enrich extensively'.

Nevertheless, although Blumlein did much to enhance the well-being of society, his name is not widely known to members of the general public. Attempts to write a biography were made from 1967/8 and 1971/2 but without success.

The task of writing a biographical account of Blumlein's life and work has not been easy for several reasons. EMI, pre-war, discouraged its staff from publishing papers — Blumlein's publication list comprises just two learned society papers. No collection of personal letters exists and there is a paucity of information about his private life. Moreover, most of his co-workers, friends and colleagues are now dead, many of their earlier recollections have not been available to the author, and Blumlein lived for just over half the biblical three-score years and ten; he was 38 years of age when he was killed.

Fortunately he applied for, and was granted, 128 patents from 1927 to 1942. These, together with various unpublished internal reports, memoranda and letters, published papers, family records and taped reminiscences of some of his associates, have enabled an account of the greatest British electronics engineer of the twentieth century to be written.

Much care has been taken to ensure that the material presented is accurate. Primary source documentation has been used, where possible, throughout the book and extensive reference lists have been prepared. Unreferenced biographies/histories must always be read with a degree of circumspection: people's memories are not always infallible; sometimes a person's recollection can have been influenced by what he or she has read or heard between an event and the recall; and it is possible that a person may wittingly, or unwittingly, exaggerate his or her importance in the saga of events.

Special pains have been taken on the chapters which describe Blumlein's radar work, since several unreferenced or poorly researched autobiographical books by some former radar workers have been published which contain errors. A great deal of time has been spent by the author in the Public Record Office researching its very extensive radar collections, and detailed source references have been quoted.

The work of Blumlein was, of course, highly technical in character and many of his patents are very complex. Their detailed description is not appropriate for a general account of his life and times. However, this book would not be complete without some mention being given of his contributions. These have been broadly considered in the text, but several technical aspects (which may be omitted by the non-technical reader) have been dealt with in the technical notes which accompany several of the chapters, and in Chapter 14.

Acknowledgements

Many persons have aided me in the writing of this book. It can truly be said that without their generous support my task would not have been realised. In all cases they were ever ready to provide me with materials or recollections relating to a remarkable person.

First and foremost I should especially like to thank Simon Blumlein, A D Blumlein's eldest son, for the loan of much family memorabilia including school reports, family histories and reminiscences, letters, photographs and miscellaneous papers. He has always taken an active interest, and given very helpful support to me, in my efforts to further the life and work of his father: I am indebted to him.

Many thanks, too, must be accorded to Philip Wake (the son of Mina Blumlein — A D Blumlein's sister), who has been ever willing indeed to answer my numerous questions and requests for photographs and documents.

Special mention has to be made of the efforts of the late Mr J A Lodge who, after his retirement from EMI's Central Research Laboratories, was responsible for their archive collection. His extensive knowledge of the archives, his wide experience of the work of the Laboratories, his excellent memory, and his generosity, were of very great assistance to me. He never demurred when I asked him to search for particular items of information, and visits to his room at CRL, Hayes, were always occasions for friendly and informative discussions.

Of the other members of EMI's Archives, and formerly EMI Music Archives, I should like to thank Mrs R Edge, the Archive Manager and Chief Archivist, Mr R Willard, Mr L Pett, and Mr W Bailey. It has been a joy to spend time researching EMI's excellent archives and to appreciate the helpful and able service of Mrs Edge and her staff. EMI Archives' splendid new accommodation and facilities are a model for other companies and institutions to emulate. Mr W P Lucas, formerly Administration Manager of CRL, generously provided me with some photographs of the Laboratories. To all these persons I am most grateful.

Sadly, many friends, colleagues and associates of A D Blumlein, who kindly invited me into their homes or offices, to let me have their personal memories of the great man, are no longer alive. They were Mr I L Turnbull, Mr E A Nind, Mr E A Newman, Professor J D McGee, Professor F C Williams, Mr H Melling, Mr D C Birkinshaw, and Professor H Miller. Other persons from whom I received very valuable information, recollections from meetings and correspondence include Dr E L C White, Mr I J P James, Mr T B D Terroni, Sir Samual Currans, Mr B J Benzimra, Mr R Clayden, Mr H C Spencer, Mr R B Spencer, Mr E B Callick, Mr M G Harker, Mr W Turk, Mr A S Radford, Mr B Greenhead, Mr F Charman, Mr W S Bland, Dr E E Pawley, Mr J M M Pinkerton, Mr D H Tomlin, Mr R Stevens, Mr P Houchin, Mr J E Day, Mr S J Preston, Dr L F Broadway, Mr J G Davies and Dr B J O'Kane.

No book which purports to be an accurate biography/history can be written without extensive use of good libraries and record centres. Since a vast amount of my time and effort were spent in these places, I should like to express my gratitude, for their courtesy and support, to the staffs of the libraries and archive centres of the Institution of Electrical Engineers, the National Information Reference Service, the British Library, the British Lending Library, the Public Record Office, the Nottinghamshire County Council, the University of Nottingham, Nottingham Trent University, the BBC Written Archives Centre, the Post Office Archives and Records Centre, the Historical Archive of Nortel Technology, the College Archives of Imperial College of Science, Technology and Medicine, the library of DERA (Malvern), Swiss Cottage Public Library, the library of the RAF Museum (Hendon), the library of the Imperial War Museum, the Greater London Record Office, the library of Swimming Times Ltd, the private papers of Mrs W R B Foster and the National Sound Archives.

The latter establishment contains many taped interviews between members of its staff and persons associated with sound recording/reproduction. Of central importance to me in my work were the recordings made of the reflections and recollections of Mrs D Blumlein, and J B Kaye, A D Blumlein's wife and best friend respectively. These were quite invaluable and povided a fascinating insight into the personality of A D Blumlein.

The photographs considered for the book came from several sources: Mr S Blumlein, Mr P Wake, Mr W P Lucas, the late Mr J A Lodge, the late Mr P D Leggatt, the late Mr T B D Terroni, the late Mr H Melling, the Defence Evaluation Research Agency (Malvern), the Imperial War Museum, the BBC, Mr D H Tomlin, the Historical Archive of Nortel Technology, Highgate School, Abbotsholme School, Mr M Smith and the RAF Museum. To all the persons who responded magnificently to my requests for illustrative material I am especially grateful.

Abbreviations

ACAS	Assistant Chief of the Air Staff
ADEE	Air Defence Experimental Establishment
ADGB	Air Defence of Great Britain
a.f.	audio frequency
AGC	Automatic Gain Control
AI	Air Interception (radar)
AMRE	Air Ministry Research Establishment
ARP	Air Raid Protection
ASDIC	Allied Submarine Detection Investigation Committee, or Anti-Submarine Division(ics)
ASE	Admiralty Signals Establishment
ASV	Air-to-Surface Vessel (radar)
AT&T	American Telephone & Telegraph Company
AVC	Automatic Volume Control
BIR	Board of Invention and Research
CD	Coast Defence (radar)
CH	Chain Home (transmitter antennas)
CHL	Chain Home Low (radar)
c.p.s	cathode potential stabilisation
c.r.t	cathode ray tube
CSSAD	Committee for the Scientific Survey of Air Defence
CTE	Controller of Telecommunications Equipment
CVD	Communications: Valve Development Committee
c.w.	continuous wave
DC	Director of Contracts
d.c.	direct current
d.f	direction finding
DRP	Director Radio Production
DSR	Director of Scientific Research
ECL	Emitter Coupled Logic
e.m.	electromagnetic
e.m.f.	electromotive force
EMI	Electric and Musical Industries Ltd

FIU	Fighter Interception Unit
GCI	Ground Control Interception (radar)
GL	Gun-Laying (radar)
HB	Holman and Blumlein, sometimes Hells Bells, or Hot and Bothered
h.f.	high frequency
h.f.d.f.	high-frequency direction-finding
HMV	His Master's Voice
h.t.	high tension
IFF	Identification of Friend or Foe
ISEC	International Standard Electric Corporation
IT&T	International Telephone & Telegraph Company
IWE	International Western Electric Company
MAC	Merchant (ship) Aircraft Carriers
MAP	Ministry of Aircraft Production
MTI	Moving Target Indicator (radar)
OB	Outside Broadcast
p.p.i.	plan position indicator
PFF	Path Finder Force
RADAR	Radio Detection and Ranging
RAE	Royal Aircraft Establishment
RDF	Radio Direction Finding
RESB	Royal Engineers Signals Board
r.f.	radio frequency
RFC	Royal Flying Corps
r.p.m.	revolutions per minute
RPU	Research Prototype Unit
SLC	Search Light Control (radar)
STC	Standard Telephones and Cables Ltd
u.h.f.	ultra high frequencies
v.h.f.	very high frequencies
WE	Western Electric Company
WT	Wireless Telegraphy (Board)

Chronology relating to Semmy Joseph Blumlein and Alan Dower Blumlein

1863	Semmy Joseph Blumlein born on 9th September in Offenbach, Germany.
c. 1877	Arrived in England; resided in Liverpool, in the care of B Newgass. Worked for Newgass.
c. 1881	Sailed for South Africa. Landed in Durban, Natal.
c. 1883	Obtained employment in a stationer's shop in Kokstad, East Griqualand.
1886	New Year's Day; introduced to the Rev. W Dower, a Scottish missionary, and his family. Met Jessie Dower, his daughter.
1887	Friendship of SJB and Jessie.
1888	November: SJB left Kokstad to become Manager of Griffen's shop in Pietermaritzburg.
1889	28th March: SJB and Jessie married in the Griqua Church, Kokstad.
1890	17th January: Mina Philipine born in their first home, a bungalow in Long Market St., Pietermaritzburg. February: moved into a large flat above W D Griffen's store at corner of Church and Chapel Streets.
1891	Introduced to E F Bourke and T Bourke who had interests in the Rand mines.
1892	SJB became Managing Director of the E F Bourke Trust in Pretoria. End of year: SJB and family left Pietermaritzburg for Pretoria, Transvaal Republic.
1892–1899	Resided at 'Hillcrest', Sunniside, Pretoria.
1899	May: Jessie and her sister left South Africa for the UK. Stayed in Hastings where Mina was at school. Later, SJB departed from South Africa. Renewed his association with Newgass. Became Manager of B Newgass and Company.

1900   From December 1899 to May 1900 the Blumleins lived at 20 Hilgrove Road, Hampstead, London; and from May 1900 they lived at 23 Albion Road, Hampstead.

1902   March: address now 31 Netherfield Gardens, Hampstead.

1903   Alan Dower Blumlein born at 6.30 p.m. on Monday, 29th June. Christened on Sunday, 13th July at Marlborough Road Presbyterian Church.

1904   SJB severed association with Newgass. Obtained employment with a firm of South African merchants. About December: Manager of Swaziland Corporation Ltd.

1905   31st January: SJB naturalised.

1906   February: SJB sailed for South Africa. JB and ADB stayed at Miss Smallwood's boarding house, Belsize Park Gardens. About May: JB and ADB embarked for South Africa. On arrival, lived in a house in Park Town, Johannesburg.

1907   29th June: SJB resigned from his position. The family left Johannesburg and stayed with ADB's maternal grandparents in the Manse attached to the Union Church, Port Elizabeth.

1907–1908   Returned to the UK. Stayed at Miss Smallwood's boarding house.

1908   Spring term: ADB began school. Family resided in the Misses Greig's house, 12 Hilgrove Road, Hampstead. SJB worked on a prospectus for the formation of the Kyshtim Corporation Ltd. Autumn: new company floated; SJB appointed Managing Director. Visited Russia during winter of 1908–1909.

1908–1909   Winter: furnished a house at 11 Glenmore Road, Belsize Park.

1909   Furnished a house at 66 Boundary Road, St John's Wood. 25th September: Mina married. Summer term: ADB enrolled at Miss Chataway's South Hampstead Kindergarten and Preparatory School, 43 Belsize Road.

1911   L Urquhart became Managing Director of Kyshtim Corporation Ltd. SJB remained a Director.

1912   Autumn term: ADB at Loudon House School, Hampstead.

1913   September: ADB admitted to Abbotsholme School, Uttoxeter, Derbyshire.

1914   Tuesday, 28th July: death of SJB. Funeral held on 1st August.

1915   16th January: ADB a boarder at Ovingdean Hall School, Sussex. February: JB sailed for South Africa. Summer: JB back in the UK.

1918	January: ADB became a pupil at Highgate School. JB rented a flat belonging to Miss Smallwood.
1919	2nd January: JB now at 19 Albion Road, South Hampstead. ADB matriculated and entered the VIth form in the autumn term.
1920	July: ADB awarded the Bodkin Prize (General). He swam for the School. Home at 20 Lyncroft Mansions, Lyncroft Gardens, Hampstead.
1921	October: ADB a student at the City and Guilds College with a Grosvenor Scholarship.
1923	Awarded first-class honours BSc degree. Became an Assistant Demonstrator.
1924	Postgraduate research at City and Guilds College. Joined IWE, which became ISEC.
1925	Paper submitted to the IEE; awarded an IEE premium. February: a member of Collard's group. Various periods in Switzerland.
1926	Investigated crosstalk in loading coil equipment.
1927	Awarded a bonus of £200 for work on loading coils.
1928	15th October: obtained pilot's licence. Worked on confidential company problems.
1929	March: recruited by Shoenberg of the Columbia Graphophone Company to work on a new electric recording system.
1930	Met Doreen Lane. Had digs at 57 Earls Court Square.
1931	Formation of EMI. Awarded a bonus of £200 for work on a new electric recording system. Worked on stereophonic recording and reproduction.
1932	Moved to 32 Woodville Road, Ealing, and then to 7 Courtfield Gardens, West Ealing.
1933	22nd April: married Doreen Lane at St John's Church, Penzance.
1934	Lived at 32 Audley Road, Ealing. November: awarded gold watch by EMI. With Shoenberg gave evidence to the Television Committee.
1935	26th March: first son born but died soon afterwards.
1936	11th May: Simon John Lane Blumlein born.
1937	Moved to 37 The Ridings, Ealing.
1938	21st February: David Anthony Paul Blumlein born. ADB awarded an IEE premium for paper on television.
1939–42	Worked on various radar systems of vital importance to the integrity of the UK.

1942   7th June: killed in an air accident. His death considered a
            'national loss' and a catastrophe'. 13th June: funeral service.
            19th June: memorial service.
1958   Awarded a posthumous citation by the Audio Engineering
            Society for work on stereophonic recording and reproduc-
            tion.
1977   1st June: blue plaque unveiled at 37 The Ridings, Ealing.

Chapter 1

Early life

Shortly before A D Blumlein was born, Rudyard Kipling wrote a short poem titled 'My six serving men'. The poem begins:

> I keep six honest serving men,
> They taught me all I knew,
> Their names are What, and Why, and When
> And How, and Where, and Who.

Kipling's men well serve a writer who endeavours to delineate the life and work of a great person such as Blumlein. When and where was he born? What were the influences which affected him during his formative period? Where was he employed? With whom did he work? How did he undertake his creative activities? What were his contributions to society?

Certainly Blumlein's original contributions to knowledge and to the well-being of mankind were very great. As Isaac Shoenberg, the Director of Research at Electric and Musical Industries Ltd (EMI), said after Blumlein's tragic death at the age of just 38 years: 'There was not a single subject to which he turned his mind that he did not enrich extensively'. Blumlein's work embraced the fields of telephony, electrical measurements, monophonic recording and reproduction, stereophonic recording and reproduction, high definition television, antennas, cables, electronics generally, and radar. When he was killed in June 1942 the Secretary of State for Air, Sir Archibald Sinclair, described his loss as a 'national loss', and the *Daily Telegraph* wrote of his death as a 'national disaster'.

There can be no doubt that Blumlein's efforts in the Second World War were of central importance in the defeat of the German night bombers during their Blitz attacks on the United Kingdom, the defeat of the U-boats in the Battle of the Atlantic, the effectiveness of the British bombing offensive over Germany, and the British night-time torpedo attacks by aircraft on Axis shipping supplying the German forces in North Africa.

Blumlein has been described as a genius. Of course, to say that a person is a genius is to invite several questions. What is the definition of genius? How can it be assessed or measured? This topic is considered at length in the final chapter, but here it suffices to say that genius is used to designate creative ability of an exceptionally high order as demonstrated by actual achievement. It differs from talent in both quality and quantity. In assessing genius it is necessary to bear in mind that genius involves originality, creativeness and the ability to think and work in areas not previously explored, so as to give mankind something of worth it would not have otherwise.

Returning now to Kipling's six serving men, the questions: When and where was Blumlein born? What were the factors which influenced his early life? Where was he educated? How did he respond to these factors and his education? Who were his mentors?—and others—form the basis of this chapter.

Alan Dower Blumlein was born at 6.30 p.m. on Monday 29th June 1903 at 31 Netherhall Gardens, Hampstead, London. He was the only son of Semmy Joseph Blumlein, a financial merchant, and his wife Jessie Edward Blumlein (née Dower) and was their second child, Philipine Blumlein, the Blumlein's only daughter, having been born at 10.00 a.m. on 17th January 1891 in Pietermaritzburg, Natal.[1]

Very little is known of the very early life of S J Blumlein (Figure 1.1). He was born on 9th September 1863 in Offenbach, Germany. Neither Semmy nor his sisters could explain the origin of his first name. His father Joseph B Blumlein was Jewish and was born on 23rd March 1840 at Hessdorf, Bavaria. Until late 1862 he was a 'merchant of Wurzburg'. Of his wife the only information available, from early records, is that her maiden name was Philipine Hellmann. Anecdotal recollections have described her as being a very beautiful French woman. S J Blumlein appears to have inherited her goods looks, for he was 'a most handsome man'.[1]

He had six sisters—Marie, Ida, Anna, Tilla, Lina and Emma—and two brothers—Otto and Victor—but did not maintain a close relationship with his parent's family. He was educated in Aix-la-Chapelle, rather than a German school, because of his father's 'bitter antagonism towards Germany' and his 'detestation of the German military system'. At the age of about 14 years Semmy was sent to Liverpool, England, in the care of an old family friend, Benjamin Newgass (or Neugass). Conscription into Germany's National Service was thus avoided.

Figure 1.1 Semmy Joseph Blumlein (1863–1914)

Source: Mr P Wake

From the very sketchy reminiscences of his teenage years it seems that SJB worked for Newgass. A sister of SJB's future wife has written[1]:

'The only thing I ever heard about this period of his life was his friendship with a lady who was an ardent church woman and he attended her church. She presented him with a Book of Common prayer when he left Liverpool. [He took] this to South Africa.

'He went to London, I believe, with the same employer. Anyway he became the trusted Secretary of his employer, and was very young — probably only 17 — when he was trusted to go to Mexico to put through a business transaction for the firm.'

When SJB was 18 years of age he left the United Kingdom to seek success elsewhere. It seems that he was required to undertake a business transaction of which he did not approve. He resigned from his post and sailed for South Africa.

'I do not know how he financed this voyage — probably worked his passage out, as many another man and youth were doing. [At] Madeira, he received a cable from his employer, offering to double his salary, and pay his return passage money back again, and begging him to return. But he refused and continued his voyage.'

He landed at Durban, Natal, and set off on foot for the North. After a trek of more than 200 miles through East Griqualand he arrived in 1883 at Kokstad, the capital.

'What his employment was [for the first two to three years] I [Mary Dodgshun] do not know. But I do most vividly remember his stationer's shop in 1886, and even though I was only eight years of age, I was impressed by the neatness and order of the books, notepaper, inks and fancy goods. . . . That impression was a right one, for this sense of order was one of his outstanding characteristics, which impressed me again in his private offices in Pietermaritzburg and Pretoria, and in his little study in his home in Hampstead, London.

'Shortly before he left Kokstad in 1888 I saw him and was served by him in H Griffen's store, the employment which proved to be the stepping stone to his position in H Griffen's establishment in Pietermaritzburg, Natal.

'I judged his arrival in Kokstad to be 1883 because of an incident that took place concerning my sister Jessie [born 11th February 1869] as she left for school in England, in September in 1884. After [her engagement to Semmy] he told her that the place, where he was working at the time, was quite near to the home of a Miss Ogle to whom she went for music lessons. He watched out for her coming and going, and wished so much to meet her.

'And our father [the Reverend W Dower] told of an incident in February 1884. He was caught, with several others, in a severe storm of rain in Kokstad, and took shelter in a small library, and there they were entertained by a young handsome man who played an instrument he, father, had neither seen nor heard before — a zither. He made a mental note to ask the young man to the Manse. But the rain ceased as suddenly as it came, and everyone rushed off to their homes, and it passed from his mind until they met on New Year's Day of 1886.'

At this point it is necessary to introduce William Dower and his family[2,3]. He was born on 15th November 1837 at Banchory Ternan, Kincardineshire, Scotland, and was the only son of a local builder John Dower and his wife, Jean Forbes. Various legends have associated the Dower family with the Heads of the Gordon clan, to which clan they undoubtedly belonged. William attended the Mechanics Institute of Aberdeen, and Aberdeen Grammar School before entering Edinburgh University. He studied at Theological Hall, Edinburgh, and was ordained in that city on 27th June 1865. He married Jessie Edward (Figure 1.2) (born 16th September 1838), also of Banchory, and on 22nd August 1865 they left Scotland for South Africa.

A few months after their arrival in September, William, in January 1866, began his work for the London Missionary Society at Hope Dale,

Figure 1.2   The Reverend William Dower and his wife Jessie (Edward), with their daughters Mary Isabella, Jessie Edward and Alice Elaine

Source: Mr P Wake

Cape Colony. He was 'a great linguist' and soon learned Dutch and 'many native languages'. In 1868 Adam Kok, the Chief of the Griquas, rode to Grahamstown to ask the Governor of the Cape Colony to obtain a missionary for his people. Several missionary societies, including the London Missionary Society, and the Glasgow and the Wesleyan Methodist Missionary Societies, were active in the Cape Provinces at this time and had set up missions to introduce Christianity to the native African populations. Among these societies, the London Missionary Society, which had arrived in South Africa in 1799, had established mission posts at Bethelsdorp and at a number of other places in the Eastern Cape in the Orange river valley, and in Namaqualand.

Among the various religious faiths in South Africa, the Dutch Reformed Church believed in predestination, the religious theory that God has decreed forever that part of mankind shall have eternal life and part shall have eternal punishment. The Boers, as Dutch followers of the Church, considered themselves to be a superior race by birth and the native, non-white, races to be servile races who could never enjoy the status which befitted the Boers. For them the Africans were an inferior people, to be used to carry out menial tasks and to perform hard labour. They were not, in the Boers' view, the equal of the white man, and as such were despised by the Dutch colonists. The Boers extolled the principles of 'equality', 'fraternity', and 'liberty' insofar as they could be applied to themselves in particular and to other 'whites' in general, but there could be no thought whatsoever of extending these noble principles to non-whites or natives[4].

To the missionaries of the London Missionary Society, the Roman Catholic Church and the Moravian Brethren, all men were equal and entitled as of right to fair and humane treatment without regard to their colour. The noble principles had to apply to all and many missionaries fought for the rights of the natives, thereby creating discord between themselves and the Boers.

Among the missionaries of the London Missionary Society, Dr John Philip worked among the Griqua, Khaikhai and Bantu soon after his arrival in South Africa, in 1819. His work and his exposure of the whites' attitude to the blacks led to the 1828 enactment of the 50th Ordinance which restored civil liberties to non-whites. The non-Dutch Reformed Church missionaries gave comfort and protection to any non-white who was the victim of Boer injustice, and so the mission stations became havens of peace and sympathy to the oppressed. Moreover, the missionaries endeavoured to provide a rudimentary education for the coloureds. This was resented by the colonists. They believed that such education would lead the Africans to have aspirations which were discordant with their station in life.

The Boer policy of white supremacy could not co-exist with the introduction of legislation which removed the shackles which bound the Africans and non-whites to their place in society. But in 1833 the emancipation of the slaves by Great Britain was applied to all parts of the Empire, including South Africa. The law was viewed with great dismay and resentment by the Boers and became one of the most important sources of disquiet between the Boers and the British administration.

Thus, the various native tribes, such as the Griqua of Griqualand, saw the British missionaries as friends who would defend their rights, particularly their land rights, and who would treat them charitably.

In 1820 the total Griqua population was about 3000 and there were three Griqua villages, namely, Griquatown at Klarwater, Campbell, and Philippolis whose Chiefs were Andries Waterboer, Cornelius Kok and Adam Kok. They subsisted at a basic level for many years until 1867 when the discovery of diamonds in Griqualand West led to a huge increase in the value of their land. Though the actual sites where the finds were made were legally an integral part of the Orange Free State (a Boer Republic), this fact was disputed by Waterboer. He was encouraged by an Englishman, D Arnot, to press his claim that the land was under his (Waterboer's) jurisdiction and asked for British support. The United Kingdom accordingly entered the discussions, and in 1871 Griqualand

West was declared a British territory. It was incorporated into the Cape
Colony nine years later. Both events caused much fury and antagonism
towards the British in the Orange Free State, and relations between the
two white communities in South Africa further worsened.

Following Kok's visit, the London Missionary Society asked the
Reverend William Dower to take up an appointment among the Griquas.
With his wife Jessie and their two children, Mina (3 years) and Jessie (15
months), and Mrs Dower's sister Margaret Edward, the Reverend Dower
left Hope Dale on 17th May 1869 and trekked to Noman's land (later
Griqualand East). They reached Mount Currie laager (camp) on 19th
August and returned to Hope Dale in October 1869. Dower agreed to stay
among the Griquas provided they built a new township. This was accepted
and on 7th March 1870 the family once again arrived at Mount Currie.

With a Mr Edward Barker, William Dower planned the town of
Kokstad, afterwards assisting the Griquas in their move from the Mount
Currie laager, and constructing the first building in the new town. After its
establishment Barker became the Clerk to the Chief Magistrate and
Postmaster.

In 1870 Margaret Edward, a teacher of the London Missionary
Society, and Edward Barker, a widower with three sons, married. The two
families had many interests and from their first Christmas in the new
Mission House the Barkers always dined with the Dowers, and vice versa
on New Year's Day. The Barkers and the Dowers also provided hospitality
for any lonely men and women who worked in Kokstad.

Dower's association with the London Missionary Society ceased in
1878 when he assumed the Pastorate of the Griqua church at Kokstad. His
parsonage soon became a meeting place for numerous notable persons
who visited the area.

On New Year's Day 1886 the Barkers included Semmy Blumlein in
their guest list. He was a great success and entertained the company with
his zither.

'From that very first meeting, we children adored him. He was so different to
anyone else we had ever met, so jolly, and friendly but a great tease (this to the
end of his days).'[5]

In December of that year a Dower family reunion took place. Jessie,
who had been in England for 27 months at a boarding school
(Walthamstow Hall, Sevenoaks) for missionaries' daughters, and her two
brothers, Will and Jim, who had been away from home for 22 months at a
boarding school in the Cape Colony, all returned together for Christmas
at the Manse. They and their father were met just outside the town by
twenty to thirty of the townsfolk, including Semmy Blumlein who had

become a very welcome and frequent visitor to the Manse. On that day he wrote in his diary: 'Today I have met my future wife.'

Throughout 1887 Semmy and Jessie saw much of each other and their friendship grew apace. They were both fond of music — Jessie played the piano — and together they played duets and arranged concerts. After her return from England she became the organist at the European Congregational Church which her father had founded in Kokstad and which Blumlein had begun to attend following his introduction to the Dower family. Also his friend Tom Coulter, who had trekked with him from Durban to Kokstad, in 1883, had become an official in the church and this added to Blumlein's interest in it. Horsemanship was another common interest of Semmy and Jessie.

In November 1888 SJB left Kokstad to become manager of a new department of Griffen's, in Pietermaritzburg, which had been set up to deal with the business which had arisen from the rapid development of the gold mines on the Rand. His career was progressing and on 28th March 1889 he married Jessie in the Griqua Church, Kokstad (Figure 1.3). Her father officiated at the ceremony.

Just under two years later, on 17th January 1891, their first child and only daughter, Mina Philipine, was born in their first home — a bungalow in Long Market Street. A month later the Blumlein's moved into the large flat above W D Griffen and Company's store at the corner of Church and Chapel Streets. Among their guests was Jessie's sister, Mary. She stayed with them from March until June 1891. Many years later she recalled[5]:

'I . . . got to know Semmy very much better, and to love and admire him more. They [Semmy and Jessie] enjoyed their somewhat brief stay of nearly four years in Pietermaritzburg before they moved on to Pretoria, capital of the Transvaal [a Boer Republic].

'Henry Griffen, the head of the Griffen establishment, lived about two to three miles out of the town in a large bungalow type of a house surrounded by a broad verandah, and standing in large well kept grounds. Tennis courts and croquet lawns were the special attractions for the many folks young and old who gathered there, chiefly on half-holidays, etc. W H Griffen, the son, was not interested in these games, and so Semmy was asked to be MC. And these visits on Saturday afternoons, and on holidays became a regular feature in their lives in Pietermaritzburg.

'Henry Griffen's two daughters, Eleanor and Jessie, had married two brothers, E F Bourke and Tom Bourke, who had great interests in the Rand mines. They were amongst the early prospectors in and near that area and were large share owners. Here Semmy met E F Bourke of Pretoria on his frequent visits to Pretoria. The result was that near the end of 1892 Semmy became Managing Director of the E F Bourke Trust in Pretoria.

Figure 1.3   S J Blumlein and J E Dower, photographed at the time of their marriage in 1889

Source: Mr P Wake

'They [Semmy and his family] had to travel to Pretoria by stage coach (the only means of public transport) a distance of about 500 miles in stages where accommodation at some of these stages was not far removed from the primitive. In consequence of this Jessie contracted typhoid fever, and was seriously ill for the first months in Pretoria, which affected her general health for many years.

'The year 1892 was a momentous one of change for the Dowers and the Blumleins, and most appropriately they spent New Year's Day together, six years to the day after they met Semmy at the Barkers. Just before Christmas a portion of the Dower family (parents, two sons and two daughters) paid a farewell visit to the Blumleins on their way from Kokstad, East Griqualand, to their new life in Port Elizabeth. [William Dower had been appointed Minister of the coloured congregation of the Port Elizabeth Union Church in 1891. In 1897 he became Chairman of the Congregational Union of South Africa.] They left Durban by sea for Port Elizabeth on 17th January (Mina's first birthday) and the Blumleins left Pietermaritzburg for Pretoria, Transvaal Republic, on their new venture, at the close of that same year.'

They resided there from 1892 until 1899 when the conflict between the British and the Boers obliged them to evacuate from South Africa. The Boer War which commenced on 12th October 1899 drastically changed the lives of the Blumleins and many other families. Jessie and Mary returned to the United Kingdom in May 1899 and stayed in Hastings where Mina was at school. Semmy's departure from Pretoria was not without some cause for anxiety[5]:

'When details were available, we heard that Semmy had come across, in examining a sideline of the business, a huge bit of fraud concerning the medical supplies to the Boer Defence Force, and involving some influential Boers. In this circle a desire for revenge for their possible exposure was rampant.

'So a few of his Boer friends begged him to leave Pretoria. In fact, one friend, though closely connected with the government, told him he was a marked man and that a plan had been made by those fraudulent men that immediately the war, they expected, was declared, he would be shot in the back, with the excuse that he was a spy, and they could prove it.

'So, with the knowledge of only a small number of his faithful friends, he left Pretoria for a destination unknown. He left his home ['Hillcrest' in Sunniside, then the latest and most modern suburb of Pretoria] as it stood, with the assurance from these friends that they would care for removable articles. This was done by them, and some years after the Boer War ended the Blumleins received, in London, their table silver and silver articles, a few ornaments, household linen and other fabrics, and some such articles as the carpet (I believe Persian) that . . . was on their verandah. But in the conflict the other furniture and other household possessions disappeared, including the bedroom suite made for them by our father [William Dower].'

During the British occupation of Pretoria 'Hillcrest' became the residence of Lord Kitchener, Second-in-Command of the British forces. Tom Bourke's house, which was about fifty yards away from 'Hillcrest' on the opposite side of the road, was taken over by Lord Roberts, Chief of Staff of the Army. Later, after the Boer War, this house became the British Residency.

On returning to the United Kingdom in December, Blumlein renewed his association with Newgass and became Manager of B Newgass and Company, London.

Blumlein's immediate task was to find accommodation for his family and himself which was, first, conveniently situated near the City of London to enable him to commute to the offices of the company, secondly, close to a good school for Mina, and thirdly, situated in a pleasing part of

London. Of the various choices, the attractive borough of Hampstead had, at the turn of the century, well developed bus and train services, and was well endowed with many schools. Under the Education Act 1918 the numbers of private schools recorded included 24 with 1270 boys and girls, 5 with 337 girls, and 4 with 165 boys. Another 32 schools had applied for inspection and recognition[6].

From December 1899 to May 1900, the Blumleins lived at 20 Hilgrove Road, Hampstead; and then until March 1902 they resided at 23 Albion Road, Hampstead. In that month they moved to 31 Netherhall Gardens, Hampstead, a street close to Finchley Road and the nearby Finchley Road station. Their house was built about 1888/89.

Netherhall Gardens is situated in an affluent area of the borough. Initially called Netherhall Terrace, Netherhall Gardens and Way were named in 1877 after a Maryan Wilson property in Sussex. The street attracted 'the nicest sort of resident': several GLC (Greater London Council) commemorative 'blue plaques' mark the houses where notable personages lived for a time[7]. A blue plaque on No. 10 records the fact that it was the first home of the social reformers and historians Sidney and Beatrice Webb (1816–1898, 1858–1943) after their marriage in 1892. And on No. 51 a plaque notes that it was the home of John Passmore Edward (1823–1911) from 1908 until his death. The plaque describes him as 'Journalist, Editor and Builder of Free Public Libraries', but does not mention his wide philanthropy. He founded more than 70 public institutions (including homes and hospitals as well as libraries), and 25 free libraries; erected 11 drinking fountains; placed 32 marble busts of eminent men, by important artists, in public buildings—and refused two knighthoods. Of the other distinguished properties, No. 7 was for many years the home of Louis Sinclair (1861–1928), a Member of Parliament (Romford Division 1897–1906), who became so prosperous from his 'commercial pursuits' in Australia that he was able to retire from business at the early age of 25; and No. 59 which was the home, during the Great War years, of the Irish tenor Count John McCormack.

The greatest Englishman to have lived in Netherhall Gardens was possibly Sir Edward Elgar (1857–1934). He and his family occupied No. 42 during the period 1912–21, a fact recorded on a blue plaque. Sir Edward named his house 'Severn House' to show his Worcestershire background. According to one account Elgar chose the Mansion (which had been designed by Norman Shaw) as a status symbol to show that composers in general, and himself in particular, could live in the grand manner. It seems to have been a burdensome symbol, however, for he had to accept commissions for unsubstantial theatrical works to maintain his property.

Shortly after their arrival in London the Blumleins began to attend St John's Wood Presbyterian Church[8]:

'Dr Monroe Gibson was the Minister there—a Scot with an outstanding personality and a good preacher. As a pastor, in London, and in those conventional days, he was somewhat unusual in the way he linked up people in his Church.... [One day he called when] Jessie only was at home. He told her that he had taken the liberty of asking her neighbour next-door, Mrs Hamish McCunn, a member of his Church, to call on her. And, with her permission, he wished to do the same with three other families, he thought she would like to meet and know, namely the Misses Grieg, the Misses Glen, and Mrs Wallace. That action started friendships which lasted a lifetime of all those people mentioned.'

At 6.30 p.m. on Monday 29th June 1903, at 31 Netherhall Gardens, Alan Dower Blumlein was born. He was christened on Sunday, 13th July, by Dr Monroe Gibson at Marlborough Road Presbyterian Church[9]. (See Figure 1.4.)

Figure 1.4   A D Blumlein and his sister Mina photographed in 1904

Source: Mr P Wake

Notwithstanding the loss of many of their possessions in South Africa, the Blumleins seem to have enjoyed a fairly comfortable lifestyle in Hampstead. SJB's remuneration with B Newgass and Company is not known, but when he became the Managing Director of the Kyshtim Mining Corporation in 1908 his salary was £600 per annum. To put this figure in some context: in the period 1901–10 a labourer's weekly wage was about £1.10, rent for a working-class house was about 35 pence per week, school fees in London were about £9.00 per term, a two-bedroom house could be bought for £300, coal was £1.25 per tonne, and an ounce of tobacco and a gallon of beer were 2p and 6p respectively[10]. Consequently the Blumleins were well be able to employ a housemaid (Ada Pullen), a cook (Ida Pullen) and a nurse (Florence Jeeps) to care for Alan. The nurse was given the name Didi by her young charge.

An illustration of Blumlein's affluence and generosity has been recorded by Jessie Blumlein's sister Mary[11]:

'In February 1902 my younger sister Elaine (two years my junior) was married to Sandy Selkirk which left me as the only member of the family in the old home of my parents. Shortly after the birth of her first child ... I received a surprising letter from Semmy. He began by saying he thought it was essential that a member of the Dower family should come to England to see their son. I appeared to be the only one now free and unattached to do so.... He invited me to come that year as soon as possible, at his expense, and with the assurance that he would be responsible for an adequate allowance while I was with them, at least for a year. As I was teaching at the time and earning, what was considered a good salary in those days, this last assurance gave me the needed feeling of security to make this adventure possible, for I certainly had not saved very much.

'The invitation coming at that time showed a perceptive understanding of my feelings. Elaine, until she was married the previous year, had been my inseparable companion all my life....

'So, I accepted and arrived in England in May 1904 and spent with [the Blumleins] ... ten and a half months, amongst the most memorable of my life....

'Of course, the very first thing I did on arrival at Netherhall Gardens was to go to the nursery at the top of the house to meet my first nephew about ten and a half months old. He was a lovely baby, and the picture of health with his rosy cheeks, light brown hair and big eyes and such a solemn expression. He was an unusually serious baby, and one felt quite rewarded when one got a smile or laugh from him.'

At about this time Semmy Blumlein severed his association with B Newgass and Company (Merchants) of 7 Lothbury, London, the firm of his father's old friend, which had given him his first training and experience of business when he (SJB) had acted as Newgass's Secretary in their travels in America and on the Continent. Now, in 1904, Blumlein was with the firm of 'L... and M...' (South African merchants) which appears to have had control of Swaziland Corporation Ltd. It is known that in December 1904 SJB was the Manager of Swaziland Corporation Ltd since in that month he presented a Memorial[12] to the Home Office for a Certificate of Naturalization. This was granted on 31st January 1905[13] and the Oath of Allegiance was taken on 2nd February 1905. Among the five persons who provided statutory declarations in Blumlein's support there were Dr Gibson and two relatives, R S Dower of Ilkley, Yorkshire, and C J S Atkinson of Wharfedale, Yorkshire, who were a cousin and a cousin-in-law of Jessie Blumlein.

Little is known of Blumlein's work with the Swaziland Corporation. It seems to have been established in 1898 and owned more than one million acres of land in Swaziland, then a British dependency of about 4 million acres, with a population of about 85 000, situated between Natal and Mozambique. The main activity of the corporation was the mining of tin.

Blumlein's work necessitated him revisiting and working in South Africa. He sailed in February 1906 with 'only a few days notice.'

'Soon [there] came a cable from Semmy to say he was in Johannesburg and that we [Jessie, Alan and Mina], with the maids and all the furniture, were to follow him. . . . I [Jessie] arranged to move out of the house as soon as possible and the furniture was taken to the City to be packed and shipped to Durban. . . . I found a boarding house with a private sitting room — Miss Smallwoods, Belsize Park Gardens — and from there got through business arrangements, shipping and packing. Didi was with us at Miss Smallwoods . . . [but] she was not coming with us — only Ida Pullen and her sister Ada, who were to meet us at the boat-train at Waterloo. . . .

'We sailed in a intermediate boat — slower than the mail boats carrying only first class passengers and a few steerage. On Alan's third birthday we anchored at Las Palmas (Canaries).'[14]

Alan's nurse was now Ada Pullen, whom he called Dudu, the Swahili word for insect. (See Figure 1.5.)

'Semmy came to Cape Town to meet us and we travelled with him to Johannesburg. . . . [He] had taken a house in Park Town, a spacious comfortable house with a big garden, stables and backyard . . .

'We were only one year at Talana house — a very happy year in Alan's life. . . . [But] from a business point of view all was not well. Semmy was known to be not only a very able man, but a man of the highest integrity with a great reputation behind him for his work in Pretoria. L ... and M ... could not bend him to condone their crooked ways and in the end it was L ... and M ... who were broken. Semmy was a man of such kindly attitude to everyone, and the

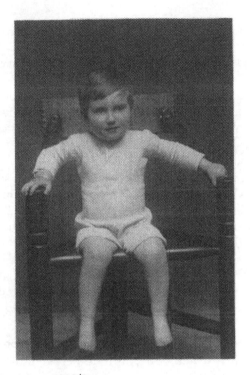

Figure 1.5   A D Blumlein, c. 1906/7

Source: Mr S J L Blumlein

discordance of the life was abhorrent to him and he just had to end it. Thus he tore up the contract he had made with them and came out of the entanglement with no prospects and very little money—in fact only our passage back to England. We left Johannesburg on 29th June 1907—Alan's fourth birthday . . . [and] went by train to Port Elizabeth and stayed with my parents.'

From Port Elizabeth the Blumleins and Dudu sailed in the *Llandovery Castle* to London. Here they spent the winter months of 1907–08 at Miss Smallwoods, and Alan began school early in 1908. In the spring, with the impending visit of Jessie's parents, the Blumleins decided to move into a larger furnished house

'where [they] could entertain father's and mother's friends and make our pre-parations for the wedding [of Mary, Jessie's sister]. Our friends, the Misses Greig, let us have their house [at 12 Hilgrove Road, Hampstead] furnished while they were travelling in Italy, and we were very lucky to be so comforta-bly housed. There was a garden at the back with a wall and on the other side of the wall were very many rails running from Euston station to Scotland and Ireland. Lots of big express trains rushed past day and night. It was a great joy to Alan . . . [Railways with all their operational devices and systems would remain a life-long attraction for him. His interest stemmed from the time when he was a baby at 31 Netherhall Gardens and could see, from his nursery window, the railway lines of the Midland Railway which ran out of a tunnel under Netherhall Gardens.]

'From the time of his return to England Semmy had been at work on a scheme concerned with mining interests in Russia — strictly speaking it was in the Urals. There were rights, concessions, holdings and ownership, and the entanglement had baffled everyone who tried to report on it, or to bring it to the stage where a company could be floated. Brown Shipley held big interests and Semmy's friend James Leigh Wood (he gave Alan his silver christening mug) asked him to tackle the difficult job. During the winter he [SJB] had worked almost day and night on it and that summer [actually autumn 1908] saw the prospectus ready and the Kyshtim Mining Co. [actually the Kyshtim Corporation Ltd] was floated . . .'[15]

Before the onset of hostilities in 1914 Imperial Russia under Czar Nicholas II had an economic growth rate which exceeded that of any European country. Much foreign capital, especially British capital, financed various Russian enterprises, and from the Ural mountains to eastern Siberia British registered syndicates and corporations began to develop, particularly after 1902, the mineral resources of the vast country[16].

Among the British entrepreneurs, Leslie Urquhart was an engineer who had had extensive experience of Russia. Of Scottish descent, he was born in Russia in 1874 and had studied mining engineering at the Universities of Glasgow and Edinburgh[17]. From 1896 to 1906 he had worked in the Baku oilfields in the Caucasus, becoming manager of four British oil companies at the age of 30, but had been forced to flee following an attempt by revolutionaries to assassinate him.

On his return to London, he promptly organised the Anglo-Siberian Company to undertake the exploitation of the huge mineral deposits of Russia. A year later another company, the Perm Corporation, which had been set up for the same purpose, succeeded in negotiating a deal with the Kyshtim Mining Works, a Russian company established in 1900 with limited liability, operating in the Urals. The Anglo-Siberian Company obtained control of the Perm Corporation and on 29th October 1908 the Kyshtim Corporation Ltd was registered to acquire from the Perm Corporation the entire share capital of the Kyshtim Mining Works. This capital consisted of 34 440 shares of 200 roubles each fully paid, having an equivalent sterling par value of £725 000[18].

On its formation C F H Leslie and S J Blumlein were appointed Chairman and Managing Director respectively. Blumlein's salary was £600 per annum[19]. The Board of Directors comprised L Urquhart, Baron

Meller-Zakomelsky and five other directors. Urquhart was also appointed Managing Director of the Kyshtim Mining Works.

The Kyshtim estate in the Ural mountains near Ekaterinburg belonged to a distant branch of the Romanov family, then headed by Baron Meller-Zakomelsky. The estate covered an area of 2198 square miles (over 1 406 000 acres) and contained rich deposits of copper, gold and iron, as well as ironworks which had been producing specialised products (such as iron castings) for more than 150 years. The local population was about 76 000.

However, the estate needed modernising and developing. Capital, meaning foreign capital, was necessary and this the Kyshtim Corporation Ltd was prepared to channel into the Kyshtim Mining Works. The new company's plans were ambitious and made provision for £8000 to be spent on the ironworks, £35 000 for a branch railway to the copper mines, and £80 000 for a large modern copper smelter, workers' houses, and other requirements. In all, £260 000 had to be raised. For this purpose a large share and debenture issue was offered in November 1908.

Blumlein's first visit, with Urquhart, to Russia was made in the autumn of 1908. They stayed in the imposing palace at Kyshtim and began the task of reshaping the backward and impoverished Kyshtim enterprise[20]. By late 1909 a transformation was being effected. 'Thirty miles of railway connecting the town of Kyshtim with the mines were complete; at Karabosh Lake preparations were well underway to erect in 1910 a smelter capable of producing 5000 tons of copper per year; a new electrolytic refinery was in place; and exploration and development of the mining areas [had] discovered several substantial deposits of copper ore.'[21]

Towards the end of 1909 Herbert Hoover, a US mining engineer, who later became President of the United States of America, became associated with the Kyshtim organisation. From this time the history of the enterprise becomes quite complex. Hoover's claims for the part he played in remodelling the Kyshtim estate led, in 1951, to indignant letters being written to Hoover by Urquhart's widow. She accused Hoover of grossly exaggerating and misrepresenting his role in advancing the prosperity of the Kyshtim concern and of various inaccuracies in detail. Fortunately this history and contention are not relevant to this book. Blumlein continued to be the Managing Director of the Kyshtim Corporation Ltd until 1911/12 when L Urquhart replaced him. The Company file in the PRO shows that Blumlein remained a Director of the Corporation until his death in 1914.

While Blumlein was wintering in Russia, Mrs Blumlein and Alan moved to a furnished house in Glenmore Road, Belsize Park. Yet another move was made early in 1909 when Blumlein returned from Kyshtim.

Their new abode, at No. 66 Boundary Road, was, according to Alan's mother, 'the best of all the many houses' the family lived in.

'There, Mina became engaged and her wedding was there—a glorious summer day [25th September 1909] — and we entertained the guests in the garden. There Semmy died and with him much of my life died too. 66 Boundary Road had a large room level with the garden. This was Alan's playroom or, as he called it, his workroom.'[22]

At the beginning of the summer term 1909 Alan was enrolled at Miss Chataway's South Hampstead Kindergarten and Preparatory School, at 43 Belsize Road. The Blumlein's knew of this school from the Hobsons; Mrs Hobson was a daughter of the Greigs in whose house the Blumleins had stayed during part of 1908. Gothic House, the name by which the school was usually known, would have several links with the Blumlein family. Mina, Alan's sister, taught there when a member of the staff became ill; both of her sons, Robert and Philip, became pupils. at the school; and in 1930 Alan began courting Doreen Lane, one of the kindergarten teachers.

He attended Gothic House from the beginning of the summer term 1909 until the end of the summer term 1912. His school reports, which are still extant, consistently affirm his intelligence, but he had one weakness—reading.

'Alan is working well, though is quite capable of doing still more. His handwork is good. He is most intelligent, and a great help to the class.' (Summer term 1909)
'Alan is working very satisfactorily and has made some progress in reading and writing though these subjects still need effort. We find him very intelligent.' (Spring term 1910)
'Alan has made decided progress with all his work, he is very intelligent and has an excellent memory. His reading still needs effort.' (Autumn term 1910)
'Alan has at last made progress in reading, though his spelling is still weak. He has good ability for arithmetic, and is keenly interested in history and geography.' (Spring term 1911)
'Alan's spelling and reading have decidedly improved, though still weak. He could make still more effort. All his other work is most intelligent and interested [*sic*].' (Autumn term 1911)

On Alan's early school days his mother has written:

'Dudu used to help Alan with his homework. He always hated reading and dodged it whenever possible as he could always remember so well anything he heard. He used to listen to Dudu when she read his history and geography to him, and usually once was quite enough to fix the facts in his mind. Thus he acquired habits of concentration which stayed with him. And, of course, as he rarely read, he did not learn to spell! I remember Mary walking in from the garden and looking at a drawing he was making of an engine which was his own idea and which consumed its own smoke to save fuel. Beside the plan of the engine was a picture of a railway engine propped up. Mary asked him about the idea he was trying to work out and added: "I suppose you are using that picture of an engine to copy from?" "Oh! No, certainly not, it's my own invention entirely." "Then why have you got that picture there?" "Only to see how to spell 'engine'." '

Alan's spelling was certainly his Achilles heel throughout his short life. He would spell 'rotten' as 'wrotten' and his wife would say to him: 'No "w".' He would reply: 'Well you spell wrong with a "w" why can't I spell "wrotten"?' Kipper was 'kippa'. 'I [had] kippas for breakfast', he would write. In his last letter to his wife (sent two days before he was killed), there were nine spelling mistakes in two pages.

From Miss Chataway's school Alan moved to Loudoun House School run by Mr Stephen Cox Newton, MA. The only report which has survived (for the autumn term 1912), mentions that Alan's spelling 'is very weak, but has improved considerably of late'. He was ranked first in a class of seven boys.

Through some friends, the Wyalls, the Blumleins were introduced to Dr Cecil Reddie, an educationalist who in 1889 had founded Abbotsholme school[23]. The school was the earliest of the new progressive schools which were to have such an appreciable influence on education not only in the United Kingdom, but also on the Continent of Europe. Reddie was the founder of the movement which embraced the Leitz Schule in Germany, and which led to the establishment of Bedales, Gordonstoun, and many other progressive schools in Great Britain. The introduction led to the Blumleins sending their son to Abbotsholme.

Reddie's object in opening his school, for boys aged between 10 and 19 years, was to provide an education of a thoroughly modern and practical nature; namely, to teach the boys 'the laws of health and the exercises enjoined by them, and [the] habits of gentleness and justice, and to prepare him for the calling by which he [was] to live'.

To further this object, Reddie structured the school's curriculum so that:

'The whole life at the School [was] planned to develop harmoniously all the faculties, physical, artistic, intellectual, and moral; and for this purpose, the mornings [were] devoted to study, the afternoons to athletics and manual work, and the evenings to arts and social recreation.'

The timetable (Figure 1.6a) from the school's first prospectus indicates how the daily routine was organised to accommodate these various activities.

TIME TABLE

6.15	A.M.	Rise (in winter at 7.)	*Every boy will have a light meal*
6.30	''	Musical Drill	*immediately after rising.*
6.45	''	First School.	HOURS A DAY
7.30	''	Chapel.	Mental work 5
7.45	''	Breakfast. Dormitory	Athletics and Manual
		Parade.	work.......................... 4½
8.30	''	Second School.	Social and Artistic
10.45	''	Break for Lunch.	occupations................... 2½
11.15	''	Third School.	12
12.45	P.M.	Singing.	Sleep 9
1.0	''	Dinner.	Meals and Free Time 3
1.30	''	Organ or Piano Recital.	24
1.45	''	Games.	The Dormitories are open (*i.e.*
4.0	''	Workshops, &c.	not divided into cubicles). This
6.0	''	Tea.	is considered more wholesome
6.30	''	Singing. Recitations.	in every way. Boys of different
		Music &c.	ages will occupy different
8.30	''	Supper and Chapel.	rooms
9.0	''	Lights out.	

No lesson, except Writing, Drawing, Practical Chemistry, &c., which involve comparatively little mental effort will exceed three-quarters of an hour. Good results have been generally obtained by curtailing both the number and length of the usual lessons.

There will be no class work on Sunday. There will be Early Morning and Evening Chapel, and at mid-day the boys will be able to attend the Parish Church, which is about two miles from the School.

Figure 1.6   (a) The timetable of Abbotsholme School, c. 1889/90

Source: 'A history of Abbotsholme School, 1889–1989', by D Sederman (Abbotsholme School, 1989)

Dr Reddie seems to have been a martinet. He produced an enormous amount of written administrative directives and school rules, (See Figure 1.6b). By way of illustration, his document No. 7, on 'How is the life in dormitory organised?', contains 18 sections on administration, obedience, tone, sitting up, visiting other dormitories, leaving the dormitory for hot baths or for cabinets, visiting cabinets at night, undressing, ventilation in dormitories, temperature, illumination, punctuality, management of beds, the proper position in bed, disturbed sleep, getting up, dormitory parade, and clean clothes and dirty clothes[24].

Each of these sections contained precise instructions. Rule 14, for example, on 'The proper position in bed' mentions:

'The sleeper should lie on the side, right or left (perhaps alternately for the sake of resting all muscles); and turn over upwards. He should never sleep on his back or face; both postures, particularly the former, induce bad dreams, and

A few copies of Monographs, Rules, &c., very beautifully hand-painted at The Abbotsholme Press (intended for framing):–

No.	Sheets	Sides		Price.
				s. d.
1	1	1	How is the Beginning of Term Organised?	5 0
2	1	1	What are the Rules regarding Unpacking at the Beginning of Term?	5 0
3	1	1	How is the End of Term Organized? ...	5 0
4	1	1	What are the Regulations regarding Packing and Dispatch of Luggage at the End of Term?	5 0
5	1	1	How are the Earth-Cabinets to be used?	2 6
7	1	2	How is Life in Dormitory Organized? ...	5 0
8	1	1	What are the General Aims as regards Clothing?	5 0
9	1	2	What are the Rules for the Boys regarding Clothes?	5 0
10	1	1	What are the Rules regarding Boot-Cleaning?	5 0
11	1	2	What are the Rules for the Boys regarding, Boots and Shoes. Boot-Rooms. Lavatories. and Boot-Cellars?	5 0
12	1	1	How is the Sick-Department Organized?	5 0
13	1	1	What are the Rules regarding the Use of Set-Rooms?	5 0
14	1	1	What are the Duties of the Set-Room Officers?	5 0
15	1	1	What are the Instructions for the Boys regarding Examinations?	5 0
16	1	1	These are the Instructions for the Staff *re* Examinations	5 0
17	1	1	What are the Methods for Indicating and Correcting Mistakes. etc.. in Written-Work. used by all Boys and Masters at Abbotsholme?	2 6
18	1	2	What are the Educative Merits of Bathing?	5 0
19	2	3	What are the Educative Influences of Haymaking?	7 6

Figure 1.6 (b) School rules, 1900–1919

Source: 'A history of Abbotsholme School, 1889–1989', by D Sederman (Abbotsholme School, 1989)

the latter interferes with proper breathing. The legs should be slightly bent to keep the body steady. Other details will be given to each boy individually.'

But as the school's historian has written: 'What else can they be?'

Under Dr Reddie's headmastership and strict control the school thrived during the first decade of its existence. The early years in many ways were 'glorious in their happiness' and in the pioneering of a new educational system. By 1895 the school had 50 boarders and was alive with activity. Much development work was carried out by the pupils. Their achievements ranged from the building of wooden dovecotes, beehives, and henhouses, to ambitious projects such as the laying of Football Lane, the construction of the piggeries, the cricket pavilion, the boathouse, and the wooden bridge over the River Dove.

All of this attracted the attention of learned educationalists and numerous visits and inspections were made by them to study at first hand Reddie's philosophy and methods. These were embraced enthusiastically by his staff. In particular J H Bradley was so taken with Reddie's approach that he left Abbotsholme in 1896 to found Bedales School; and Dr H Leitz returned to Germany to establish the Landerziehungsheime schools. Later Kurt Hahn introduced and developed the liberal educational method at Gordonstoun, where later Prince Charles was a pupil.

Sadly, the prosperity and happiness of the school declined during the second decade of the school's existence. Dr Reddie, for more than ten years, had worked untiringly to create a school modelled on his progressive methods, but now, in the first few years of the twentieth century, the stress of all his responsibilities and activities began to exact a toll. He suffered several nervous breakdowns: some of the staff became disenchanted with his personality and his direction of the affairs of the school and left; and school numbers fell from a peak of 61 in 1900 to just 33 in 1903, though they increased again throughout 1904 and 1905.

Further difficulties arose in 1904 when two teachers ventured to suggest that Reddie's mental illness made him incapable of performing his job competently. He took an extended leave of absence, travelled to Russia in the summer of 1906 and then to the USA in November of that year, and did not arrive back at Abbotsholme until October 1907.

On his return he found that two of the staff had written to all the boy's parents and had asserted that Reddie's character was inappropriate for his position as headmaster. They invited the parents to withdraw their children from Abbotsholme and enrol them at a new school which they would establish on Abbotsholmian lines. Of course some parents became alarmed and about half the children were removed and sent to other schools: the school was never full again during Reddie's headmastership.

A graphic account of life in the school during these unhappy years has been given by an Old Abbotsholmian[25].

'Reddie was always a complex combination of devil and saint, the devil now took command. His temper was ungovernable. He shouted, stormed and raged. He seldom came into a class without a cane. Teaching, what there was of it, was thrown completely to the winds, and instead we suffered tirades against the English — [Reddie was Scottish], against women, and against public schools.

'Only Germany was extolled although at that very time German bullets were tearing Old Abbotsholmians to death. His classes became reigns of terror, and rested like a dead weight upon the happiness of the boys. Even bathing was spoilt with Reddie always present, shouting at us to do this and do that ... The freedom and gaiety of the school disappeared. Instead there was oppression and failure. Each term more boys left.'

None of this could have been known to the Blumleins, for in September 1913 Alan was enrolled at the school. Surprisingly, in view of the staff's disquiet about Reddie, Alan, according to its mother, was very happy at Abbotsholme. He loved the country surroundings of the school and enjoyed the freedom which it offered. Certainly, the school with its many facilities and its own grounds of 133 acres in rural Derbyshire, on the edge of what is now the Peak District National Park, must have appeared a paradise for those boys brought up in the soot-stained industrial towns and cities of the Black Country and elsewhere.

After Blumlein had been at Abbotsholme for approximately one year his parents began to have some doubts about whether his leanings towards science and technology could be vigorously developed by the progressive methods espoused by Dr Reddie. Whether the Blumleins ever became aware of the tensions pervading the school is not known.

During the last week of July the Blumleins were making arrangements to holiday in Looe, Cornwall. They planned to travel there on Friday 31st July but for several days previously Semmy Blumlein had felt ill. He consulted his doctor on Monday 27th July and, being assured that there was no cause for anxiety, went to his City office as usual the following day. At about three o'clock on the Tuesday he returned home in a distressed condition. The doctor was called but was unable to effect any easement or improvement. Ninety minutes later Semmy Blumlein was dead[26].

Semmy Blumlein was a much-loved father, husband, friend and colleague. His children adored him.

'He was the perfect "host" — thoughtful, considerate and imaginative. And what impressed me [Mary Dodgshun] ... was his attitude to all those who served him in all sorts of ways. He was so courteous to all, and yet, at the same time, quite positive and firm in his requests and demands and in seeing them executed. And people like railway officials, porters, waiters, cabbies ... seemed to sense this quality, and would respond, so that he seemed to get service where other men failed to procure it. ...

'People interested him, and he made them feel this was the case, very often people whom others would ignore. ... He had the wonderful ability of piercing right through the exterior of men and women and finding ..., the very essence of their personality. Even when this was immature or distorted he recognised it, and gently drew it out.'[27]

Tuesday, 28th July, was a day Mrs Blumlein would never forget. Not only did she lose her beloved husband after 25 years of loving marriage, but also it was the day when the evening newspapers carried the dreaded news that Austria had declared war on Serbia. Life for the Blumlein family and much of the world would never be the same again.

Dr Reddie was informed by telegram of SJB's death and was asked to convey the sad news to Alan. 'He was quite shattered. It broke his little life.' He returned home the following day. The funeral was held on 1st August 1914.

Alan's closest male relative was now his maternal grandfather, the Reverend William Dower, the former 'grand old pastor' of Kokstad. He sent Alan a letter which expressed his deep sympathy and his advice for the future.

'Since hearing of the sad event which made you fatherless I have thought much of you as well as of your mother and Mina. Your mother has told us of how bravely you stood by her in her hour of sorrow. Her heart will cleave to you now more than ever and I sincerely trust — and expect that you will be to her a comfort and a joy and grow up to be worthy of the name you bear. Your mother will naturally look to you and lean on you the more as you grow in years. Now more than ever you need to apply yourself to those studies which will fit you for the serious business of life. Though your dear father has been taken away from you while still so young you must allow the remembrance of his care and affection to be to you a restraint and inspiration.

'The best proof of affection for your mother and respect to the memory of your late father is to devote yourself with all fidelity to your school work.

'Above all remember your Creator in the days of your youth. Seek ever Divine guidance and grace for daily duty and victory over temptation. This is the real secret of success in life and may it be yours to win it.

'Your grandmother and I have felt very sorry for you and have asked God to help and comfort you.'

Soon afterwards Alan's mother wrote to Dr Reddie and informed him that Alan would have to leave Abbotsholme at the end of the autumn term. Suspecting that this request had been made because of some domestic financial constraints, Dr Reddie generously offered to reduce Alan's school fees. However, the decision to withdraw Alan from the school had effectively been taken prior to his father's death. Although Abbotsholme is now a very fine school indeed, from just a few years after the turn of the century the school, as noted above, had been drifting towards eventual closure. Fortunately such a disastrous event was averted when in the spring of 1927 Mr Colin Sharpe was appointed headmaster following Dr Reddie's retirement. Sharpe was a 'visionary with courage'. He had to be so endowed. When he arrived at Abbotsholme the school comprised the headmaster's secretary and just two boys.

Following the funeral and a holiday at Bexhill, the most important question which Alan's mother had to resolve concerned Alan's future education. She consulted Miss Chataway for advice and was told that the Reverend D H Marshall ran a good preparatory school for boys at Ovingdean, near Brighton, on the Sussex coast.

Miss Chataway's knowledge of the Reverend Marshall stemmed from the prestige of the schools which he and his wife ran in the Belsize Park area of London (which was where her school was situated). In 1898 the Reverend Marshall purchased, from the Reverend F J Wrattesley, Belsize School, which he had founded in 1889, and which was situated at 18 Buckland Crescent[28]. Five years later the Marshalls took over an adjoining house and in that year the school had an enrolment of 58 boys, including ten boarders. Further expansion occurred in 1905 when Marshall bought the Allen–Olney girl's school situated in Crossfield Road. This was continued at 18 Buckland Crescent by Mrs Marshall, the Reverend Marshall having moved the boys to Crossfield Road and renamed his school The Hall. It prepared boys aged 5 to 13 years for entrance to public schools and achieved much success from the award of many scholarships to its boys. In 1909, when the roll was over 100, Marshall sold the school to G H Montauban.

Miss Chataway wrote to the Reverend Marshall and soon Mrs Blumlein was invited to meet the headmaster and his wife at Ovingdean Hall school (Figure 1.7)[29].

'They were very kind and offered a reduction of fees. When I [Mrs Blumlein] responded with a repudiation of my intention to ask for this Mrs Marshall said: "Oh no, of course not. The people who ask for reduced fees come in a Rolls Royce with a chauffeur and a footman!" It was settled that Alan [would] begin at Ovingdean after Christmas 1914 and Mrs Marshall gave me advice about the necessary [school] outfit. The school blazer was rather pleasing in black flannel bound with a $\frac{3}{4}$ inch fold of rose pink; the monochrome DVS — [actually OHS] — on the cap and pockets in the same pink. Black stockings with pink stripes at the top [completed the uniform]. . . .

'Mr Marshall was a classical scholar and an excellent headmaster. But it was to Mrs Marshall that Alan owed so much of what he gained at Ovingdean School. She took the mathematics [classes] and she was a most able teacher.'

Ovingdean House, as Ovingdean Hall was called prior to 1891, was built, in the spring/summer of 1792, by a local wealthy landowner,

Figure 1.7   Ovingdean Hall School

Source: Mr J W M Smith

Nathaniel Kemp, on his 350-acre estate[30]. When he died in 1843 the house was let until Elliot MacNaghten, the retired Chairman of the East India Company, bought the Kemp estate around 1858. MacNaghten died in 1888 and the property and land were then purchased by F Chasley. He opened a 'young gentlemen's school' in 1891. At that time the house contained '13 principal and four servants bedrooms' but as these did not provide sufficient accommodation for the school, Chasley soon had additional buildings constructed. Most of the fields of the estate were sold and from 1910 Ovingdean Hall School stood in 'a park of 20 acres, with a wall and trees all round. It [was] situated on a slope of the down, about 120 ft above sea, which [could be] seen from all parts of the park'. The total site comprised 40 acres, of which 4 acres were devoted to garden produce.

An early, about 1908, school brochure[31] enthuses on 'the great and essential advantages' which accrued from the situation of the Hall. First, there was 'the enjoyment of a complete country life, with the additional recommendation of proximity to Brighton and easy access thereto'. Secondly, 'the entire community [was free] from risk of infectious disease. This [was] owing to the comparative isolation of the Hall, the wonderfully pure air, and the absence of such channels of illness, a church, a public laundry, bath, etc.'

The Reverend D H Marshall, whom Mrs Blumlein met in 1914, was the third headmaster of the school: his predecessors were the Reverend F W S Price (1891–1906), and the Reverend C F S Wood (1906–09). They had modelled the preparatory school on classical lines, since the objective of the school was to prepare boys for all the public schools and for Osborne. The 1908 brochure lists the subjects taught as scripture, Latin, Greek or German, French, mathematics, and English in all branches, and it is probable that Marshall continued this tradition. No mention is made of any science teaching. Extra-curricular activities included violin or piano tuition (at £2.10 per term), dancing (£2.10 per term), carpentry (£1.05 per term), swimming (50p per annum), and drawing (£2.10 per term).

Among the facilities offered by the school there were the swimming baths (where a swimming competition was held every year), the gymnasium, the library (which adjoined the headmaster's study and where 'quiet [was] preserved'), the sanatorium (staffed by an experienced nurse), the playgrounds (of 12 acres), and the chapel (where 'the services [were] thoroughly hearty and moderate in character'). The school games included cricket, football, tennis, croquet, hockey and fives. Altogether the school seems to have provided a most pleasant environment for young boys.

Alan joined the school on 16th January 1915 when he was $11\frac{1}{2}$ years of age (Figure 1.8).

Figure 1.8 A D Blumlein, c. 1915

Source: Mr S J L Blumlein

Shortly after he was enrolled he fell ill with a 'bad attack of quinsey' and was moved to the sick bay. A telegram was sent by the school to his mother, who was at Tilbury about to embark on the SS *Kenilworth Castle* for the journey to South Africa, but she was unable to visit him. Her passage had been booked some weeks previously for 13th February and the arrangements could not be changed. Fortunately the school matron soon had Alan fit and well, but it must have been a miserable time for him.

Alan was one of nine boys in his class; he was 16 months older than the average age of the other boys. Presumably the Reverend Marshall had not been impressed by Alan's previous education at Abbotsholme and recognised that he was behind in average ability for his age. For the term ending July 1915 his reports showed that he was 'very good' in geometry, arithmetic and Latin; 'good' in algebra, composition, drawing and history; 'very fair' in literature; 'fair' in scripture, geography and grammar; and 'poor' in French: his positions in the class, based on the term's marks and the examination marks, were seventh and third respectively. By the end of December 1915 these positions had improved to second and first, in a class of 11, and the headmaster was able to write on Alan's report: 'Excellent. Spelling still atrocious.'[32]

Mrs Marshall seems to have been an important influence during the young Blumlein's formative period. She taught arithmetic, geometry, algebra and trigonometry, and for these subjects Alan consistently received 'very good' gradings. His French was 'weak'. Literature, too, was not a strength — 'fair' being the usual grading.

As an aside: when a Greater London Council (GLC) commemorative 'blue' plaque, in memory of Blumlein, was unveiled at a ceremony held on 1st June 1977 at 37 The Ridings, Ealing, where he had lived, the press made much of the oft-quoted statement that he did not learn to read until he was 12. Sometimes this statement is given as: 'He could not read at 12 but he knew a lot about quadratic equations.' The story is apocryphal rather than anecdotal, as his school reports confirm. Miss Chataway's reports show clearly that Alan's reading was 'good'. Presumably he did not enjoy reading but he did enjoy mathematics.

In 1916 the Marshalls entered Alan for the entrance scholarship examination for Aldenham's school. Although he passed and was offered a House Scholarship, Mrs Blumlein refused the offer, much to the Reverend Marshall's displeasure. Aldenham was a good, traditional, old school but was viewed by Alan's mother as being of the classical type. For several years Alan had shown an interest in science and inventions and now, in 1915, had expressed a wish to become an engineer. Aldenham was considered unsuitable for this purpose. Enquiries were made by Mr Marshall and 'the result was that failing Oundle, Highgate seemed the school best fitted to help Alan'. He was entered for Highgate School in August 1917[33].

'He had to go to the school and live in one of the school houses for about a week and, no doubt, his manners and ability to be a good mixer played as important a part as the examination papers he wrote. Soon, from the Headmaster and Governors at the school, came the news that he had been awarded a Living Scholarship of £50 per annum and extra for teaching and books. This was great good news for Alan . . .'

He joined Highgate School in January 1918 and was placed in Fitzroy Lodge, with Mr C A Evors, MA (late scholar of Jesus College, Cambridge, 2nd Class Classical Tripos), as the housemaster.

Highgate School[34] was, and still is, one of the great schools of London. On 29th January 1565 the Great Seal of Queen Elizabeth I was attached to Letters Patent authorising Sir Roger Cholmley to found 'a grammar school . . . for the good education and instruction of boys and young men' in Highgate, and 'in any other convenient manner to provide for the relief and support of certain poor in the said town or hamlet'. Three months later, on 27th April, the Bishop of London granted to Sir Roger the site of the chapel and the surrounding land at the top of Highgate Hill for the development of the school.

For more than two and a half centuries the school was governed according to its statutes. The nadir in its fortunes occurred during the early 1800s. A Parliamentary Commission found a lamentable state of affairs in the school. 'We cannot but observe', wrote the Commissioners, 'that this school does not appear to have kept pace in the progress, either with the intention of the founder, or with the gradual improvement of its funds, and the necessities of the neighbourhood in which it is placed. As a grammar school it has fallen into complete decay.' School numbers fell alarmingly and in 1838 only 19 boys were on the register. Disaster was averted by the appointment, in 1838, of the Reverend J B Dyne, DD, as headmaster. Under Dyne the school flourished; the numbers of boys enrolled steadily increased and the school's reputation grew. After Dyne's headmastership (1830–74), the Reverend C McDowall, DD, (1874–93), and the Reverend A E Allcock, MA (1893–1908) further enhanced the well-being and standing at the school.

When Allcock resigned in 1908 the Governors showed great wisdom in choosing Dr J A H Johnston MA, DSc, to succeed him. At that time, most headmasters of public schools were clerics and classicists: it was a well established nineteenth-century tradition that public school headmasters should be in holy orders. Highgate had had clerics as its headmasters for nearly 200 years; but, now in 1908, a scientist was appointed headmaster. Five years previously, when Marlborough School had selected Mr F Fletcher, a non-cleric, as its new headmaster, the *Times* had described him as 'one lay apple in the clerical dumpling'.

Dr Johnston had studied mathematics and natural science at Edinburgh, his birthplace, before going up to Pembroke College, University of Cambridge where he became the fourteenth wrangler. For a year he was a Professor of Physics at the Royal Agricultural College in Cirencester, and then for ten years was an assistant master taking science and mathematics at Tunbridge School.

One of Highgate's historians, in assessing Johnson's influence on the school, has written: 'Without doubt Johnston will be judged Highgate's most important Head-master of the twentieth-century. Underlying everything he did for the school was his enthusiasm for mathematics and science'[35].

Like Dr Dyne, Doctor of Divinity, who greatly improved the status of the school, Dr Johnston, Doctor of Science, was a natural administrator and a shrewd businessman with an aptitude for educational reform. School numbers rose from the 295 which prevailed when he was appointed, to 480 in the autumn of 1913, and then to 603 in May 1921, the year Blumlein left Highgate to enter the City and Guilds Engineering College.

Johnston was a formidable figure: the boys found him alarming. 'He went through the school "like a tornado". He would enter chapel in his gown, carrying his mortar board, from a doorway beside the organ, and come down the aisle to his seat like "a ship in full sail".... He was an overpowering sight, no question of that.... Mrs Johnston was a charming woman. People used to wonder how a woman like that could marry such a horror.'[36]

Another former pupil has recollected that Johnston was known as 'the Pate' because of his shining bald head. 'He had a flowing blond moustache and a red face and strode up and down Hampstead Lane four times a day — to and from the School House in Bishopswood Road — with billowing silk gown and mortar board and umbrella. He looked very fierce but was probably quite mild. He was fond of singing operatic arias.'[37]

Mild he was not. On one famous occasion when some boys in assembly prolonged the final 's' of the word 'fundatoris' in a new song the consequence was probably predictable[38]:

'"It was a very successful hissing", wrote an old boy. Johnston [stood] up and [glared] at the back rows, his face livid. All who had hissed were told to report to him and perhaps 50 did — (over 100 according to some boys), thinking there was safety in numbers. But Johnston caned every one of them. When the first wretched people went in there, they got three each; there was a rush to be last — we thought he'd be weakening. But then somebody got four, and there was a rush for the doors'.

When Johnston was appointed, Highgate School was a school where the classical side was pre-eminent, and the modern side was a concession to parental pressure. Johnston soon changed all that. Under his direction the school developed a strong science stream, with science fifth and sixth forms, based largely on chemistry, biology, physics (including electricity and magnetism and optics) and astronomy[39]. In his 1913 report to the Governors he mentioned that in the development of the school curriculum Highgate was leading the way, and there were indications that other public schools were about 'to follow in our footsteps'.

The difficulty which Johnston faced was a lack of suitable laboratory accommodation. Science during the Great War was taught in 'a ramshackle building' which included biology, physics and chemistry laboratories of a 'primitive sort'. This did not accord with Johnston's wishes and year after year, in his reports to the Governors, he exhorted them to extend and remodel the science building. 'I would beg you again to consider the urgent needs ... for wider and more modern science accommodation', he wrote in January 1915. 'The demand for science teaching is growing greater everyday. I have enlarged on this question in every one of my Annual Reports for six years' (January 1920).

And so, when Blumlein entered Highgate in January 1918 he joined a school in which a modern science stream had been created and in which the headmaster was totally committed to the furtherance of science education.

Johnston's enthusiasm for science never ceased. When, in 1925, the Governors finally acceded to his many appeals, he was able to say: 'It is with feelings of very profound relief that I view the advent of the new Science Building, so long needed and desired'. The 21st July 1928 was 'The most notable public occasion in [the school's] history' according to Johnston; for that was the day when the new science building was formally opened. In his address Johnston declared: 'This is no mere traditional home for physics and chemistry'. In the new building 'the budding engineer [would] study the engines themselves, their principles of action, as well as their construction, and these are not merely the machines that rule the land and water, but, in response to a growing and natural demand, the engines and machines also which have achieved the conquest of the air'.

Sir Samuel Hoare, Minister for Air, opened the new science block. He seems to have been chosen because Johnston was keen to develop the teaching of aeronautics in the school. He believed that schools should provide pupils with an education relevant to the age in which they were living. Consequently they should be 'air-minded'. To further his belief Johnston equipped the school with a 'Snipe' aeroplane and five aeroplane engines which were accommodated in the woodworking and metal-working rooms respectively. Another aircraft, an Avro 504 K, was housed in a hangar, specially built for it, which was situated at the top of the new building.

None of these developments aided Blumlein, of course, for he left Highgate in 1921. However, they illustrate the great stress which Johnston placed on the teaching of science. Highgate was certainly in the front rank in its progressive attitude to the disciplines of physics, chemistry, biology, astronomy and aeronautics. It would seem most unlikely that Blumlein could have received a public school education which embraced more science teaching. Certainly, Blumlein benefited enormously from his time at Highgate. When he was admitted to the City and Guilds Engineering College, his knowledge of the fundamentals of science was so advanced that he was able to enter directly the second year of the engineering degree course.

Blumlein was $14\frac{1}{2}$ years of age when he was admitted in January 1918 as a boarder to Fitzroy Lodge. His form master was Mr A H Ardeshir, BA (late scholar of Trinity Hall, Cambridge; 2nd Class Historical Tripos). It has been insinuated that because of his late entry he was lonely in school. The evidence does not support such a view. Alan was one of 32 pupils who were enrolled in that month. Twelve of the boys were of the same age as himself. In addition seven boys of his age had joined the school in the previous September, and a further five similarly aged boys had been admitted in April 1918. Hence Blumlein was one of 24 new entrants, all born in 1903, whose names were entered, during the academic year (1917–18), in the school rolls[40].

Alan did not make a favourable start at Highgate[41]. He probably under-estimated the amount of effort required for good progress. At the end of the first term 1918 his form master's general report noted: 'Not a satisfactory term. He must work harder. Capable of taking a much higher place [than 21st in the class]'. The house tutor's report mentioned: 'Conduct in house excellent. Must take to heart this report'.

These comments were taken to heart and at the end of the summer term 1918 Alan's class position was thirteenth and his form master was able to say: 'Started inauspiciously but made a great effort and recovered much lost ground'. One term later Ardeshir could write: '[Alan] continues to do sound work, and is making good progress'. His class position was now third.

One of Blumlein's contemporaries, R G Macbeth (1917–22), who became the Director of the Department of Otolaryngology, Radcliffe Infirmary, Oxford, has said[42] that life in School House (one of the six school houses) was 'like being in prison'.

'For breakfast there was bread and scrape (margarine or dripping) with only occasionally a bit of bacon or an egg. For lunch there was often lentil pie. It was really galling to eat this stuff and see the Headmaster necking into roast chicken. . . . It was always said that he had a gastric ulcer and that therefore he had to have a special diet. Most of us believed that as much as we believed any other fairy story.'

Another contemporary, C P Fox (1915–22), who subsequently followed a career in electrical engineering, has recollected[43] that 'boarders, despite organised games, [had] perforce a number of hours of idleness to fill each day and week':

'Alan was, I well recall, a voracious reader, and [it] seems that this could well have been his defence against the eternal small cricket and the like which left him so cold.

'I can see him now, sitting with his nose in a novel, completely oblivious of the bear-garden around him in the House common room. And I can well re-call too his quick impatience, softened with a smile when I once offered him a book. A glance, then: "Oh, I have read that." — and he was gone, no doubt in search of something more attractive.

'Manual Instruction, as it was termed, offered some practical assistance for a few boys to fill in the hour between school and lunch. The school had a well-equipped workshop presided over by a congenial despot named W B Bryce.

'Mr Bryce taught carpentry as well as drawing to the younger boys at High-gate for 28 years [1900–28]. He must have been something of an odd-job man too, for he repaired broken desks and benches. More important, he was be-loved. He took pains in helping the hasty and the inept with their sawing and chiselling. Alan was neither hasty nor inept. He saw the Manual Instruction hours as a welcome outlet for his practical abilities as well as a happy alterna-tive to aimless boredom. He seemed to need little instruction I remember, and while other boys were starting model boats or book shelves but seldom finishing them, he designed and built for himself a balance cabinet of quite professional excellence.'

Physics (including electricity and magnetism) was taught by C A Carus-Wilson, MA (Pembroke College, Cambridge), an assistant master (1916–22). He was a 'gentle gentlemen who must have been as delighted with a boy like Alan Blumlein as he was frustrated by the majority of his flock. . . . It is likely that [Carus-Wilson] was [Alan's] greatest single influence during his school days, and the boy could not have been in better hands'.

Other memorable masters of Blumlein's period included the mathemati-cians J Ll Thomas, MA (1915–47), and H M Sylvanus, MA (1915–40). Thomas 'was such a gentle person in the classroom, probably too gentle . . . he was very easily taken for a ride, and it was only if you really wanted to learn that you could learn enormously'. Sylvanus initially taught a junior form until his abilities were discovered by Johnston, who then promoted him to take sixth form mathematics.

A well liked science master was A Izard (1905–41) who coached three boys for the first medical examinations attempted from the school. 'He took it very seriously, and got us through, too. It was considered rather a triumph for the science staff.' Whether Blumlein received coaching from Izard is not known, but they did have one common characteristic. Izard[44] was 'a great comedian'. 'The sort of thing that he would do was to take up [a piece of] rubber tubing and use it as a catapult, and if he saw a boy on the other side of the room fooling about, this chap would as likely as not find this piece of tubing coming at him.' Blumlein, too, had a great love of fun. Dr Dutton, of Electric and Musical Industries, who was one of Blumlein's friends, has been quoted as saying that Blumlein with a paper clip and an elastic band could 'ping a bloke's ear at 25 ft'[45].

On the general atmosphere which pervaded the school, Blumlein's contemporary F Forty, MB, BS, FRCS, (January 1918—July 1923) has written[46] that Highgate 'was not particularly warm and friendly—it was impersonal—[and] the regime was not strict'.

> 'I was a bit of a non-conformist and broke most of the rules, but was never severely rebutted. It took me some time to realise that this was really due to indifference. Nobody really minded what you did or did not do. "Laissez-faire" was the general attitude. There was the school and you made of it what you would or could. On the whole, I think it was a good place, there were no theories about "education" and no fanaticism about anything. If there was merit there was nothing to prevent its flourishing, as evidently happened in the case of A D Blumlein.'

(Forty was in the modern languages sixth form.)

Blumlein matriculated in the Second Division in 1919 and entered the science sixth form at the beginning of the third term 1919. It was a small form—just six to ten boys depending on the subject being studied—and Blumlein followed classes in chemistry, physics, extra physics, mathematics and extra mathematics. He started well and at the end of the first term his form master, J B Finter, was able to write: 'Has made much progress throughout the term'. A 'fair' spring term's work led to success in the summer term. Alan's position at the end of this term was first and Finter noted: 'Has improved considerably all round. Should do very well next year'. Individual (subject) comments were:

Extra maths	Has made a sudden dash forward and is practically fit for the Intermediate exam now.

Chemistry	Has done remarkably well this term. Is very keen on his work.
Physics	Very useful term's work. Has made considerable progress.
Special mathematics	Disappointing in exams but has made real progress all round.

These results led to the award of the Bodkin Prize (General) to Blumlein in July 1920.

Interestingly, his house tutor's report for the spring term recorded: 'Conduct satisfactory; but quite a negligible quantity: does not possess the sporting instinct'. This comment appears to have stimulated Blumlein to participate in some non-curricular activities. He swam for his school and in July 1921 took part in the Public School's Team Race of the Bath Club. The Club had presented, in 1910, to a Joint Committee of the Oxford and Cambridge Universities Swimming Clubs, a silver Challenge Cup as a perpetual trophy, for competition among certain of the public schools of England. In that year and the two following years the race was held over a distance of 150yd—St Paul's winning each time—but in 1913, when Harrow won, it was changed into a team race of 200yd. St Paul's again won in 1914, and in 1920, when the race was restarted after the Armistice, Blumlein, J M S Zambra, and C W Schoedelin formed the Highgate School team for 1921 but did not win their heat. Later Blumlein swam for Imperial College and for the London Otter Club[47].

Also in 1921 he was one of three courtiers in a school play taken from *A Midsummer Night's Dream*. In the play:

> 'Theseus, Duke of Athens, is celebrating his marriage with Hippolyte. He is informed by his Master of Revels that certain poor Athenian craftsmen have written and rehearsed a play, which they desire to perform before their ruler. The Master of Revels makes it clear to Theseus that neither the play nor the actors are of the highest quality. Nonetheless, Theseus replies:
>
> > "I will hear the play;
> > For never anything can be amiss
> > When simpleness and duty tender it." '

Blumlein continued to make good academic progress in 1921: 'Has made [in] this term [first term 1921] a most gratifying spurt. The result is very evident in his work'.

Blumlein left Highgate School in July 1921. In October 1921 he was admitted to the electrical engineering degree course at the City and Guilds College, which formed the engineering section of the Imperial College of

Science and Technology. The college was a school of the University of London in the Faculty of Engineering.

Imperial College[48] evolved from a series of changes in the provision of science and technology education in London. The series commenced with the establishment in 1845 of the Royal College of Chemistry. Six years later the Government School of Mines and of Science Applied to the Arts was created; then in 1853 these institutions were combined to form the Metropolitan School of Science Applied to Mining and the Arts. Subsequent developments were as shown in Figure 1.9

When Blumlein joined the City and Guilds College, the three constituents of Imperial College of Science and Technology comprised the following departments:

The Royal School of Mines   Mining, Metallurgy and Geology
The Royal College of Science   Biology, Meteorology, Physics,
                               Mathematics and Mechanics, Chemistry,
                               Chemical Technology and Aeronautics
The City and Guilds College   Civil Engineering, Mechanical
                               Engineering, Electrical Engineering and
                               Mathematics and Mechanics*

(*This department was effectively a sub-department of Mathematics and Mechanics in the RCS.)

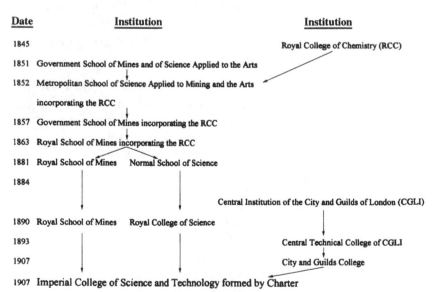

Figure 1.9   The antecedents of the Imperial College of Science and Technology

Source: 'The University of London', by Negley Harte (Athlone Press, London, 1986)

The Department of Electrical Engineering, in 1921, was very small by present-day standards and comprised just five teaching staff above demonstrator grade, namely[49]:

Professor	T Mather, WhSch, MIEE, FRS
Assistant Professors	S P Smith, DSc, AMICE, MIEE
	E Mallett, MSc, AMIEE, AMICE
Lecturers and Instructors	A Rushton, MSc, AMIEE
	B Hague, ACGI, DIC, MSc, AMIEE
Demonstrators	F E Meade, AMIEE
	H E Park
Assistant Demonstrators	G W Sutton, BSc
Draughtsmen and Design	J W Sims, AMIEE, AMIMechE
Office Assistants	D W Hopkin
Workshop Instructors	F W Andrews, AMIEE
	W Diamond

From the 1921 prospectus[50] 'the work of the Department [was] specially designed to meet the requirements of persons who [desired] to obtain a technical knowledge of the application of Physics, either to enable them to rightly understand the processes carried on in certain industrial operations, or to fit them to become technical teachers'. The degree course of three years in duration was of a traditional character and was founded on a common first-year programme of mathematics, physics, chemistry, engineering drawing, engineering workshop and laboratory classes.

Students who wished to enter directly into the second-year course, and those who wanted to compete for a Governors Entrance Scholarship, were required to take written and practical examinations in both physics and chemistry. In the written examinations candidates were expected to show some knowledge of practical laboratory work, and in the practical examinations to show a knowledge of theory.

The excellent education which Blumlein received at Highgate School now manifested itself. He not only passed the examinations which enabled him to be admitted into the second year, but also he was awarded one of only six Governors Scholarships, worth £60 each per annum[51]. These could be renewed for a second or third year on the recommendation of the Engineering Board of Studies. The records show that Blumlein held his scholarship for the two years he was an undergraduate student.

All second-year engineering students followed a common course for 27 of the 30 hours per week, namely:

Strength of Materials, Mechanics, Machine Drawing, Design, and the Steam Engine	11 hours per week
Electrical Technology	5
Mechanics and Mathematics	8
Chemistry	3
Optional Subjects	3

(For electrical engineering students the 3 hours per week allocated to optional subjects were spent in the electrical laboratories.)

J M Turnbull, a Kitchener Memorial Scholar, was another direct entrant into the second year. He has described how he met Blumlein for the first time[52]:

'[It was] on the steps of the old dynamo room in City and Guilds. We were the odd men out because the ex-first year students had already paired-up [for laboratory classes]. I ambled over to Blumlein, little knowing what I was letting myself in for, and suggested that we form a partnership. Blumlein said OK, but it did not work out. I could not keep up [with him] and after about three or four months the partnership was not going too well and broke up. [The reason was] while I was still reading the instruction notes, Blumlein [had read the notes and] had just about finished the experiment! [He] showed his brilliance [even] then.'

Turnbull, who later worked, as did Blumlein, at the Columbia Graphophone Company and later at Electric and Musical Industries Ltd, remembered him as being 'quite popular' and 'mixing pretty well', although he 'did not suffer fools gladly'. He was a very good swimmer, represented Imperial College in swimming events and received the college's colours for this activity.

The second-year electrical technology syllabus covered the fundamentals of electrical engineering and comprised lectures on alternating current theory, the generation, transmission and storage of electric energy, arc and glow lamps, and electric traction. This programme was supported by the laboratory course which was given in the six electrical and magnetic research laboratories, in the arc lamp room, the optical test room, and in the dynamo and motor laboratories. Besides experiments on dynamos, motors, transformers, accumulators, capacitors, arc and glow lamps, and so on, the work embraced the measurement of the strength of magnetic fields, the magnetic properties of iron, coefficients of self and mutual induction, and the calibration of various commercial meters.

During Blumlein's second year Professor Mather retired and from August 1922 his chair was occupied by Professor C L Fortescue, a wireless expert who had previously been at the Royal Naval College, Greenwich[53].

The third-year course for electrical engineering students consisted of:

Electrical Engineering 23 hours per week (average)
Mathematics 4
Mechanical Engineering 2
Physical Chemistry 1

and in electrical engineering the following subjects were considered: telegraphy, telephony, radio telegraphy and wireless telephony, electric machine principles and design, electrical theory and measurements, transformer design and construction, electrical engineering practice, and practical photometry.

Following completion of the three-year course students who had satisfactorily passed their several examinations, including the matriculation or the entrance examination, were awarded the diploma of Associate of the City and Guilds Institute (ACGI). Additionally, students could sit, either as internal or external students, the examinations necessary for the award of the BSc (Eng) of the University of London. Blumlein sat these examinations as an external student since he had not accumulated sufficient time, three years, to be considered an internal student at the university. Turnbull, on the other hand, had been a student at Battersea Polytechnic (an internal school of the university), prior to going to the City and Guilds College, and was able to take the examinations as an internal student. He has stated that because the external students were not so well known to the university as the internal students, the examinations for the externals were much stiffer than for the internals. Blumlein took the final year BSc (Eng) degree examinations at the age of 20 and was awarded a first-class honours degree: his examination subjects were electrical technology, electrical machine design, electrical generation, distribution and utilisation.

Blumlein, during his formative period, was much aided by an excellent memory. Not only could he process information at a quicker rate than most of his contemporaries, he could retain the knowledge without difficulty. 'Certainly', according to his mother,[54] 'he did not spend long hours grinding up his stuff'. He studied rapidly and with intense concentration. This ability appears to have been developed when he was a small boy. Blumlein, then, was a lazy reader and would always ask his nurse, Dudu, to read aloud to him. After reading a passage once through she would ask: 'Have you got it all?' 'Now and again the reply would be "Just go over again the points about so and so" —some incident in history, or the chief towns, or exports of the country. His mind fastened round what he had heard and he did not need to re-read it, or "swot" at it.' Such a facility led to the impression among some of Blumlein's student friends that he did not work. One of them told Blumlein's mother that it was generally said of him: 'If Blumlein worked you would not see him for dust'[55].

Some of Blumlein's attributes—those of his powerful intellect, endless patience, and fine teaching ability—produced a useful financial gain during his days as an undergraduate. He coached several students for their examinations. 'They came up to the flat for their coaching whenever Alan could fit them in and as they paid well Alan was glad to have the money.'

It may be thought that his excellent academic progress—the awards of scholarships to Highgate School and to the City and Guilds College, and the award of a first-class honours BSc degree, after just two years' study, had been the consequence of an unremitting regime of study and hard work and that leisure pursuits had not featured in his student days. Such a view would be wholly incorrect. Blumlein was not a dull, unsociable swot who closeted himself away from others. The speed at which he could assimilate knowledge allowed him time to experience some of the joys of student life. John W May has written:

> '[Blumlein] was one of my greatest friends from his arrival at Fitzroy Lodge [Highgate School] until just before the war when we lost touch.
> 'Throughout the 1920s we were always out together. I can remember him as a terrific practical joker and a great fun lover—Chelsea Arts balls, weekends together in Paris, bridge, parties, anything that was going.... What I can chiefly remember is a lot of fun together in the "roaring" twenties.'

The Chelsea Arts Club Balls were a particular delight for Blumlein. 'For sheer exotica, extravagance, and opulence of display, exhibitionism and glorious fun' there was almost nothing which could approach the exuberance of these events.

The Club had been formed in 1890 by James McNeil Whistler with a foundation membership of 40. Whistler (born in 1834) was an American painter and graphic artist, who after having studied art in Paris had settled in Chelsea, London, in 1859. At first the Club, together with the many artists who frequented this suburb of the capital, organised Mardi Gras Carnivals which were held in Chelsea Town Hall. Later as their popularity increased they were moved to Covent Garden and then, in 1910, to the Royal Albert Hall. Renamed the Chelsea Arts Ball, the RAH function retained the carnival atmosphere of its predecessors[56].

Blumlein was introduced to the pleasures of the ball in 1921 when some associates of the Blumleins moved to a flat near their accommodation. Daisy Marks had been a friend of Blumlein's sister Mina, when they were at South Hampstead Girls School, and had a younger sister Enid who was the same age as Blumlein. Enid Marks, after leaving Rodean School, Brighton, in the summer of 1921, joined the Slade Art School which like the City and Guilds College was situated not far from the Albert Hall. Inevitably Enid and Alan travelled by the same bus to and from South Kensington and so saw much of each other. And being students they could participate in the various social and other activities organised by the two institutions.

Each year the Chelsea Arts Ball adopted a theme which the floats and many of those who attended endeavoured to epitomise, though any form of fancy dress costume, however varied and fanciful, was permitted. Many of the floats were provided by the arts schools. They vied with one another to exceed in invention, beauty of costume and decoration, and brilliance of impact, the tableaux of the previous years. The rewards for the art students, who did most of the work in fabricating the floats, were the practical realisation of their ideas and artistry and a nominal entrance fee to the balls.

Blumlein attended several of the Chelsea Arts Balls with Enid (Figure 1.10). Mrs Blumlein, his mother, has described[57] how on one occasion Alan went to an event dressed as a devil.

'[The theme] was that everything was getting so bad that only as devils could the world survive. There was a lurid mouth of hell with black, yellow and red figures rushing in and out, giving the effect of flames. The dress was of black, fitting tight[ly], with bat-like wings, lined red or yellow, fastened down the

*Figure 1.10   A D Blumlein at a Chelsea Arts Ball with Enid Marks. He can be seen imme-
diately below the clock with Enid Marks, on his right, leaning out into the
aisle. She is sitting next to the man with a moustache*

Source: Mr S J L Blumlein

outside of the tight black sleeves. The head dress [was] a Mephisto fitting cap with red or yellow horns — rather like carrots — sewn on!

'When the dancing at the ball came to an end in the early hours of the day the students [would] adjourn to some place in the college where they could get a warm meal and await the early workmen's trains.

'One morning, early, I was awakened by someone knocking at the front door, and when I opened it there was Alan with a bunch of violets — he had bought for me at Covent Market — in one hand, a bottle of milk in the other, his black cap with red horns on his head, his ordinary dark overcoat [on], but hanging down behind he showed a tail with a red arrow at its tip! He said: "I wondered why everyone stared!"...

'Another fancy dress I made for Alan was a black and yellow Harlequin costume — very effective with very wide baggy legs narrowing down to very tight from [the] knees down, with very full white ruffles at the neck and sleeves, and, of course, a Harlequin cap. A favourite fancy dress for Enid was a Golliwog — she had very curly hair, quite black — and could make it stand on end, just like a Golliwog.'

After graduating, Blumlein had to decide whether to pursue postgraduate research for a higher degree, or enrol on one of the department's postgraduate courses, or seek some paid employment.

Research in the City and Guilds College was not strong in the early 1920s. In a 'Report of an inspection on research, teaching and equipment', the inspector, J A Ewing, observed: 'The general impression received in a most interesting visit was that the work of the college is energetic, sound and effective for its purpose. One would have been glad to see more activity in research ...'.

When Blumlein graduated, postgraduate research in Professor Fortescu's department seems to have been almost non-existent. Only three postgraduate students are mentioned in the Annual Reports of the Imperial College of Science and Technology for the period 1918–27. M G Say who had been working on electric traction for railways, under the supervision of Dr Parker Smith, had followed Dr Smith to Glasgow when Smith had obtained a post in the University of Glasgow; and A H Reeves, who had been investigating the interaction between valve-maintained oscillating circuits, had left to join the Western Electric Company. One year later, in 1924, G F Dutton began a 'very promising' investigation on the electrical reproduction of sound. The Annual Reports note that 'a research on the properties of certain insulating materials was begun and will be continued during the coming session in close collaboration with the British Electrical and Allied Industrial Research Association'. There was a snag, however. Although the work was of 'fundamental importance', it was 'remarkable that there was at present no equipment in the laboratory for carrying out tests of this kind which students [could] use. It [was] very much to be desired that this deficiency may be made good by some means in the near future'.

The position in the department with regard to postgraduate courses was rather better: regular postgraduate courses were offered in 'Electrical machinery and transformers', and in 'Electric traction and railway electrification', and Professor Mallett ran a successful special advanced lecture course on 'Wireless telegraphy and telephony'.

A major consideration for Blumlein in 1923 was the urgent need to obtain some remuneration. From 1914, following the death of his father, he had been supported by his mother and two scholarships, but now he probably felt that the time had arrived when he should seek a salaried post. Fortunately two vacancies for assistant demonstrators existed in the department and Blumlein decided to apply for one of the posts. He was successful and was soon assisting Mallet in running the course on 'Wireless telegraphy and telephony'.

E A Nind[58] was one of the students who was helped by Blumlein. His recollection was of a 'delightful man', 'very human indeed', who had a 'good sense of humour' and was 'very good at explaining anything'. 'He did not get exasperated if you did not understand and would go through a point again and again.' He had 'plenty of inexhaustible patience' and had a 'great facility for converting quite complicated mathematics into very simple circuit elements'. 'Blumlein would ask you a question and this question would absolutely be the "nub" of the question that was puzzling you and so you answered your own question. He was expert at this and could go straight to the point that was [causing difficulty].'

When Nind left college in 1926, he too, like Turnbull, went to the Columbia Graphophone Company, which became part of Electric and Musical Industries Ltd (EMI) in 1931, and once more experienced Blumlein's outstanding ability.

Blumlein spent one year researching, under Professor Mallet's supervision, a 'new method of measuring HF [high frequency] resistance'. The work led to a paper[59] which was submitted in 1924 by Mallet and Blumlein, to the Institution of Electrical Engineers for possible publication in the journal of the Institution. Following the usual practice the IEE sent the paper to several referees. They reported: 'The paper appears to have been somewhat hurriedly put together; the authors should be asked to go carefully through it, to rectify errors and improve the English'. This was undertaken and on 17th January 1925 an abstract of the revised paper was read before a meeting of the Wireless Section. Later Blumlein and Mallet were awarded one of the IEE's premiums for their work: Blumlein was just 21 years of age, an exceptionally young age for such a presentation.

References

1 DODGSHUN, M.: 'Semmy Joseph Blumlein', an unpublished manuscript, 57pp, private collection
2 LEVERTON, B. J. T.: 'William Dower', the 'Dictionary of South African Biography', Vol. III, p. 238.
3 See entry no. 631 on 'William Dower' in the Register of L. M. S. Missionaries, p. 85
4 DAVENPORT, T. R. H.: 'South Africa, a modern history' (Macmillan Press, London, 1977)
5 Ref. 1
6 ELRINGTON, C. R. (Ed.): 'The Victoria history of the County of Middlesex', Vol. IX (Oxford University Press, Oxford, 1989), p. 165
7 THOMPSON, F. M. L.: 'Hampstead. Building a borough, 1650–1964' (Routledge and Kegan Paul, London, 1974), p. 58; see also 'Who was who'
8 Ref. 1
9 BLUMLEIN, J. E.: 'Granny's story of Alan Dower Blumlein made for her grandsons', an unpublished manuscript, 1946, 29pp, private collection
10 PRIESTLY, H.: 'The what it cost the day before yesterday book, from 1859 to the present day' (Kenneth Mason, Hampshire, 1979)
11 Ref. 1
12 Statement of Semmy Joseph Blumlein, 21st December 1904, HO/144/778/124981, PRO, Kew, UK
13 Certificate of Naturalization to an Alien, No. 15078, 31st January 1905, HO/334/39, PRO, Kew, UK
14 Ref. 1
15 Ref. 12
16 NASH, G. H.: 'The life of Herbert Hoover. The engineer, 1874–1914' (W.W. Norton and Co., New York, 1945), chapter 21, pp. 426–446
17 See entry for J L Urquhart in 'Who was who, 1929–1940'
18 Report in *The Times*, 4th November 1908, 22c
19 Ref 16, p. 430
20 Ref 16, p. 429
21 NASH, G. H.: 'The life of Herbert Hoover. The engineer, 1874–1914' (W.W. Norton and Co., New York, 1945). Chapter 21, pp. 429
22 Ref. 17
23 SEDERMAN, D.: 'A history of Abbotsholme School, 1889–1989' (Abbotsholme School, Uttoxeter, 1989)
24 Ref. 23, pp. 27–32
25 Ref. 23, p. 18
26 Ref. 9, p. 14
27 Ref 1, pp. 48, 49, 51
28 DAVIES, J. G.: Notes on 'Ovingdean local history', private collection
29 Ref. 9, p. 15
30 Ref. 28
31 ANON.: 'The Ovingdean Hall School', *c.* 1908, 17pp.

32 School reports, spring term 1915 to autumn term 1917 (inclusive) of Ovingdean Hall School, private collection

33 Ref. 9, pp. 21–22

34 HUGHES, P., and DAVIES, I. F. (Eds.): 'Highgate School register, 1833–1988', 7th edition (Castle Cary Press, Somerset, 1988/89)

35 HINDE, T.: 'Highgate School, a history' (James and James, London, 1993), p. 81

36 Ref. 35, p. 83

37 Letter: F Forty to R W Burns, 23rd July 1991, personal collection

38 Ref. 35, pp. 84–85

39 FOX, C. P.: 'Blumlein at Highgate', personal collection

40 Ref. 34

41 School reports, 1918–1921, of Highgate School, private collection

42 Ref. 35, p. 86

43 Ref. 39

44 Ref. 35, p.92

45 HADDY, A.: taped interview, National Sound Archives, London, UK

46 Ref. 37

47 Miscellaneous papers in the author's collection

48 HARTE, N.: 'The University of London' (The Athlone Press, London, 1986)

49 Prospectus of the City and Guilds (Engineering) College, August 1921

50 Ref. 49, pp. 430–432

51 Letter: Registrar, City and Guilds (Engineering) College, to A D Blumlein, 30th September 1921, private collection

52 TURNBULL, J. L.: taped interview with the author, personal collection

53 Prospectus of the City and Guilds (Engineering) College, August 1922

54 Ref. 9, p. 26

55 Ref. 9, p. 27

56 THACKRAH, J. R.: 'The Royal Albert Hall' (Terence Dalton Ltd, Lavenham, 1983), pp. 111–115

57 Ref. 9, pp. 25–26

58 NIND, E. A.: taped interview with the author, personal collection

59 MALLETT, E., and BLUMLEIN, A. D.: 'A new method of high frequency resistance measurement', *Jour. IEE*, 1925, **63**, pp. 397–414

Chapter 2

Long lines

Until his death in 1922, at the age of 75, Alexander Graham Bell's fertile imagination and interests ranged over many topics. Apart from multiple telegraphy, the telephone and the photophone, Bell's experimental activities embraced spectrophones, phonographs, telephonic probes, kites, aeroplanes, hydrofoil boats and air-conditioning, among others. He took a keen interest in sheep breeding, longevity, eugenics and the National Geographic Society. 'There cannot be mental atrophy', he wrote[1], 'in any person who continues to observe, to remember what he observed, and to seek answers for his unceasing hows and whys about things.'

Shortly after Bell had demonstrated his telephone on 10th March 1876 he wrote a document[2] which became remarkable for its prophetic vision of the future:

'At the present time, we have a perfect system of gas pipes and water pipes throughout our large cities. We have main pipes laid under the streets communicating by side pipes with the various dwellings, enabling the members to draw their supplies of gas and water from a common source.

'In a similar manner, it is conceivable that cables of telephone wires could be laid underground, or suspended overhead, communicating by branch wires with private dwellings, country houses, shops, manufactories, etc, etc, uniting them through the main cable with a central office where the wires could be connected as desired. Such a plan as this, though impracticable at the present moment, will, I firmly believe, be the outcome of the introduction of the telephone to the public.'

Bell's vision required funds for its implementation and therefore companies had to be established. While Bell was working on his multiple, or harmonic, telegraph system, T Sanders, a leather merchant and friend of Bell, made a verbal offer to Bell to help finance his endeavours in return for a share in whatever patent rights might accrue. Soon afterwards, G G

Hubbard, a Boston attorney and Bell's future father-in-law, made a similar offer. These offers were accepted and on 27th February 1875 the Bell Patent Association was formed. In several stages this organisation became the American Telephone and Telegraph Company (AT&T) which was incorporated on 3rd March 1895[3].

Prior to 1878, all of the telephone apparatus for the Bell organisation had been constructed by T A Watson, initially at the shop of Charles Williams, Jr., a manufacturer of telegraph instruments at 109 Court Street Boston, and then in a workshop which Bell set up in January 1876 at 5 Exeter Place, Boston. Watson's tasks soon increased to the stage where he could no longer fabricate all the equipment required, so he engaged a number of manufacturers, including Williams, to assist him. Contemporaneously the Western Union Telegraph company was having some of its telephone devices made in its own workshops, and some by the Western Electric Manufacturing Company of Chicago. This caused consternation in the Bell organisation, since it considered that its patents were being infringed, and in the summer of 1878 a lawsuit was brought against the Western Union Telegraph Company[4].

When agreement between the parties was reached in November 1879, Western Union relinquished its telephone activities, and the Western Electric Manufacturing company was licensed to manufacture telephones for the Bell organisation. Two years later, on 26th November 1881, the American Bell Telephone Company purchased a controlling interest in the Western Electric Manufacturing Company. The company was renamed the Western Electric Company and on 6th February 1882 it officially became the manufacturing unit of the Bell System, thereby providing a dependable source of instruments and apparatuses which would be coherent in system use and also be of consistently reliable quality[5].

Before its incorporation into Bell, Western Electric had for a number of years been interested in the sale and manufacture of communication equipment abroad. Its overseas business continued after the merger and branches of Western Electric existed throughout the world by 1918. At that time, its foreign interests were so extensive that it was considered desirable to concentrate them into a new subsidiary, the International Western Electric Company. This organisation grew rapidly. At the same time a huge expansion of the telephone network in the USA was taking place and the Bell management felt that it would be commercially desirable to avoid any possibility that these foreign activities would interfere with the domestic (US) business. Hence, in 1925, the International Western Electric Company was sold to the International Telephone and Telegraph Company (IT&T) and its name changed to International Standard Electric Corporation (ISEC). Neither of these companies had any further corporate connection with the Bell system.

When Blumlein applied in 1924 to join the International Western Electric Company the head of the European Department was R A Mack. He had received details of Blumlein's ability from Professor Mallet but, according to J B Kaye[6], '[Mack] wouldn't look at Blumlein. He had a Jewish name and I was told at the time that the Western Electric Company in America would not engage Jews — why I have never known. [Anyway] it was not correct...'. Mack came under some pressure from Mallet, who told him that if he did not appoint Blumlein he would lose one of the best men on telephony that they had ever had at the City and Guilds Engineering College. With such praise Blumlein became a telephone engineer with IWE.

At IWE Blumlein soon met J B Kaye. Kaye subsequently became Blumlein's closest friend; he was best man at his wedding in 1933, and became godfather to Blumlein's son, Simon, who was born in 1936. Consequently a few words of introduction are necessary. In June 1923 Kaye obtained a not particularly good degree in science, which consisted of physics, chemistry and geology, from the University of Cambridge and applied for a post with IWE. The company wanted engineers and 'did not seem to have any openings for anyone like myself [Kaye]. But ... Mack said he would take a chance and turn me into an engineer. So I had a full course of all types of work — fieldwork, in the factory and so forth — and then I started at Connaught House [in Aldwych, London (Figure 2.1)] on the 15th September 1924. That was the date on which Blumlein — that's how I addressed him, I was always JB, he was always Blumlein — [started work]'. Kaye's weekly wage was three pounds per week but Blumlein, being an engineer, received four pounds 15 shillings per week.

Kaye has recorded some quite delightful and fascinating reminiscences of his association with Blumlein while at IWE and ISEC. Since no other anecdotal histories exist, much of what follows is based on Kaye's recollections. In 1924 IWE recruited many '4th year men from the City and Guilds' who had qualified in light current electrical engineering.

'Now [IWE] was a bit puzzled as to what to do with us, because they couldn't quite make out Blumlein ... [he was] a little difficult to assess. So they put us together as a pair in the Transmission Laboratory. That was a lab. which dealt with all the cable work, measurements, instrumentation and all that was required to fortify people in the field. These were the engineers dealing primarily with all the main cable routes in Europe, where the company had contracts, [viz.]: Paris – Strasbourg, Milan – Genoa, Paris – Madrid [and so on].

'They put us under the control of Roland Webb. Well, we must both have had a sort of funny reputation because the betting was that Roland Webb would either cope with us or he would have a nervous breakdown. Where our reputation came from I've never been quite clear. . . . We both took to one another, but

Figure 2.1   Connaught House, Aldwych, London

Source: Historical Archive, Nortel Technology

[were regarded] as a bit of a puzzle. The result was that we worked on all sorts of strange things — odd testing stuff off the normal routine.'

Among their disparate tasks Kaye and Blumlein tested the first samples of permalloy tape. This had been brought to the UK from the USA by F Gill, the European Chief Engineer, following his visit to the laboratories of AT&T. Kaye vividly recalled the occasion since 'it was the only time when [he] beat Blumlein to the punch'. In the 1920s the standard method of testing a sample of new ferromagnetic material was to wind two windings onto a toroidal core of the material and then use a ballistic galvanometer connected to one winding to observe the changes of magnetic flux in the core as the current in the other winding was reversed. By varying the magnitude of the current changes, the hysteresis curve of the material could be obtained. On the above occasion, while Kaye and Blumlein were plotting these results:

'Blumlein exclaimed: "Oh, good [gracious] we've forgotten one thing. This stuff has a very high permeability and therefore it [will] easily saturate in the earth's field. We haven't taken account of the earth's field. I'll go and work it out." So he sat down at the desk, and nearly half an hour later he came along with his results and told me: "It's perfectly all right, [the earth's field] doesn't affect it because it's a toroid." I said: "Well, as a matter of fact I found that out within 10 seconds of you sitting down to start working that out." He [asked]: "How did you do that?" I said: "I simply took the coil in my hand and twisted it smartly." The earth's field was there of course cutting the coils . . . [but] . . . it didn't deflect the galvanometer one iota: it [had] no effect at all. So that was it. But that was the only time in my recollection when I ever beat him, and I beat him by half an hour!'

On another occasion the pair tested one of the first low-voltage cathode ray tubes to incorporate a Wehnelt cathode (an indirectly heated cathode). This cathode ray tube had been designed by J B Johnson[7] of the Engineering Department, Western Electric, and was intended to replace the high-voltage, hard vacuum, continuously pumped cathode ray tubes which were a feature of cathode ray oscillography in the first 25 years of the twentieth century.

By 1924 the technical enquiries of the Bell system had increased so much in scope and size, and in the number of personnel employed, as to suggest the formation of a single new organisation to deal with most or all of its activities. Such an organisation was established on 27th December 1924, and commenced operations on 1st January 1925 as the Bell Telephone Laboratories, Incorporated, but sometimes referred to simply as the Bell Labs.[8]. The Labs. had a dual responsibility — to the American Telephone and Telegraph Company for fundamental researches and to the Western Electric Company for the application of the results of these researches to designs suitable for manufacture.

At the date of its incorporation, the personnel in the Labs. numbered approximately 3600, of whom about 2000 were members of the technical staff and comprised engineers, physicists, chemists, metallurgists and mathematicians. The Bell Labs. occupied space in the building at 463 West Street, New York City: this was to remain one of the major locations of the Labs. for more than 40 years.

Kaye, in his recollections, was most complimentary about the 'marvellous people' in the Bell Telephone Laboratories; 'They would try out various things without any thought of commercial return'. With their vast resources, the Laboratories could afford to undertake researches for which possibly there would be very little financial return.

One of the products of their work was the artificial larynx, a device which was later used by thousands of people who had lost their vocal cords following a laryngectomy operation for throat cancer. With the artificial larynx the normal passageway for the air passing from the lungs through the vocal chords was replaced by tubing attached to the windpipe which terminated in a hole in the front of the neck. The airflow from the lungs then activated a vibrating metal reed which injected sound vibrations into the mouth, where they were modulated by the same cavity resonances as in normal speech. The artificial larynx was due to R R Riesz, who joined Western Electric in 1925 from the physics department of Wisconsin University[9].

One of these devices was sent over to IWE. 'Again [IWE] didn't know what to do with it so they gave it to Blumlein and me and we tested it: one [of us] was blowing and the other was modulating—I can't remember which. It sounds very unhygienic, but [we proved] the thing could work. After that we heard nothing more; [presumably] I imagine because so many people just didn't survive.'

Much fundamental research work on speech and hearing was undertaken by staff at the Western Electric Laboratories prior to the incorporation of the Bell Telephone Laboratories. There had been numerous investigations, over a 50-year period, of hearing sensitivity, beginning with those of Toepler and Boltzmann in 1870 and Rayleigh in 1877 but, as Fletcher and Wegel of Western Electric showed in 1922, the results were too widely disparate to be of practical use. There had been no appreciation of the great variation in the sensitivity of hearing of people with supposedly 'normal' ears. This variation could be as much as a thousandfold in terms of acoustical energy. One of the difficulties which the early workers faced was the lack of suitable instrumentation for their measurements. Fortunately, following the necessary development of radio communications during the First World War, valves and radio components became easily attainable from about 1920. Electronic oscillators and amplifiers enabled more accurate and reliable measurements to be obtained, and when an oscillator and amplifier were combined with a capacitor microphone, a thermal receiver and a calibrated attenuator, a new instrument, the audiometer, became available for hearing tests.

Using the audiometer, Fletcher determined the sensitivity versus frequency characteristics (audiograms) of a large number of subjects who were thought to have normal hearing. He produced a pair of curves, which were first published in September 1923, showing the threshold of audibility and the threshold of feeling for the average human ear, as related to frequency[10]. The curves have great significance in the reproduction of speech and music.

Bell Telephone Laboratories sent one of their audiometers to IWE: it was passed to Blumlein and Kaye for assessment. They determined the ear characteristics of 'everyone [they] could find' and found the characteristic for an average ear.

'From that Blumlein then did something that was absolutely typical. He calculated a network that would respond [in the same way] to the average ear. . . .
The great feature was this, and this is what he was after. All measurements in
those days were done primarily by listening [using] a pair of head phones. But
with one of these things [the electrical analogue of the ear] on a cable system
which was being tested [an engineer] could check miles and miles of cable . . .
and do measurements which gave much more accurate results. I don't know
how far it was used but that was the principle and of course the principle is used
to this day in many fields.'

On their roles in the partnership Kaye has commented:

'I might mention that Blumlein had the ideas, he thought of things, he had a
different outlook. My pleasure in life was to make them work — to construct
the bread-board of the circuit, to check it, to do the measurements and things
of that nature. . . . We were complementary. He had the brains and I had patience to test these things out, although that's not denying that he hadn't got the
patience either; he would keep at it. But I had no mathematical outlook, I was
largely experimental. It was . . . the day of the experimental physicist and
mathematics hadn't really come into the telephone engineering world in a big
way.'

In the intervals between testing and assessing various pieces of
equipment Blumlein and Kaye attempted to inject a 'little lightness' into
Connaught House. (See Figure 2.2.)

'Blumlein was a natural clown and I am told I was. . . . Blumlein found he
could play the opening bars of Pier Gynt on the oscillator. But we always came
to a gap [in the music] and we never got over that gap. He would be working on
the opening bars and I would be trying to cover the gap, but I didn't have three
hands and couldn't quite get the frequency jump that was required.
  'Also [we tried out] one or two experiments in dynamics. The labs and the
offices were all equipped with those desk chairs which rotate and, unlike the
modern ones, these would really spin. You could adjust the height with a great
big screw thing at the bottom. So we would lower [the seat], then persuade
some innocent victim to come and take part in the experiment. The victim
would be persuaded to sit in the chair, stretch his legs out forward and his arms
out sideways as far as they would go. Then we would give him a spin. We would
spin it very hard. And he [the victim] had been instructed that at a signal from
us he should draw in his arms and legs quickly; and when he did that he had
not realised what would happen of course. The speed of rotation accelerates
tremendously. I forget which law of dynamics covers the situation — [it is the
conservation of angular momentum] — but the speed [increase] was quite
dramatic. And in at least one case the occupant of the chair was flung out and
landed across the labs. That was quite interesting!'

Figure 2.2   Connaught House staff, c. 1924–29. A D Blumlein is the third person from the left in the back row

Source: Mr T B D Terroni

Not all of Blumlein's and Kaye's tests followed a predictable course of action. On one occasion they were very close to being embarrassed by causing damage to a valuable resistance bridge. During their work on determining the sensitivity of people's ears Blumlein and Kaye had a requirement for some resistors having accurate resistance values of only a few milliohms. For this purpose they borrowed a Post Office type Wheatstone bridge, which was under the control of E K Sandeman. The bridge was connected to the laboratory low voltage supply of 24 V and work commenced on the manufacture of the required resistors. It seems that these comprised lengths of copper wire which could have their resistances altered slightly by filing. However, the bridge set-up was insufficiently accurate.

'Blumlein pondered the problem and said:

' "More volts, that's what we need". So we slapped on another 12 V, pressed the bar button—[a switch]—and there was a loud "poof".... Then smoke started to emerge from this expensive and celebrated bridge. We hastily disconnected everything, and carried it out through the swing doors onto the fire escape. There was a light-well at the back of Connaught House and the other half of the light-well was occupied by the Air Ministry. So, they had a vision of two very young engineers emerging through the fire escape doors with a very expensive looking box that was puffing out smoke. Well, they were a bit shaken.

'We let everything cool down and then we unscrewed the lid. There was a funny smell coming out of it as well as smoke by that time. The resistors were covered with what looked like carbon; it had been fabric textile insulation. But luckily, they were the resistors against which we were balancing a resistance of milli-ohms and they [too] had very low resistances, which meant that they were constructed of pretty good thick wires. We could hardly believe it when we had cleaned the bridge up inside, and then quietly taken it away from observation and measured the standards in the bridge, by means of another bridge of equivalent accuracy, and found it hadn't been damaged at all. We put the lid on and screwed the screws down and that was it. It caught us out badly at that time.'

Although Blumlein commenced working at the International Western Electric Company in September 1924 he did not apply, either personally or jointly, for a patent until November 1927. Whether he wrote any reports on the various projects on which he was engaged with Kaye is not known.

Clearly, Blumlein and Kaye were receiving a good training and experience in various aspects of telephone engineering but, initially, it would appear that their disparate projects were of a fairly minor nature. Kaye's recollections make reference to tests carried out by them on equipment which originated in the Bell Telephone Laboratories, but the recollections do not include any description of major projects executed by Blumlein and Kaye. It may be that they prepared reports on their tasks and assessments which were then sent to the Bell Labs. and the findings incorporated in later equipment designs of the Labs.

An interesting example of a possible technology transfer concerns the type of capacity microphone which was invented by Wente in 1917[11]. Bell Labs. sent such a microphone to IWE and as usual Blumlein and Kaye were asked to test it. In use, the stretched diaphragm of the microphone, onto which sound waves impinge, vibrates and causes the capacitance between a fixed electrode and the diaphragm to vary. If the microphone is included in a series circuit comprising a high resistance and a d.c. supply, the changes of capacitance produce voltage changes across the resistance. In the 1920s these changes had to be amplified by a valve amplifier. Blumlein immediately observed that a cable could not be employed to connect the microphone to the high input impedance of the first amplifier stage. Ideally, he reasoned, the first stage should be housed in the microphone casing, and the stage used to convert its high input impedance to a low output impedance for use with the necessary cabling, without causing any undue loss of signal or distortion.

> 'So what we did was to use a Western Electric peanut valve. These [1.4 V] valves were only about one and a half inches long and probably just half an inch in diameter. They had a barium coated cathode, directly heated, of course, as distinct from the bright emitters. They had a very small glow, and they were very good indeed. . . . We mounted one of these inside the microphone housing, and as the valves were microphonic we suspended them on ordinary rubber bands . . . just the ordinary rubber bands found in the office drawer. . . . And that was it, and it worked. It was battery fed. . . . As far as I know it was used by the BBC on the Birmingham transmitter for some time. Of course, it had a very good top [high frequency] response. That was a very good example of Blumlein's approach to problems. He seemed able to get right down to the basic root of a problem. Nobody had ever said get the grid of the first valve right up against the microphone. In these days [the 1970s] it sounds silly that no one had thought of that.'

Interestingly page 183 of 'A history of engineering and science in the Bell system' shows the capacitance microphone and associated amplifier, about 1928, housed together. Whether this was based on the work of Blumlein and Kaye is not stated.

In February 1925 Blumlein was transferred from his post in the transmission laboratory and joined Dr John Collard's group, which was concerned with interference tests and acceptance tests on long distance inter-city telephone lines in continental Europe. It would seem that the period, September 1924 to February 1925, had been a probationary or training period for Blumlein. He had with Kaye tested and assessed—and made some minor contributions to—apparatus which had been sent over to IWE by Western Electric and later Bell Telephone Laboratories, but now he was to engage in specialist work which would occupy his intellect for at least three years.

On Collard and Blumlein, a former member[12] of the company has stated that: 'There is no doubt that in Collard and Blumlein we had two greats in the profession who were competent, complementary and in complete harmony'.

After joining Collard's group, Blumlein was very much concerned with the investigation and prediction of noise interference in telephone lines, caused by the power lines and the overhead conductors of electric railways, since in some European countries, particularly in Switzerland, the telephone cables were laid alongside or were parallel to the power lines. Part of this investigation entailed measurements of earth resistances. These were taken at Pickering in Yorkshire and at Herne Bay and so embraced two types of soil. For this purpose a search coil was designed and constructed. By moving the coil to different distances from the interfering source, and by comparing the signal induced in it with that from a standard noise unit (a buzzer of American design), the effect of the unwanted signal on speech communication over a telephone line could be predicted. This search coil test data was then processed by means of Blumlein's 'funny functions' and 'queer functions' (families of graphs) to determine the minimum distance, from the power source, for laying a cable[12] (Figure 2.3).

Throughout his working life, Blumlein had a penchant for inventing names, not only for functions, but also for pieces of equipment, and units. While at ISEC, Kaye and Blumlein encountered a member of staff who was 'very slow'. 'He was one of the old brigade [and] was so slow at getting any action going that it was almost unbelievable. For some reasons he had been nicknamed Beelzebub—quite a decent bloke [really]. So we had a unit called the 'beelze' [which was the] ratio of the time it took Beelzebub to do a job to the time it took any normal [person] in the office to do it. But it was such an enormous unit [that] for all practical purposes we all had to work in decibeelzes.'

When Blumlein was transferred to Collard's section, Kaye, who was now in the loading coil section, continued to see his close friend. They nearly always lunched together when Blumlein was not abroad and, since they usually had some spare time before returning to Connaught House for the afternoon session, began to take an interest in pistol shooting. Kaye had found a shooting gallery, in Villiers Street (just off the Strand by Charing Cross station), with a 'wonderful selection' of hand weapons.

Figure 2.3   In 1927, Collard (standing outside the car) and Blumlein carried out search-coil tests on the level of interference in Switzerland, following the route of power lines and electric railways

Source: Mr B J Benzimra

Villiers Street had a rather notorious reputation at night, because of the transactions which were conducted there, but fortunately the street's notoriety did not extend to the daylight hours. The shooting gallery itself was located underneath an arch of Charing Cross station. Various firearms were available, including Smith and Wesson, Colt, and Webley pistols. Both Kaye and Blumlein thoroughly enjoyed the sport — 'it was great fun'. Usually they practised target shooting with .22 Colt automatics because the ammunition was much cheaper than that for .38 or .45 weapons. Blumlein became 'really proficient'. Several years later, about 1937, when travelling by liner to the USA he won a pistol shooting competition and was awarded a cup.

Blumlein joined International Western Electric at an opportune time because major developments were taking place in the provision of European long distance telephony. Prior to 1924 the expansion of the long distance telephone system on the Continent of Europe had been on a slower scale compared to that of the United States. The prime reason for this state of affairs was the radical difference in the political conditions of the two geographical regions. Long distance communications across the national boundaries were dependent upon the provision of a proper organisation and the adoption of certain practices. More particularly, it was necessary to standardise the methods of construction, upkeep and operation of the international circuits and to ensure their security.

In the July 1921 issue of *Annales des Postes Télégraphes et Téléphones* a paper[13] written by Monsieur M G Martin was published entitled 'Long distance telephony in Europe'. Martin made the suggestion that an association should be established by the various European telecommunications administrations for the purpose of constructing, operating and maintaining long international telephone circuits. A similar proposal was put forward, independently, by Mr F Gill of the International Western Electric Company (later the ISEC) in his 1922 Presidential Address[14] to the Institution of Electrical Engineers (UK) when he spoke on 'Electrical communication—telephony over considerable distances'.

As a result of these pronouncements Monsieur P Laffont, the Under-Secretary of State for Posts and Telegraphs in the French government, convened a meeting, early in 1923, in Paris, of the representatives of the telephone administrations of Belgium, France, Great Britain, Italy, Spain and Switzerland. Their discussions ranged over the whole ground of the problem and a number of recommendations were adopted unanimously and unreservedly. These recommendations[15] were later approved by the administrations of the six countries and the stage was set for practical progress. The recommendations were grouped under six headings, namely:

1 Administration
2 Transmission–engineering construction
3 Engineering maintenance, including removal of faults
4 Traffic
5 Programme of immediate additional construction work
6 Preliminary programme of further construction work

On administration, the delegates agreed to the formation of a permanent 'International Consulting Committee for International Telephone Connections' and further recommended that:

1 The greatest attention should be given to the selection, to the installation, and to the maintenance of apparatus and installations used for the equipment of trunk circuits intended for long distance international telephony.
2 Telephone administrations should equip themselves with the necessary measuring apparatus required for the proper supervision and for the proper maintenance of the installation.

Blumlein would make major contributions in both these areas of activity. From early 1925 to early 1928 he spent much time on the

Continent of Europe carrying out interference and acceptance tests on long distance telephone lines. A brief chronicle of his activities for this period is as follows:

1 February 1925, engaged on the acceptance tests of the Lausanne–Geneva cable;
2 June 1925, placed in charge of a small group of engineers making interference tests on the open-wire lines in France which it was planned to use in connection with the new Paris–Madrid circuit;
3 July 1925, engaged on the acceptance tests of the Berne–Lausanne cable;
4 December 1925, in charge of the group carrying out the acceptance tests of the Winterthur–St Gallen cable;
5 January 1926, in charge of the interference tests which were carried out in Wales with Post Office engineers;
6 May 1926, at Euston railway station during the General Strike;
7 September 1926 to March 1927, placed in command of a special group to investigate and find remedial measures for crosstalk in loading coil equipment;
8 December 1926, responsible for the completion tests of the Zurich–St Gallen telephone cable;
9 April 1927, supervised a new method of manufacturing loading coils;
10 July 1927, in charge of the theoretical and mathematical development of the search coil testing theory;
11 September 1927, responsible for the completion tests on the Winterthur–Shaffhausen cable;
12 Remainder of 1927, in charge of the engineers on the Altdorf cable, where special balancing tests had to be carried out to reduce the very severe interference experienced by this cable circuit; also in charge of the group testing the St Gallen–Oberst cable, and the Winterthur–Schaffhausen cable;
13 November 1927, testing new method of cable jointing in Switzerland;
14 December 1927, responsible for the completion tests of the St Gallen–Oberst cable;
15 February 1928, responsible for the theoretical side of the company's interference investigation;
16 1928 to end of employment with the company, in charge of a group dealing with one of the company's more confidential problems.

For the European telephone network it was intended that telephone speech signals should be transmitted, without frequency translation, by quadded cables and terminal equipment arranged so that independent signals could be sent over the side and phantom circuits of the telephone multiplex system.

The first proposal for a telephone multiplexing system was made by F Jacob in 1882. He put forward the idea, illustrated in Figure 2.4, whereby using a Wheatstone bridge arrangement an additional telephone signal C could be superposed on a pair of circuits A′ and B′, known as the side circuits, without the additional signal interfering with the signals A and B

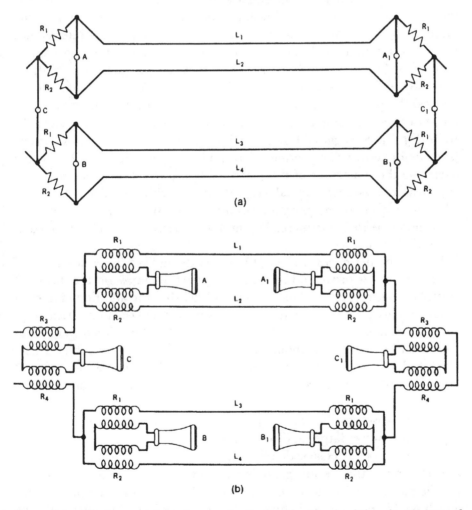

(a)

(b)

Figure 2.4   Development of the phantom circuit: (a) phantom circuit based on concepts of F Jacob, intended for the simultaneous operation of three telephone circuits over two metallic circuits; (b) phantom circuit (scheme of J J Carty) using induction or repeating coils instead of resistors. A, B, and C are telephone sets

Source: 'A History of Engineering and Science in the Bell System,' M D Fagen (Ed.) (Bell Telephone Laboratories, Inc., 1975)

carried by the side circuits. Effectively the two conductors of circuit A′ and the two conductors of circuit B′ act as the two conductors of another circuit, called the phantom circuit, which can be used for the transmission of an extra signal.

Referring to Figure 2.4, the current from C divides into two equal parts by means of the resistance bridge and each part is transmitted over one wire of the pair used for A. A similar arrangement is used to combine the two components at the far end and to provide a return path over the pair of wires employed for B. Since the phantom currents flow in the same direction over the two wires of the pair, there is no mutual interference between the phantom circuit and the side circuit. Thus, the two pairs of wires of the quadded cable can actually carry three independent signals.

Jacob's use of resistors for the derivation of the phantom circuit allowed the phantom signal current to be switched into two equal parts, but introduced considerable attenuation into the transmission system. In 1886, J J Carty suggested a solution based on the use of transformers (sometimes known as repeating coils) in place of the resistors. His scheme overcame the major loss of the Jacob configuration, but difficulties were experienced in attempts to make repeating coils with satisfactory balances. Thus the concept of the phantom circuit, though simple, was not immediately implemented commercially, and for many years it remained a scientific curiosity.

During the period 1873–1901, O Heaviside had published a series[16] of papers in the *Electrician* and in the *Philosophical Magazine* on the theory of electrical transmission and had shown in 1893 that the addition of a series inductance to a transmission line could lead to a decrease in its attenuation constant. Stone, as early as 1894, had proposed the utilisation of additional continuously distributed inductance for the same purpose, but it was Campbell and Pupin who, independently, in 1899 advanced the notion of lumped loading, with series connected loading coils, as being more practical[17].

The coils used in the early experiments were of the air-core, solenoid type. They were large and unsatisfactory for commercial telephone practice because, as Heaviside remarked, 'inductance coils have resistance as well, and if this be too great the remedy is worse than the disease'. Furthermore, because of their size, the coils established magnetic fields which could interfere with other telephone circuits. By April 1901 the toroidal, or doughnut-shaped, coil had been developed. This has the property that the magnetic field produced is mostly confined within the core material. Initially the coils were wound on cores made up of many miles of very fine, lacquer-insulated iron wire, but by about 1916 this type of core had been replaced by a powered iron core. This was manufactured from insulated iron granules which were compressed into toroidal rings. Later cores consisted of special heat-treated iron alloys, 'the permalloys', having larger permeabilities.

The development work on loading coils led to improvements in the construction of repeating coils and the application of phantoming. However, its widespread utilisation did not come about until coils could be made for loading the phantom circuits as well as the side circuits. This was accomplished in 1910 with a set of three coils, one for each side circuit and one for the phantom. By 1911 the basic problems of loading and of phantoming both open-wire lines and underground cables had been solved, though refinements were needed to achieve a high degree of balance (Figure 2.5).

It was soon realised that the application of loading could yield large economies on long distance cable circuits, and by the end of 1907 about 60 000 loading coils had been installed on about 86 000 miles of cable in the US telephone network. Further expansion led to the use of about 1 250 000 coils on 1.6 million miles of cable circuits and 250 000 miles of open-wire lines in the Bell system.

With the growth of the long distance lines in Europe, following the establishment of the International Consulting Committee for International Telephone Connections, it was essential that good telephone engineering practices should be implemented. Loading and phantoming were adopted, and the star-quad cable was extensively used. This cable had the plane of the two wires of one side circuit at right angles to the plane of the wires of the other side circuit. Provided this configuration was rigorously

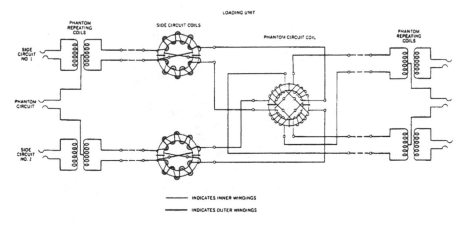

Figure 2.5 *Bell System standard method of loading phantom circuits and their side circuits*

Source: 'A History of Engineering and Science in the Bell System', M D Fagen (Ed.) (Bell Telephone Laboratories, Inc., 1975)

maintained there was theoretically no coupling between the pairs. In practice the cables had to be balanced.

With any telephone circuit it is, of course, essential that the speech signals transmitted by a given circuit should not be heard in adjacent circuits; that is, the 'crosstalk' between telephone circuits should ideally be zero. Crosstalk between telephone circuits in close proximity is due mainly to differences in the resistances of the two wires of a pair and to electric and magnetic inductive effects[18]. The former effect is important only in connection with crosstalk between side circuits and the associated phantom and, whereas for non-loaded open-wire circuits the two inductive effects are approximately of equal importance, in adjacent cable circuits the electric inductive effect predominates. An accurate indication of the magnitude of this effect may be readily obtained from capacity unbalance measurements[19].

The distributed capacitances of a quad (excluding the direct capacitances between the two wires of a pair), may be represented as shown in Figure 2.6a. This circuit arrangement can be transformed to give the configuration of Figure 2.6b from which, using the Wheatstone bridge principle, it is apparent that crosstalk will not occur from one pair of coils,

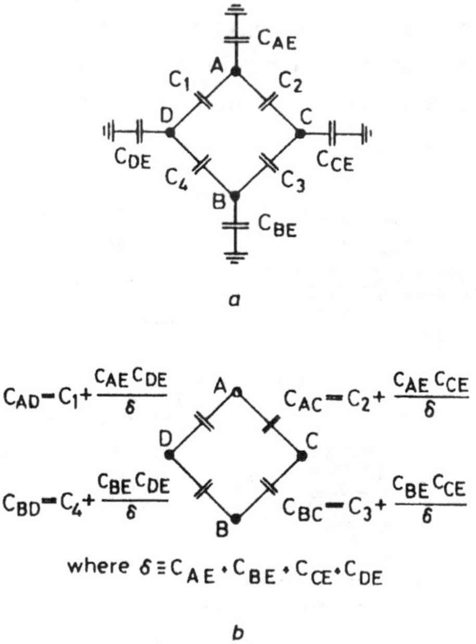

a

$$C_{AD} = C_1 + \frac{C_{AE}C_{DE}}{\delta} \qquad C_{AC} = C_2 + \frac{C_{AE}C_{CE}}{\delta}$$

$$C_{BD} = C_4 + \frac{C_{BE}C_{DE}}{\delta} \qquad C_{BC} = C_3 + \frac{C_{BE}C_{CE}}{\delta}$$

where $\delta \equiv C_{AE} \cdot C_{BE} \cdot C_{CE} \cdot C_{DE}$

b

Figure 2.6   (a) *The distributed capacitances associated with a quad; (b) the equivalent circuit*

Source: the author

say AB, to the other pair, CD, when C_{AD}/C_{AC} is equal to C_{BD}/C_{BC} , and *mutatis mutandis*, for the elimination of crosstalk between a side circuit and the phantom circuit. Hence the measurement in the field of the capacitances of a quad, and its subsequent balancing, was of great impact in the commissioning of the long telephone lines of the European network.

As noted earlier, from September 1926 to March 1927 Blumlein was in charge of a group investigating, and finding remedial measures for reducing, crosstalk in loaded telephone circuits. His first patent[20], dated 1st March 1927, with J P Johns, a loading coil engineer, was based on this investigation. They accurately analysed the problem of balancing the windings of a loading coil and put forward proposals for the design of a new coil, which was used for many years. Blumlein also devised the manufacturing methods for the construction of the new low crosstalk coils and evolved a new bridge for determining capacitance. According to P E Erikson, the European Chief Engineer of ISEC (after F Gill retired): 'This work has borne fruit in all our loading business and our standard loading-coil equipment is now being manufactured in accordance with these principles' (i.e. those contained in the patent). The patent made a 'very great contribution' to long distance cable telephony. 'Before the invention, the loading coils were the major source of crosstalk but I [Kaye] think I am correct in saying that afterwards they made the smallest contribution.'

On Blumlein, Erikson once said to Kaye: 'You've got to get up very early in the morning indeed to catch that young man out'.

The number of loading coils utilised by ISEC is not known, but in the United States of America in 1974 the number of loading coils manufactured was in the order of 14.5 million. In Kaye's view the 3-coil loading unit was 'really a tremendous step forward'. He thought it was the greatest advance in the field of loading coils which had taken place since the publication of the Fondiller and Shaw patent, which covered the loading of the phantom circuits, as well as the side circuits, in a long distance quad cable.

E L E Pawley[21] was one of the engineers of ISEC who was engaged in work on long distance lines. He joined the Corporation in 1926. 'My first assignment was to carry out tests on telephone cables in Switzerland. Blumlein came out to take part in the tests on the Winterthur–Schaffhausen cable, because it was the first to incorporate his invention for reducing cross talk by transposing the connections of the windings of the loading coils so as partially to cancel out capacity unbalances between them. The tests took over a month and included 10,000 measurements of cross talk—a most tedious process because each measurement required the adjustment of an attenuator until two complex tones were judged to be equally loud.... During the whole month, Blumlein concentrated on calculations—even doing sums on the tablecloth during meals. Our only relaxation was an evening at the Tonhalle in Zurich to hear a recital by Paderewski. Some Swiss people were amused because Blumlein spoke no German in spite of his name.'

Melling also worked in Switzerland with Blumlein[22]. 'I have in my files a 35 page memorandum [written by Blumlein and dated] 30th December 1927 on "Balancing to sheath". [It was] the product of a hard winter's work at Goschenen, on the Altdorf–Gotthard cable, the main telephone and rail route over the Alps into Italy and a classic test route for telephone interference, under very severe climatic conditions even in this Alpine approach to the Gotthard tunnel. I learnt here at first hand from Swiss engineers of all grades in the Cortaillod and government organisations what they thought of this brilliant young English scientist engineer. They were almost as much impressed by his physical determination and endurance as by his technical genius and tenacity.'

ISEC certainly appreciated the importance of Blumlein and John's contribution and gave a bonus to each of them[23]. Blumlein was only 23 years of age at the time. His future wife has recalled: '... [the company] knew he wanted [a car] as he always used to go around the countryside on his motor bike, Cuthbert, much to his mother's disgust. So they gave him £200 to £250 and he went out immediately and bought a second hand car, but he could'nt drive. He told the people from whom he bought the car to take him to Regent's Park. [There] he drove [the car] round and round ... until he could drive it home to Hampstead ...' The car was a bull-nosed Morris open tourer.

The Morris gave Blumlein much joy. As with anything practical, he was keen to carry out experiments under unusual conditions. On one occasion he told Kaye he had discovered an interesting way of skidding. Using his method, Blumlein claimed he could get around the Marble Arch island quicker than anyone else. The technique, which only worked on a wet surface, was to position the off-side wheels in the gutter of the island and then accelerate to as high a speed as possible without the wheels losing their contact with the depressed surface of the gutter.

> 'He took me out to demonstrate it around Russell Square. It was quite fascinating [but] was a bit nerve-racking for the passenger. I gather that one wet night he was using this technique in Wigmore Street — I think he was [trying] to turn [into] one of the side streets. Anyway he became detached: he broke away, went across the road and [got] stuck in the plate glass window'.

Blumlein's cornering technique was not the only aspect of his motoring which could unnerve his passengers. While driving he would sometimes 'sketch out' a circuit on the windscreen, with his finger. During daylight hours this mannerism did not cause much alarm, but in the dark, late at night after a long and tiring day in the lab, some apprehension would be felt as to whether careful attention was being paid to the road conditions ahead[24].

It seems that Blumlein was a skilful driver:

'He was coming into the office one morning — [travelling] down the hill that leads from Hampstead towards Camden Town – and was cruising down at pleasant pace [when] suddenly he saw a wheel pass him, a motor car wheel. He was still carrying on normally and said: "Huh, somebody's lost a wheel, what fun." This wheel went careering on and to his horror he looked in the mirror and saw that there was no other car on the road behind him that could possibly have shed a wheel. He came to the logical conclusion that the wheel had come off his bull-nose Morris, which it had. It shot down the road, struck the curb, went straight up in the air and landed in someone's garden. Meanwhile Blumlein with some considerable skill — he could think very, very quickly on all occasions — brought that car to rest by careful braking and steering.'

Fun was never far from Blumlein's mind. One morning Kaye and Blumlein had set off for the office and were travelling along the Thames Embankment when Blumlein mentioned that he needed some petrol. 'Suddenly a policeman stepped out and [made] the usual statement of those days: "You have been timed over the measured furlong and you have exceeded the legitimate speed. You travelled at the speed of [so much]." And the reply the policeman got from Blumlein, with a broad grin [on his face] was: "If I hadn't run out of petrol in the middle of that furlong I would have done it much quicker."

The year when Blumlein was given increased responsibility, 1926, was the year of the General Strike[25]. The dispute which led to the strike commenced in the spring when the government withdrew cash subsidies paid to the coal industry. Mine owners, anxious to safeguard profits, reduced the miners' wages and so precipitated a bitter argument. On 3rd May the Trades Union Congress agreed to support the miners, in what appeared to be an unresolvable conflict of interest, and workers in the iron and steel industries were called out, together with employees engaged in the press and transport sections of industry. The General Strike resulted in a national paralysis and volunteers were invited to maintain some essential services such as railways and telephone communications.

Blumlein and Kaye, D L B Lithgow and J S Low of ISEC responded to the call and offered their services. They reported to an office which had been set up in the Savoy Hotel and said they were prepared to work in the field of communications. Following acceptance, the four engineers notified the head of the engineering department in ISEC of their intentions. Gill was not pleased, and told the four that ISEC was an international company and that consequently they had no nationality and their only loyalty was to the company. Furthermore, he intimated that they should not expect to have a job waiting for them after the strike was settled. Gill was not threatening the four with the sack: rather, since the outcome of the strike and its effect on industry was uncertain, he wanted to make the point that if the four engaged in some voluntary activity, any redundancy would be as a consequence of their actions.

The four ignored Gill's advice and were directed to keep the telephone exchange at Euston station in working order. They worked 12-hour shifts from 8 o'clock in the evening to 8 o'clock the following morning when young female staff took over the operation of the switchboards. The exchange was not large, about four or five full-size boards, and had not been well maintained. Some of the cord circuits were broken and the four effected repairs. Towards the end of the strike 'the switchboards in the exchange were working at a level of efficiency, technically, that had never previously been achieved'. At times the telephone exchange was very busy and every calling light would be on. The four also had to tend to the telephones in the offices of the 'top brass', and as a result of the four's activities Euston station's telephone exchange was fully operational for the duration of the strike.

The strike ended when a compromise deal was negotiated, under severe pressure from the government and the TUC, between the mine owners and the miners. Initially the miners had refused to accept the deal, but after the TUC withdrew its support on 12th May the General Strike was over.

When the four returned to ISEC they were 'received with open arms'. The railway company had told the firm what had been done and it 'was tickled pink' because of course there was the prospect of it receiving lucrative railway signalling equipment contracts. The four young engineers were handsomely rewarded for their efforts. They were paid wages by both the railway company and ISEC, they received a bonus from the railway company, and in addition the company presented each of the four with a medal.

During the 1920s ISEC employed four different types of capacitance unbalance measuring sets[26]:

1 American 1—A capacity unbalance set which allowed phantom-to-side, and side-to-side unbalances to be measured;
2 American 2—A capacity unbalance set which was similar to the 1-A set;

3 EMS 476—a capacity unbalance set which enabled side-to-side, phantom-to-side, phantom-to-earth, and side-to-earth unbalances to be determined; and

4 the London capacity unbalance set, which was similar to the EMS 476 set.

About September 1924 various modifications were made to the EMS 476 set, because of unsatisfactory reports on its behaviour when used for unbalance measurements on the Paris–Strasbourg cable and on cables in Switzerland, and the improved set was known as the EMS 813 capacity unbalance set. Later in 1926 a new American capacity unbalance set, the 3-A, became available, but this was not adopted as the standard set for capacity unbalance measurements in Europe as phantom-to-earth and side-to-earth measurements could not be measured and these were required.

All of these bridges were the subject of criticism, from time to time, by telephone engineers working in the field, and in July 1928, the company decided to develop a set which would supersede all the current sets and which would be 'at least equal to any of them in any particular detail and which would in general be superior to any of them'.[26]

ISEC's decision seems to have been initiated by the 35-page report, dated 30th December 1927, on 'Balancing to sheath on the Altdorf–Goeschenen cable', which Blumlein wrote following his work at Amsteg. He noted that 'the object of balancing to sheath [was] to reduce those unbalances which produce noise in the cable'.[27]

As with all his work, Blumlein carefully considered the fundamentals underlying the problem and observed: 'As soon as the three unbalances which cause noise are determined, and apparatus made to measure these unbalances, then anything which reduces the unbalances as measured will reduce the noise'. He proposed a radical, but simple, solution to the measurement problem. The resistor ratio arms of the standard type of Wheatstone bridge would be replaced by a centre-tapped coil. Blumlein wrote:

'The [centre-tapped coil known as a] retard coil is so arranged that for the generator currents the windings are opposing so that the voltage drop in the windings is practically only due to the resistance of the windings which is very small compared to the impedance of the [capacitances] in the arms of the bridge, viz: the direct [capacitance] of wires 1 and 2 to layers (2) and (1). The voltage drop through the coil will then be very small compared to the total voltage, i.e. the conductors being measured will be sensibly at the same potential as the other wires of the quad, the remainder of the layer and the sheath.

Capacities between the wires being measured and the others joined to the mid-point of the retard coil will not affect the balance.

'The windings of the retard coil will, as viewed from the telephone, be connected in series aiding and so have a very high impedance, and hence not seriously shunt the telephone. It is important that the retard coil should be very highly balanced, the impedance of the two halves being balanced to the same accuracy as the ratio arms of an ordinary bridge used for capacity unbalances.'

In these few sentences Blumlein outlined a new form of impedance bridge which was to revolutionise the measurement of the primary electrical parameters of resistance, inductance and capacitance. Prior to Blumlein's conception most, if not all, electrical bridges were of the Wheatstone form. With this bridge, the accurate measurement of resistance, inductance and capacitance can be carried out only if the bridge is augmented by an additional pair of impedances, Z_A and Z_B, connected in a 'Wagner earth' arrangement. However, the bridge then becomes tedious to use since two 'balances' must be obtained. Furthermore:

1 Z_A and Z_B must be adjusted with every change of earth capacitance, i.e. with every measurement when a series of different impedances is being measured;

2 Z_A and Z_B must, in general, be changed if the frequency is changed; and

3 since the two balances are mutually dependent; convergence may not always be rapid.

For the tests which Blumlein made on the Altdorf–Goeschenen cable no suitable retard coil was available, so he had to utilise resistive ratio arms, as in standard practice. These usually comprised 2 kΩ resistors but Blumlein replaced them by 400 Ω resistors in order to reduce the measurement errors. This caused the sensitivity of the bridge to be 'seriously reduced owing to the large shunting effect on the receiver of these low resistance arms', so he concluded: 'The use then of a retard coil in place of low resistance ratio arms would appear desirable'. He suggested that a suitable bridge for this purpose could be obtained by replacing the resistance ratio arms of a standard type EMS 813 capacity unbalance set by a highly balanced retard coil.

Blumlein's ideas for a new capacitance set based on the transformer's ratio arm bridge were adopted by the company. The set was engineered by Melling[28] in the 1920s and was field tested, in very variable weather conditions, for a period of four weeks on the Zurich–Chiasso cable in Switzerland. The tests were successful and the new set was put into production and coded the 74100-A capacity unbalance set. Figure 2.7 taken from

Blumlein's patent no. 323 037 dated 13th September 1928, shows the basic circuit diagram of the bridge.

The patent makes clear that Blumlein fully appreciated the merits of transformer ratio arm bridges, for it mentions three of the most important advantages of such bridges. First, stray capacitances between 10 and 9, or 10 and 11 (Figure 2.7) have a negligible effect on the balance of the bridge because the potentials of the nodes 9, 10, and 11 are substantially the same.

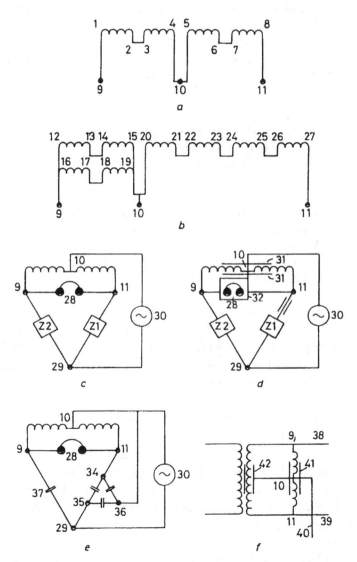

Figure 2.7   *Illustrations from Blumlein's patent no. 323 037 of 13th September 1928*

Source: British patent no. 323 037

Secondly, the bridge can measure the direct capacitance between two conductors in the presence of a third conductor. And thirdly, the ratio of the transformer arms can be quoted to a much higher accuracy than the ratio of bridge equipment in most electrical engineering laboratories. As Blumlein observed[29]:

'In bridges [of the Wheatstone type] it has been found impossible to obtain an absolute accuracy in the ratio of [the] resistance arms, and this defect is especially important when using ratio arms for measuring small differences of impedance, i.e. unbalances.'

At an IEE discussion meeting, exhibition and lecture on 'The life and work of A D Blumlein' held on 26th October 1992, A E Newman, opined[30]:

'In the early 1940s [at EMI] we had some Blumlein bridges, and they were far away the best bridges that we ever made. . . . They probably still are the best. . . . They were fast, accurate, and very, very easy to use.
   'It so happened that at the National Physical Laboratory, after quite a time, I [was appointed] Deputy Superintendent of the Applied Physics Division. When I [looked at] one of the parts of that [the part dealing with electrical measurements], and [saw] what was happening I thought that what we really wanted was a Blumlein bridge. So I saw to it that we had a Blumlein bridge, and that Blumlein bridge was just about the cheapest, most accurate and easiest to use bridge that there ever was.'

On Blumlein's personality in the second half of the decade 1920–30, Collard[31] found him to be somewhat raw at first: he was 'a difficult person to take to meetings, especially in other countries. He could be gauche and he could be very rude. He had no patience with anyone less brilliant than himself. He would work all hours himself and was quite irritated with anyone else who, for personal reasons or just for pleasure, would not join him. In the course of time, however, he lost some of his aggressiveness and acquired a certain amount of tact.'

Melling remembered Blumlein as 'enthusiastic under any circumstances'; 'nothing could stop him when he got going', he had 'no interest in hours of work'. He was 'sociable, so clever, and could come to any problem'. 'He was indeed a genius, one of the rarer ones with the ability and guts to follow up, utilise and exploit the product of that astonishing technical brain of his.'

Kaye's recollections of Blumlein during the 1920s are amusing, interesting and enlightening, since they provide an invaluable insight into the personality of a man who, by any definition, was of genius rating:

'We knew one another very well because, at times, when his mother was away, he would come to stay in my digs. We used to go out to the pictures together, dine together and so on and so forth. We used to go for a meal in Soho and we were always discussing things. Funnily enough we never touched on the subject of young women, girls and that sort of thing. Sex had no place in our friendship, or even in our lives at that time as far as I can recall. We were just a couple of young engineers, enthusiastic, meeting up with all sorts of problems, fun, and things like that.

'He was very fond of dancing, we both were. I remember he used to go to the Chelsea Arts Ball. Somehow it didn't attract me. [On one occasion, when he was staying in my digs and had been to the Chelsea Arts Ball], I got up in the morning and padded down to his bedroom. I banged on the door to make sure he was awake, and he was groaning. He said: "I do feel ill, I do feel terrible." Well, of course, he was suffering a bit of a hangover, and although I never knew him [to be] the worse for liquor; I think he was just whacked out and had had a merry evening and that was it. So I [replied]: "Well I think you had better stay in bed." [Blumlein] didn't feel like getting up [so] I said: "Right ho, leave this to me and I'll tell your boss". At that time [his boss] was John Collard, who always seemed to me a rather sad type of individual.

'I went trotting along to the office and went and saw Collard. I said: "I'm sorry but Blumlein won't be coming in this morning, he's a bit off colour." "Oh dear," [remarked] Collard, "I'm sorry about that. What is the trouble?" I [answered]: "He looked very pale to me. I think he's suffering from a stomach upset." "Oh dear, and he's in digs," said Collard. Now Collard never lived in digs, as far as I can [recollect], and therefore he had great sympathy for those of us who had to [reside] in digs and, as he thought, rough it. So I laid it on. Collard [asked]: "Is he having a doctor?" I said: "I think a little rest will do him good." I elaborated on this theme for a bit, and [somewhat] overdid it since Collard [expressed an interest in visiting Blumlein]. I persuaded him not to go. I had [just] finished saying my piece [when] suddenly the door opened and in walked Blumlein as cheerful as a lark, grinning all over his face. I slunk out — I'd done my act.

'Occasionally and in the evening, when we were both in the same digs, we would go off to the Hammersmith skating rink. I think it was the Palais-de-Dance; it was a big ice-rink then. I would be gliding along as gracefully as was consistent with my elementary talents in that respect, [when] suddenly there would be a fearful belt in [my] back which either landed me flat on the ice or I'd go fizzing off in another direction. It was Blumlein experimenting with the laws of motion. He [would tell] me that, [because] I was travelling at a [given] speed, if he hit me at a certain speed I should go in a certain direction. He'd be tickled pink if I did.

'Another thing we would do — it was really a waste of time — [was] to go and watch the dirt track racing at Stamford Bridge. On a fine Saturday afternoon we might go down to Brooklands and watch the motor racing, and see all the well-known people of that period belting around the track. It was a very nice way of spending an afternoon. We'd see the characters, both the cars . . . and the drivers were characters, and there was really very little commercialisation in it in those days.

'The cinema films [he] enjoyed most were Walt Disneys. It was an absolute delight when, I think it was the New Gallery Cinema in Regent's Street, laid on a full programme of Walt Disney. The result was we both had to—at least I had to, he intended to—sit through the programme twice.'

When Kaye, in the 1970s, recorded his recollections he intentionally mentioned many facets of Blumlein's personality and character because he was of the view that some people felt Blumlein 'wasn't human'. But as Kaye endeavoured to point out, in an anecdotal way, Blumlein was 'very human', and 'very kind' indeed.

Kaye was a keen radio amateur and loved working with his hands. Blumlein did not share Kaye's leisure pursuit, but was always ready to assist others who did. 'He was never too clever to be above that sort of thing. . . . He would never try to destroy by reasoning, mathematics, and so on, what an amateur radio character was trying to do. He would try to show how he could do it better if he could, and help.'

Kaye had his own transmitter and, as radio amateurs do, engaged in sending messages to fellow 'hams'. One of the persons he communicated with was Walter Crowe, who had been to the same school as Blumlein. Crowe subsequently lost his sight and told Kaye that Blumlein had taught him—when he was blind—to construct a wireless receiver. 'The only thing Blumlein couldn't teach him was how to solder. It says a lot for Blumlein's patience in doing a thing like that, [but] it was typical of the man.'

In 1928 Blumlein's leisure interests extended to flying. It would seem reasonable to suppose that this interest stemmed from the proximity of ISEC's Hendon site to a London aerodrome.

When the company was owned by Western Electric the principal manufacturing plants were at Woolwich and Bexley Heath, but some of the working conditions, particularly at the Woolwich site, were primitive. There was a need to acquire further accommodation. During the early 1920s Western Electric was a successful member of the engineering industry, its sales were rising and it had the financial resources to benefit from the difficulties of others.

In 1921 the firm of J Taylor and Sons Ltd, which during the First World War had commenced developing a site in New Southgate, North London, for the manufacture of lorry engines, went into liquidation. For £20 000, Western Electric in 1922 purchased from the Official Receiver the rural 27-acre site and its two-storey concrete building. The patent department of International Western Electric was transferred from Antwerp to the newly acquired building and provision was made here for the manufacture of telephone equipment for the Post Office. It was soon determined that the existing building was too small for the immediate needs of the company and so, in 1925 and 1926, short leases were taken on properties at Hendon owned by an unsuccessful aircraft constructor, the Grahame–White Company, and by London Aerodrome[32]. With a total floor space of about 400 000 square feet the single-storey buildings were sufficiently large to accommodate the manufacturing facilities of the Bexley Heath site.

Following the change of ownership from IWE to ISEC, the expansion of the European long distance cable telephone network, and the growth of international review communications, it was essential, for the continued prosperity and growth of the new company, that it should be in the forefront of technical innovation. Prior to the sale, Western Electric had had access to the expertise of the Bell Telephone Laboratories, but now there was an imperative requirement for the company to establish and develop its own laboratory resources. Behn, the new owner, was well aware of the success and prestige which would accrue to his company if it advanced its technology capability. He established two laboratories, one in Hendon and one in Paris. Effectively, in the main, the Hendon laboratory would concentrate on R&D relating to cables and associated apparatuses, and the French laboratory would deal with advances affecting exchanges and radio equipment, but the two places would work together. Behn was not concerned with the minutiae of the work which his predominantly British and French teams undertook; what he wanted were spectacular achievements and he wanted them quickly. It was to be his boast that outside the USA he had the largest R&D organisation in the world dealing solely with electrical communications[33].

At Hendon the laboratories were housed in the premises of the former London Country Club; these were across the road from the manufacturing plant (Figure 2.8). The first-floor bedrooms of the club became executive offices, the ballroom was converted into a laboratory and the oval flower bed outside the front of the building became the centrepiece of an unofficial speed track for the young staff showing off their sports cars. Possibly Blumlein demonstrated his cornering technique when the track was wet. It would seem that the new occupants enjoyed a lifestyle which was not too discordant with that of the former club members.

About 1925 the Air Ministry was keen to encourage the formation of light aeroplane clubs throughout the country because it realised that the clubs could help aviation by assisting the country to establish the large reserves of pilots which the country needed in times of uncertainty. To

Figure 2.8   The Hendon offices and laboratories of ISEC

Source: Historical Archive, Nortel Technology

further its aim, the Air Council requested the Royal Aero Club to organise the scheme for the London district. A committee was appointed, comprising Lieutenant Colonel Sir Frances K McClean, Wing Commander T O' B Hubbard, Major R H Mayo, Colonel the Master of Sempill, and Captain C B Wilson. Later the London Aeroplane Club was officially opened by Sir Philip Sassoon, the Under-Secretary of State for Air, on 19th August 1925. It was the first light aircraft club in the country[34].

The club's headquarters were based at the Stag Lane, Edgware, aerodrome of the de Havilland Aircraft Company, which was a short distance away from the Hendon Laboratories. Initially the club had two de Havilland Tiger Moth aircraft, which were bought with the £2000 grant for equipment given by the industry. Two instructors were appointed with Air Ministry approval, the chief instructor being Mr F G M Sparks and the other Mr G T Witcombe. Membership of the club was open to British subjects only, and consisted of ordinary members and associate members who were charged three pounds three shillings per annum and one pound one shilling per annum respectively for their subscriptions. The higher fee entitled the members to take instruction in flying, and for this the charge was one pound ten shillings per flying hour. Remarkably, this included the cost of instruction, oil and petrol, possible damage to the aircraft and third-party insurance. Single-seater aircraft could be hired for an all-in price of one pound per flying hour. Associate members were permitted to hire aeroplanes for passenger flying at the rate of one pound ten shillings per flying hour[35]. Demand for flights was brisk and within a month of its inauguration club members had completed 100 hours' flying, and the club had given instruction to 60 persons. Among these were several who had completed seven hours' flying and were keen to obtain their aviator's certificate.

As the membership rapidly increased, the necessity for more aircraft became urgent. The Duke of Sutherland presented the club with the third Moth and the fourth was purchased by the members themselves. The Moths were reliable and economical, their petrol and oil consumptions being just 3.5 gallons per hour and about 1 pint per hour respectively[36].

Examination of the published news items of the club shows that Blumlein received flying instruction on three occasions in May 1928, from Captain S L F Barbe and Mr F R Matthews. Further instruction was given in September and early October 1928, and during the second week of September Blumlein flew solo for the first time. The following week he successfully completed the required tests for his aviator's certificate: this he received on 15th October 1928 (Figure 2.9). Presumably, during the period between May and September Blumlein was working on the Continent of Europe. The records show that he was learning to fly at the same time as Miss Amy Johnson, who in 1930 became the first woman to fly solo from London to Australia. Indeed, three of the entries mention Blumlein and Miss Johnson as having been given training by the same instructor, so it is possible that they knew each other.

Stag Lane aerodrome was particularly conveniently situated for Blumlein for two reasons. First, since he was living at 20 Lyncroft Gardens, Hampstead, London NW6 he could easily travel on the Hampstead–Golders Green railway to Burnt Oak station, which was a few minutes' walk from the airfield. Secondly, as an employee of IECS working in the headquarters building, Connaught House, Aldwych, London he had a need, from time to time, to visit the Hendon Labs. which were adjacent to the London Aeroplane Club.

'So, on a Thursday afternoon, not once but possibly two or three times, the man in charge of Blumlein and [Kaye] would say: "Where is Blumlein this afternoon?" [Kaye would reply]: "Oh, Sir, he is up at Hendon." —which was truthful and descriptive. Blumlein was up, but he was well above the labs. not down in them!" We got away with it every time'.

Figure 2.9   A D Blumlein when he obtained his pilot's licence

Source: The RAF Museum, Hendon

Blumlein's flying experience was not completely uneventful. On one occasion — the first time he was flying a plane with slotted wings — he was approaching the landing strip and had reduced his speed.

'[He] forgot the slotted wings would do something to his progress towards the ground because, normally, if he dropped to the speed at which he was making his approach he would have spun in. But the slots prevented him spinning so that he was floating down, rather in a horizontal position, until he saw a game of tennis rather close to him down below. He opened the throttle wide, put the nose up — fortunately not too far — and did a hop and landed right in the middle of Hendon aerodrome. He told me he didn't roll at all, [the plane] just sat down. It flapped its wings a bit and he got out unscathed. Of course, [the aircraft fitters] did take the plane to pieces because they couldn't believe it could take such a bump without overstressing something.'

As with anything mechanical, Blumlein was eager to find out the characteristics and limitations of the aircraft he flew. He had a theory that the de Havilland Moths were stable — not metastable, but absolutely stable — in flight. According to Kaye, Blumlein would take off, fly to a height of several thousand feet to give himself airspace, put the aircraft into a very awkward position, such as the climb or an inversion, and then take his hands and feet off the joystick and the rudderbar. His various experiments in the mechanics of flight convinced him the machine had been designed to enable this stability to be achieved. And of course Blumlein had great trust in anything which had been properly designed. Good design formed the keystone of everything he tackled and throughout his life he took considerable pains to instil in others the importance of good design. Experimentally determined values had to agree, within fairly close limits, with the theoretically deduced values, and if they did not his staff could expect 'trouble'.

In March 1929, after having been with the IWE and the ISEC for nearly five years, Blumlein obtained employment with the Columbia Graphophone Company. The reason for this change is given in the next chapter.

Technical notes

Blumlein's work at the International Western Electric Corporation (later the International Standard Electric Corporation and then Standard Telephone and Cables) led to eight British patents, namely:

1 291 511, March 1927 (with J B Johns), on 'Phantom loading coils for reduction of crosstalk';
2 323 037, September 1928, on 'Closely coupled inductor ratio-arm bridge';
3 334 652, June 1929, on 'Telephone loading and phantoming coils';
4 337 134, June 1929, on 'Matching coaxial submarine cable to land telephone line (the circuits used having attenuation in the common mode and low attenuation in push-pull)';
5 335 935, July 1929, on 'Submarine telephone cable carrying speech, with telegraphy signals going down the unused speech return path';
6 338 588, August 1929, on 'Circuit using variable resistors to give a fixed mutual inductor the appearance of a variable mutual inductor';
7 357 229, June 1930, on 'Double shielding to reduce interference at the ends of submarine telephone cable';
8 402 483, June 1932, on 'Shielded loading coil for telephones'.

Of these patents, those on loading coils and the closely coupled inductor ratio-arm bridge are of prime importance. Blumlein in several patents, covering a period of 12 years, described various transformations, elaborations and applications of his bridge. These patents are grouped below.

1 323 037, September 1928, the basic design;
2 334 652, June 1929, an application of the closely coupled inductor
   ratio-arm bridge technique to the design of telephone loading and
   phantoming coils;
3 461 004, July 1935, on 'Compensation of a power supply having
   reactive elements, making it appear to load as a pure resistance';
4 461 324, August 1935, on 'Arrangements to tap loads into a constant-
   impedance line along its length, with proper matching';
5 475 729, March 1936, on 'Voltage regulating system based on the
   principle outlined in patent no. 461 004';
6 581 161, January 1940, on 'Low level altimeter based on earth capacitance';
7 581 164, January 1940, on 'Improvements to closely coupled inductor
   ratio-arm bridge (using two different voltages, tapped from an autotransformer, across the impedances to be compared)';
8 587 878, June 1940, on 'Improvements to closely coupled inductor
   ratio-arm bridge';
9 541 942, June 1940, on 'A bridge for comparing low resistances with
   separate current and voltage terminals'.

Closely coupled inductors were never far from Blumlein's thoughts. As
late as 21st December 1941, just a few months before he was killed, he
produced a nine-page 'Note on transformers ratio [*sic*]' which dealt with
general 3-winding transformers. Blumlein never prepared a paper for publication in a learned society journal on closely coupled inductors, but in
January 1940 he wrote a comprehensive 30-page EMI Research
Laboratories report (No. RK/20), containing 20 figures, on 'AC bridges
with two sets of ratio arms'. This report led to an IEE paper, written by
two EMI engineers, H A M Clarke and P B Vanderlyn, on 'Double-ratio
AC bridges with inductively coupled ratio arms' (*J. IEE*, 1949, **96**, Pt III,
pp. 189–202). In this paper the authors noted: 'The material in part one of
the paper is the original work of the late A D Blumlein, BSc (Eng),
Associate Member, and the present authors have used verbatim a considerable portion of a hitherto unpublished memorandum written by him in
January 1941'. During the discussion of the paper, on 11th January 1949,
Dr L Hartshorn of the National Physical Laboratory, described Blumlein's
initial idea as 'a brilliant one' and observed:

> 'The fact that the impedance of both arms is dominated by their common
> factor, mutual inductance, also accounts for the extraordinary accuracy and
> consistency of ratio obtained with this construction. No one who has tried to
> adjust and maintain two impedances at exact equality can fail to be impressed
> by the figure quoted in the paper; coils are wound consistently with a vector

error less than one part in a million. Separate impedances of any ordinary construction will certainly not remain equal, for long, to anything like this accuracy, but with this construction, although the individual coils must change, their impedances must always change together because the dominant part of each is common to both.'

In addition to the audio frequency bridge, a bridge working on the same principles was developed under the direction of Blumlein for the measurement of impedance and frequencies up to 5 MHz. It was in use from 1935 onwards.

Part 2 of Clarke and Vanderlyn's paper described Blumlein's application of the basic principles to an audio frequency general-purpose mutual admittance bridge for the measurement of capacitance and resistance. 'In 1940 a demand arose for a measuring set for the testing and adjustment of radar equipment, AI Mark 4, 5 and 6, then under development for use in fighter aircraft. Space in such equipment was at a premium, and it was considered that a bridge that would measure the resistance and capacitance between two points in a circuit, without the need to unsolder the components which might bridge these two points to a third, would prove of advantage.'

The instrument evolved was based on Blumlein's earlier work and on the principles and methods given in UK patents no. 581 164 and no. 587 878. The new bridge was fundamentally a mutual admittance bridge but, since conductance and susceptance were not the circuit constants required by maintenance engineers, the bridge was calibrated in resistance and capacitance. In terms of these, the bridge could measure impedances between 1 Ω and 1000 Ω), the phase angles of which might be between $+5°$ and $-90°$. This range of impedances made the bridge a useful general-purpose bridge for most electronic circuit components, apart from its ability to measure under the difficult conditions mentioned above. In the form which was manufactured during the war for the Ministry of Aircraft Production, the bridge was known as Impedance Bridge Type 2, A. M. Ref. 10SB/253.

The double-ratio a.c. bridge also formed the basis of a wartime direct-reading height indicator (altimeter) for aircraft. This application was suggested by Blumlein, who also produced the original design for much of the apparatus. A number of the altimeters were modified and used in valve factories during the latter part of the war to measure inter-electrode capacitances. Both of these uses of the bridge are described in Chapter 14.

In all his work Blumlein was insistent on the use of circuits which were designable and which functioned according to their design specifications even though the properties of the circuits' valves and components might be subject to wide tolerances as a consequence of the production methods used to manufacture them. Production spreads, leakage fluxes, self-capacitances, stray fields, eddy current losses, dielectric losses, impure materials and components all had to be considered in circuit designs if the experimental validation of a particular design was to reflect its theoretical basis.

An early example of Blumlein's attention to such points is given in patent no. 338 588. The patent relates to a bridge circuit for measuring self-inductance, which incorporates a mutual inductance, M. Inductance bridges incorporating a mutual inductance had previously been proposed by Maxwell, Heaviside, Campbell, and Heydweiller. Some of their configurations employed fixed mutual inductors, but in others a variation was necessary. This variation could be obtained by changing the physical relation between the coils of the mutual inductor. However, as Blumlein observed: 'A disadvantage of this type of variable inductance for certain purposes is that the capacity between the various windings changes as their physical relation is changed.'

With another type of mutual inductor, variation of M could be achieved by altering the current through the windings. The disadvantage of this arrangement for bridge use was that 'the variation of current through the inductances [was] accompanied by a corresponding variation in the resistance of the bridge arms, necessitating a readjustment of the balance of the bridge when the mutual inductance [was] varied.'

Blumlein's solution to the measurement of inductance was to employ a toroidal coil arrangement (having static windings), in which 'variation of M was obtained through the medium of a current control network shunted across one of the windings, by means of which the current flowing in the windings [was] varied. The bridge could be made direct reading by calibrating the scale of resistor R in terms of inductance rather than resistance. This embodiment of the invention led to a valuable and widely used instrument for measurements on balanced-pair carrier cables.

From an unknown date in 1928 Blumlein was in charge of a group at STC which dealt with one of the company's 'more confidential problems'. His work at this time would appear to have been concerned with submarine telephone cables, but no information is available on this part of his career other than the three patents listed earlier.

References

1 BRUCE, R. V.: 'Bell' (Gollancz, London, 1973), pp. 486–487
2 FAGEN, M. D. (Ed.).: 'A history of engineering and science in the Bell system', (Bell Telephone Laboratories, Inc., 1975), pp. 21–23

3 Ref. 2, pp. 25–26
4 Ref. 2, p. 31
5 Ref. 2, pp. 32–33
6 KAYE, J. B.: tape recording, National Sound Archives, London
7 JOHNSON, J. B.: 'The cathode ray oscillograph', *J. Franklin Inst.*, December 1931, **212**, (6), pp. 687–717
8 Ref. 2, p. 52
9 Ref. 2, pp. 954–955
10 Ref. 2, pp. 936–939
11 Ref. 2, pp. 179–183
12 TERRONI, T. B. D.: letter to the author, 28th October 1991, personal collection
13 MARTIN, M. G.: 'Long distance telephony in Europe', *Annales des Postes Télégraphe et Téléphones*, June 1921
14 GILL, F.: 'Electrical communication—telephony over considerable distances', Presidential Address, *J. IEE*, December 1922, **61**, (313), pp. 1–5
15 Report on 'European long distance telephony', *Electrical Communications*, 1923, pp. 64–80
16 HEAVISIDE, O.: paper in *Electrician* (London), 3rd June 1887
17 Ref. 2, pp. 242–245
18 Report on 'Elementary theory of capacity, unbalance in quadded cables and of capacity unbalance sets', *Crosstalk* practices bulletin, No. 407, section 6-C, AT&T Co., 24th April 1929, pp. 1–11
19 MORRIS, A.: 'Some aspects of the electric capacity of telephone cables', *Post Office Electrical Engineers Journal*, 1927–28, **20**, pp. 43–51
20 BLUMLEIN, A. D., and JOHNS, J. P.: 'Improvements relating to loaded telephone circuitry and particularly to the reduction of cross-talk therein', British patent no. 291 511, 1st March 1927
21 PAWLEY, E. L. E.: letter to Miss S Morrison, 21st February 1984, personal collection
22 MELLING, H.: letter to the author, 13th October 1971, personal collection
23 BENZIMRA, B. J.: 'A. D. Blumlein—an electronics genius', *Electronics and Power*, June 1967, pp. 218–224. See also Ref. 22
24 NIND, E. A.: taped interview with the author, personal collection
25 MOWAT, C. L.: 'Britain between the wars 1918–1940' (Methuen, London, 1955), chapter 6
26 MELLING, H.: 'Capacity unbalance sets', Report R3111/2513/HM, International Standard Electric Corporation, 22nd February 1929, pp. 1–5, personal collection
27 BLUMLEIN, A. D.: 'Balancing to sheath on the Altdorf–Goerschenen cable', International Standard Electric Corporation, File I/42, 30th December 1927, personal collection
28 MELLING, H.: 'Re-design of capacity unbalance set', Report R 3111/EE 14345/1/HM, International Standard Electric Corporation, 3rd September 1929, pp. 1–17, personal collection
29 Ref. 20

30 Unpublished typescript of the IEE Discussion Meeting, Exhibition and Lecture on 'The life and work of A D Blumlein', 26th October 1992, personal collection
31 Quoted in Ref. 23
32 YOUNG, P.: 'Power of speech. A history of Standard Telephones and Cable 1883–1983' (George Allen and Unwin, London, 1983), p. 48
33 Ref. 32, p. 60
34 ANON.: 'The London Aeroplane Club Inauguration', *Flight*, 27th August 1925, p. 547
35 ANON.: 'The London Aeroplane Club', *Flight*, 11th June 1925, p. 353
36 ANON.: 'The London Aeroplane Club', *Flight*, 14th April 1927, p. 224

Chapter 3

Monophonic recording and reproduction

The means to transmit, to record and to reproduce speech signals were invented almost simultaneously. Alexander Graham Bell's telephone and Thomas Alva Edison's phonograph were demonstrated in 1876 and 1877 respectively. Both inventions were popularly received and by the turn of the century several companies in a number of countries had been formed to manufacture and to sell telephone apparatuses and gramophone records and players. By the 1920s worldwide sales of the units produced by the telephone and record industry exceeded many millions.

Curiously, whereas much development, based on sound scientific principles and engineering practice, was the feature of the first fifty years of telephony, the position in the gramophone industry was one of stagnation. In the early 1920s the recording of speech and music in the studio was little different from that of the 1890s. Fortunately this situation changed in 1926 when J P Maxfield and H C Harrison, of Bell Telephone Laboratories, applied the findings of telephone research to the high-quality recording and reproduction of music.

The first public description of an instrument which could visually record sound waves was given by Edmund Leon Scott in a French patent of 1857[1]. His phonautograph was demonstrated at the 1859 British Association meeting held in Aberdeen, and consisted of a horn terminated by a diaphragm to which was attached a pivoted lever carrying a single bristle. This bristle lightly contacted the surface of a strip of paper, covered in lamp black, which was wrapped around a brass drum. When this was rotated by a handle a lead screw caused the drum to move axially under the bristle. Sound waves incident on the horn vibrated the diaphragm and bristle stylus assembly, thereby causing a white tracing, on the lamp-blacked paper, of the waveform of the sound to be produced.

On 30th July 1877 Edison patented[2] several improvements in apparatus relating to telephony. He mentioned the recording of sound on a continuously moving paper tape having along its length a single inverted V-shaped ridge. Transverse indentations of this ridge could be made by means of a chisel-shaped stylus attached to a diaphragm connected to the throat end of a sound collecting horn. He also referred, in the patent, to the recording of sound by the use of indentations in a sheet of tin foil wrapped around a helically grooved cylinder or spirally grooved disc.

As with Scott's phonautograph, sound signals, for example speech signals, directed into the horn led to the stylus vibrating in sympathy with the sound waves and to the production of an indented groove, the depth of the groove being a function of the intensity of the sound waves. Reproduction of the recorded signals could be obtained by traversing the cylinder a second time. The sound output was just about intelligible.

Edison's talking machine was an instant attraction as a curiosity. About 500 machines were constructed and they were shown at fairs and exhibitions where various noises and voices were recorded and reproduced through stethoscope tubes. Edison himself soon lost interest in his phonograph—he was actively engaged in the development of electric lighting at that time—and the next major development in his machine did not take place until 1885. In that year Chichester Bell (a cousin of A G Bell) and C S Tainter invented the graphophone[3]. This was basically similar to Edison's invention, but Bell and Tainter replaced the tin foil covered cylinder by a cardboard cylinder coated with wax, and the recording and reproducing sound box was traversed along the cylinder by a lead screw. The use of wax enabled an improved sound quality to be reproduced.

Edison's inventive genius was stimulated by Bell's and Tainter's innovation and he developed a new phonograph using solid wax cylinders which could be shaved several times and so permit the same cylinder to be reused. 'Hill-and-dale' recording, in which the depth of the groove was modulated, its width remaining constant, was utilised. The playing time was increased to approximately two minutes by the use of a 'floating' self-centring stylus arrangement which enabled a finer pitch groove (100 grooves per inch) to be achieved. Although the grooved disc, as an alternative to the grooved cylinder, was mentioned in the patents of Edison and of Bell and Tainter it soon became clear that disc recording had the advantage that the master disc could be much more easily replicated than a cylinder recording. Moreover it was not necessary to mount the sound box on an accurately turned lead screw since the sound box could be swung across the record on a pivoted arm. (See Figure 3.1a.)

In 1887, Emil Berliner[4], a German migrant working in Washington DC, came to the conclusion, following a study of phonautograph traces,

WHAT WILL YOU DO

IN THE

LONG, COLD, DARK, SHIVERY EVENINGS,

WHEN YOUR HEALTH AND CONVENIENCE COMPEL YOU TO STAY

INDOORS ?

WHY!!! HAVE A PHONOGRAPH, OF COURSE.

It is the FINEST ENTERTAINER in the WORLD.

There is nothing equal to it in the whole Realm of Art.

It imitates any and every Musical Instrument, any and every natural sound, faithfully:

the **HUMAN VOICE**, the **NOISE OF THE CATARACT**, the **BOOM OF THE GUN**, the **VOICES OF BIRDS OR ANIMALS.**

From

£2 2s.

THE GREATEST MIMIC.

A Valuable Teacher of Acoustics. Most Interesting to Old or Young. A Pleasure and Charm to the Suffering, bringing to them the Brightness and Amusements of the outside World by its faithful reproductions of Operas, New Songs, Speeches, &c.

EVERY HOME WILL sooner or later have its **PHONOGRAPH** as a **NECESSITY.**

HAVE YOURS NOW ; you will enjoy it longer.

Brought within the reach of every family by Mr. Edison's last production at **£2 2s.**

Send for our Illustrated Catalogues to

EDISON - BELL CONSOLIDATED PHONOGRAPH CO., LD.,

Or to our Licensees— 39, Charing Cross Road, W.C.

EDISONIA, LD., 25 to 22, Banner Street, and City Show-Rooms, 21, Cheapside, E.C., LONDON,

Figure 3.1   (a) An early Edison advertisement.

Source: 'The British Record' (The British Phonograph Committee, London, 1959)

that the lateral modulation of a record groove would have certain advantages over the 'hill-and-dale' method of Edison. Less power would be required from the recording diaphragm, because of the constant depth of the cut, and there would be less distortion in the reproduced sound. Remarkably, the mathematical basis of Berliner's reasoning was not established until the late 1930s. Initially in his phonautograph experiments, Berliner employed a lamp-blacked glass disc on which, after a recording, the trace was 'fixed' with shellac varnish and then photo-engraved onto a

Figure 3.1   (b) An early (pre-electric) recording session in progress

Source: unknown, author's collection

zinc disc by an engraver. Later, in 1888, Berliner replaced the glass disc
by a polished zinc disc, which he covered with a very thin layer of wax by
allowing a benzene solution of beeswax to dry on the disc. After recording,
the wax was etched with chromic acid to form a zinc record ready for
copying or playing.

Commercial recording effectively dates from about 1890[5]. The
American Graphophone Company had been founded in June 1887 to manufacture the graphophone of Bell and Tainter, and in July 1888 the North
American Phonograph Company had been established, by J H Lippincott,
to handle the business and sales of Edison's phonographs. However, the
industry faced three difficulties. First, the quality of reproduction was
extremely poor; only a small part of the music frequency spectrum could
be used, since recording and reproducing heads were highly resonant.
Secondly, the wax cylinders played for a maximum time of about two
minutes, which was much too short for a vast range of music. And thirdly,
around 1890, there was no method of replicating wax cylinders. Berliner's
master discs could be replicated.

His original idea for duplicating disc records was to produce an electro-
type copy of the zinc master, but this was a slow and expensive technique.
He found, after about four years of endeavour, that a satisfactory copy of

the master could be made by electrochemical methods. This work led Berliner to conceive the idea of forming the 'negative' matrix of the original disc, from which 'positive' recordings would be obtained by a thermoplastic moulding process. Unfortunately the materials available at that time were unsuitable for the purpose. Blumlein was to experience the same difficulty in the 1930s when he was working on stereophonic recording and reproduction. Berliner at first tried rubber and then celluloid, but both of these materials had poor wear properties. He next used a substance based on shellac and found that he was able to form very satisfactory disc records with this material in a press in which the matrices were first heated and then cooled before the disc was removed. His manufacturing process was capable of producing large numbers of records from master discs recorded by the leading artists of the day. Berliner commenced pressing his seven-inch discs in 1896 in Washington following the formation, with Alfred Clark, of the Berliner Company. Clark, who had been with Edison, was 'a man of character as well as a brilliant inventor and manager'.

In 1899 he came to England to assist in the establishment of the Gramophone Company. Later he became the first Chairman of Electric and Musical Industries Ltd (EMI).

During 1896, Eldridge Johnson[6], who was fabricating the clockwork spring motors for Berliner's gramophones, introduced Berliner to the notion of recording on thick wax disc blanks cast from a soft metallic soap such as lead stearate or palmitate. In use, the surface of the disc was turned in a lathe to give an optically flat and polished surface onto which the sound signals could be recorded. This surface reproduced well the fine detail of the sound waves when it was cut with a chisel-shaped sapphire stylus. From the wax masters electrodeposited negative matrices could be prepared by dusting the wax masters with fine graphite powder to make their surfaces electrically conductive: the matrices could then be used to make shellac pressings.

By 1900, 10-inch discs were being pressed in large quantities by the National Gramophone Company (which succeeded the Berliner Gramophone Company in the USA), and by the London and Hanover branches of the same organisation. The London branch became the Gramophone and Typewriter Company Ltd (HMV) in 1900, and the Hanover branch led to the formation of the independent Deutsche Gramophon Gesellschaft. The 12-inch disc was first introduced about 1902 by the Columbia Phonograph Company, a subsidiary of the Bell and Tainter organisation, from which the English Columbia Graphophone Company was established. In 1931 this company and HMV merged to form EMI at Hayes, England.

The Hayes site, initially 11 acres of open land, was acquired by the Gramophone Company in 1907. At that time Hayes was a rapidly growing village of 2000 inhabitants. It was well placed in the region's transport network and the local canal to the London docks, the Great Western Railway and the nearby roads enabled raw materials and finished goods to be imported and exported with ease. The foundation stone of the record factory was laid by Dame Nellie Melba and soon the factory was producing both 10-inch and 12-inch records. By 1908 the output per man had exceeded the figure for the Hanover factory. The number of employees was 162 and the output was 30 000 discs per month. Growth was impressive. In just 13 years the workforce expanded to about 5000 employees and in 1923/24 the number of acoustically recorded discs was 8 875 649.

Until the early 1920s the techniques of sound recording and sound reproduction were almost wholly mechanically based. Most wax blanks were still being cut on machines, driven by weight motors, in which the horns were coupled to acoustic sound boxes. Necessarily the players and artists had to be grouped very closely in front of the horn to enable the radiated sound energy to operate the cutter. (See Figure 3.1b.)

Similarly the domestic gramophone comprised a spring-driven motor wound by hand, a stylus and sound box unit mounted on a pivoted tone arm, and a large metal or wooden horn.

With both of these systems, mechanical resonance was employed to obtain adequate cutter sensitivity and sound output. Inevitably, the frequency responses of the recording and reproducing apparatuses were far from being uniform, and much distortion of the original sound resulted. At its best, the acoustic recording process by the early 1920s was limited to a bandwidth of approximately 170–2000 Hz: the music frequency range in a concert hall is 20–20 000 Hz.

A subjective assessment of the quality of gramophone records acoustically recorded was given in the August 1923 issue of the *Gramophone*. The writer averred that:

'It [the gramophone] cannot yet render satisfactorily the full volume of an orchestra or the pure tone of the piano forte. Always the orchestra has a tinny vibration — a dwarfing of the original; nearly always the piano has many notes — particularly loud notes — resembling the banjo. . . . And finally, what is to me the greatest defect of all . . . the regulation size of the 12 inch disc imposes restrictions upon the piece of music which is to be reproduced. Too often, far too often, the music has been ruthlessly cut. . . . These defects will almost certainly be diminished in the near future. They are already less than they were a few years ago. The comparative noiselessness of the needles upon the records, the reduction of "crackle" the elimination of blare and bray, provided care is taken with the needles, are all signs of an incessant search for improvement. . . . Listening to a gramophone record used to be an irksome business. Nowadays, however, it is possible to be altogether absorbed in a fine piece of music which is being performed by means of the gramophone — to be moved by it and absorbed in it as one would be in the concert hall'.

Though only very slow progress was being made in the first quarter of the twentieth century in sound recording and sound reproduction, in telephone engineering much fundamental work on the electrical properties of circuits had been undertaken, especially by the staff of Bell Telephone Laboratories and its predecessor[7].

The growth of long distance telecommunications networks in the USA and the economic need to solve problems associated with the propagation of electromagnetic waves along transmission lines led to the development of electrical measuring instruments and to theoretical investigations of, *inter alia*, loaded lines, electric wave filters and equalisers. As Maxfield and Harrison of Bell Telephone Laboratories stated in their definitive 1926 paper[8] on 'Methods of high quality recording and reproduction of music and speech based on telephone research':

'The advance has been so great that the knowledge of electric systems has surpassed our previous engineering knowledge of mechanical wave transmission systems. The result is, therefore, that mechanical transmission systems can be designed more successfully if they are viewed as analogs of electric circuits. While there are mechanical analogs for nearly every form of electrical circuit imaginable, there is one particular class of electric circuit whose study has led to ideas of the utmost value in guiding the course of the present development.'

This was a class of circuits known as filters. The results of filter theory showed how the various resonances of a mechanical system could be coordinated so that, when a proper resistance termination was used, high efficiency and equal sensitivity were obtained over a definite band of frequencies by the elimination of the responses due to frequencies outside the band.

The practical realisation of electrical recording and reproduction stemmed from the advances in radio communication which had been achieved during the First World War. From 1918 valves could be readily purchased which enabled amplifiers and oscillators, operating over wide frequency range and at varying power outputs, to be constructed. Furthermore, the essential components of electronic circuits — resistors, capacitors, inductors and transformers — were easily obtainable.

Most early experiments in electrical disc recording made use of adapted telephone receivers as cutting heads. Two British experimenters, L Guest and H O Merriman, working privately in a London garage, fabricated the first moving coil cutting head, the design of which was based on a submarine signalling transmitter. The cutter was utilised by them in their recording, on Armistice Day 1920, of the Unknown Warrior burial service in Westminster Abbey. It was the first recording ever made from a remote pick-up, the sound being relayed over a telephone line from inside the Abbey to the recorder in an adjacent building.

In the United States of America during the autumn of 1915 the research division of the American Telephone and Telegraph Company began an extended scientific investigation of the phonograph/gramophone. Until 1919 development of the recording and reproducing processes had proceeded largely on a trial-and-error basis, though in the course of this, much empirical knowledge had been acquired. However: 'A complete theory connecting the great series of disjointed facts was still lacking. Development along empirical lines had reached its utmost and the art of sound reproduction had come practically to a standstill in its progress.' The time was ripe for the art to be replaced by science.

AT&T's research programme, under the supervision of Maxfield, was carried out in the AT&T building at 463 West Street, New York. There, trial recordings, using a private telephone line, could be made of various stage entertainments being performed at the Capital Theatre.

In designing their electromechanical systems, of the band pass type, Maxfield and Harrison stated the problem was threefold:

1 arranging the masses and compliances of the mechanical elements so that they formed filter sections;
2 determining the magnitudes of these quantities so that with or without transformers the separate sections all had the same cut-off frequency and characteristic impedance; and
3 providing the proper resistance termination.

Of these parts the last proved to be the most difficult to implement because of the lack of a satisfactory non-reactive mechanical resistance. Figure 3.2 illustrates how in a mechanical system the mechanical elements corresponding to resistance, reactance, mass and compliance become, in electrical analogues, the electrical elements associated with resistance, reactance, inductance and capacitance respectively. Also the mechanical quantities of force, velocity, displacement and impedance are analogues of the electrical quantities of voltage, current, charge and impedance. Using these analogues and the methods of filter and circuit theory, Maxfield and Harrison showed how an electromagnetic recorder and a mechanical reproducer could be designed to have much-improved frequency characteristics compared to those of earlier equipments.

Analogues

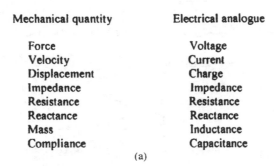

Mechanical quantity	Electrical analogue
Force	Voltage
Velocity	Current
Displacement	Charge
Impedance	Impedance
Resistance	Resistance
Reactance	Reactance
Mass	Inductance
Compliance	Capacitance

(a)

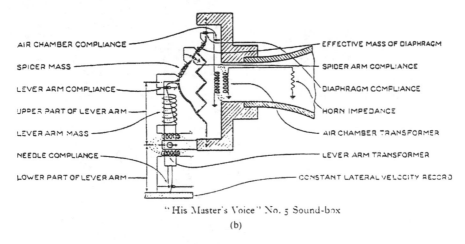

" His Master's Voice " No. 5 Sound-box

(b)

Figure 3.2   Using electrical analogues of mechanical quantities (a), an electromechanical system, such as a pick-up (b) may be modelled by an electric circuit.

Source: (a) the author, (b) the *Journal of Scientific Instruments*, 1928, pp. 35–41

In the implementation of their design of the recorder, the load consisted of a rubber rod. The cutter comprised a balanced moving iron armature system, polarised by a permanent magnet, with coils on the pole system which were connected to the output of the recording amplifier. Damping of the electromechanical system was provided by the rubber rod (nine inches in length), along which vibrations, generated by the system, travelled as torsion waves with an attenuation equivalent to that of a loaded telephone line 1500 miles in length. All the parameters of the mechanical system—the stiffnesses of the springs and the compliances of the coupling members, the masses, and the frictional resistance losses of the rubber line—were designed so that the recording velocity/frequency characteristic of the cutter was independent of frequency from 250 Hz to

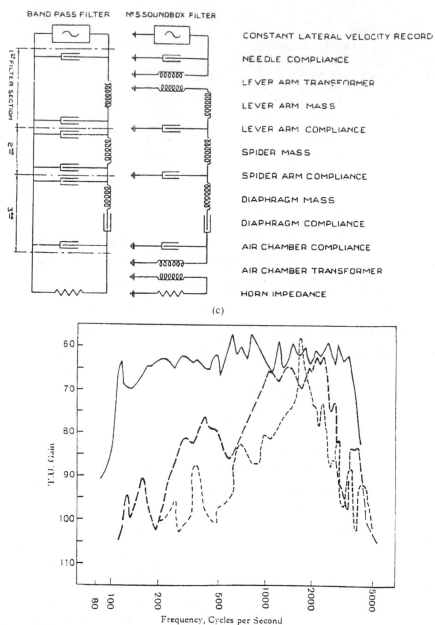

(c)

Gramophone 1897 shown - - - - Gramophone 1912 shown — ∙ —∙ — Gramophone 1928 shown —∙ ——

(d)

Figure 3.2 (continued)   *A record cutter, may be modelled by an electric circuit (c).*
Subsequent analysis and synthesis then enables an improved
system frequency response to be obtained (d)

Source: (c) and (d) the *Journal of Scientific Instruments*, 1928, pp 35–41

25 kHz, but decreased with decreasing frequency below 250 Hz. By about 1929 the Western Electric recording characteristic had been universally adopted.

Early in 1924 representatives of the Victor Talking Machine Company were invited by Western Electric to hear the new electrically recorded records and phonographs. They were unimpressed. One year later, in early February, a further demonstration was given: the sound seemed much more attractive than previously, so in March 1925 Victor and Western Electric negotiated a contract whereby Victor would have use of Western Electric's patents on electrical recording on a royalty basis[9].

On Christmas Eve 1924 Louis Sterling, the General Manager of the English Columbia Graphophone Company, obtained some test discs which had been recorded by Bell Telephone Laboratories' engineers. He was greatly dismayed. Both of the major recording companies in the UK — the Gramophone Company (HMV) and the Columbia Graphophone Company — had begun to take an interest in electrical recording systems, but neither company had engineered a cutter with the smooth response of the Western Electric recorder. For Sterling, the successful outcome of Maxfield's and Harrison's work had come too soon. The Columbia company had just completed recording, by the acoustical process, and at much expense, a very considerable number of symphonic works. With the new electrically recorded discs about to be marketed, Sterling realised that his company's investment would quickly prove to be worthless.

Sterling sailed for New York on 26th December 1924 and in January met officials of Western Electric. He learnt that the unaffiliated Columbia Phonograph Company of New York had been offered rights to Western Electric's process along with the Victor Talking Machines Company. He was told that the rights could be assigned abroad only through an American affiliate. Thus Sterling had just one course of action open to him: he bought for $2.5 million (using a loan from J P Morgan) a controlling interest in the US Columbia Company and organised it as a branch of the English Company. An agreement was reached with Western Electric and soon the Columbia Graphophone Company was using Western Electric's recording process.

The Gramophone Company's interest in electric recording dates from the end of 1923 when the Electrical Experimental Department was formed. It comprised B Mittel and G H Buckle: they were directly responsible to H L Buckle, the Works Manager. H A M Clark was employed from about February 1924, and in March/April of that year, when experimental electrical recording work began, H Hill was taken on as a draughtsman.

Investigations of electrical reproduction from records were undertaken and several pick-ups were developed. In addition a 3-valve radio receiver and a multi-range radio set were designed and built. HMV's early venture into electric recording necessitated an examination of all the available competitive commercial systems and so, at Eastertime 1925, Mittel and A Clark (of the Recording Department) visited the United States to become acquainted with the Western Electric recording process. They returned with two sets of equipment and in July 1925 the first records for sale to the general public were cut using the WE method. The Hayes experimental recording apparatus was first used at the opening of the 1925 Wembley Exhibition when the King's speech was recorded.

Subsequently the WE equipment was installed in Gloucester House, London, and in October 1925 organ recordings were made, with the aid of the Post Office's telephone lines, from St John's Church. The first public concert, of Christmas carols sung by the Choral Society at the Royal Albert Hall, was recorded in December 1920.

During 1926 six further recording sets were obtained from Western Electric and were installed in Berlin, Paris, Milan and Barcelona, and in August 1927 in India. During 1926 the electrical recording equipment installed at Gloucester House was removed and reinstalled in the Small Queen's Hall, which had been converted into a recording studio. Here the first commercial records were produced in September 1926. Towards the end of 1926 a mobile recording set had been designed and constructed and in February 1927 it was taken to Liverpool, where records were made of organ music played in Liverpool Cathedral.

The Western Electric moving iron recorder had a very good performance according to the standards of the mid-1920s. Indeed the improvement in the quality of sound reproduction from records was so great as to be almost unacceptable to some people, whose ears had become attuned over many years to the earlier records of the acoustic recording era. The correspondence in the technical press and national papers on the merits of electrical recording was so antagonistic at times that some gramophone companies queried the wisdom of adopting such a radical system.

Compton Mackenzie, the editor of the *Gramophone*, was vociferous in his editorials in condemning the 'new recordings'. In November 1925 he averred:

> 'The exaggeration of sibilants by the new method is abominable, and there is often a harshness which recalls some of the worst excesses of the past. The recording of massed strings is atrocious from an impressionistic stand point. I don't want blue-nose violins and Yankee clarinets. I don't want the piano to sound like a free-lunch counter.'

These views were supported by some of the magazines' readers, who aired their opinions in the correspondence columns.

'Mellowness and reality have given place to screaming ... peculiar and unpleasant twang ... this marvellous music is completely spoilt by the atrocious and squeaky tone ... the din is ear-splitting, a continual humming roar pervading everything'.

However, there was a simple explanation for these harsh words, as a technical correspondent pointed out.

'[A new recording] was given to me in great disgust by a friend on whose machine it sounded more like a complicated cat fight in a mustard mill than anything else I can imagine. I brought it home and put it on my own gramophone and the result overwhelmed me; it was just as if the doors of my machine were a window opening on to the great hall in which the concert was held. If it produces any less perfect result in your hands, blame your reproducing apparatus and not the record'.

Although the Western Electric recorder was a considerable advance on acoustic recorders, the electromagnetic cutter had two defects[10]. The moving iron armature operated in a magnetic field and at large deflection attitudes there was a possibility the armature would make contact with one of the pole pieces of the polarising field and 'stick', thereby putting the cutter out of action. Restoring springs, to compensate partly for this characteristic, were included in the cutter, but nevertheless the performance of the recorder at large amplitudes was not perfect. Maxfield and Harrison were aware of the possibility of this defect, but as they mentioned in their paper, there was no simple electrical analogue of the negative compliance of the magnetic field: this limitation restricted the modelling of the electromechanical system. The possibility of 'sticking' occurring due to electrical transients was always a source of anxiety to the recording engineers.

In their model Maxfield and Harrison assumed that all the mechanical elements which made up the wax cutter had linear characteristics so that straightforward linear electric circuit theory could be applied. Unfortunately the rubber transmission line was non-linear in its operation and introduced spurious resonances in the lower frequency range of the recording characteristic. Very careful adjustment of the sponge rubber packing, by which the line was supported, was necessary. This had to be manually adjusted and occupied many hours of a laboratory technician's time[10]. In 1930 Blumlein would produce a cutter design which overcame both these undesirable features.

Western Electric's cutter was quickly adopted throughout the recording industry. The first company to be licensed was the British Columbia Graphophone Company. Users, including Victor and Parlophone, had little alternative but to adopt it, and pay a royalty of between 0.875d and 1.25d on every record pressed from the masters made using the WE patents. The Western Electric system was also utilised to record large diameter 16-inch records, cut from inside to outside, for the Warner Brothers' Vitaphone sound film system.

In the UK the Gramophone Company and the Columbia Graphophone Company were each pressing about 2 million records per month by the end of the decade 1920–30 and the royalty payments to Western Electric were considerable. There was thus a strong incentive to circumvent Western Electric's patents. In the late 1920s the Columbia Graphophone Company decided to challenge the monopoly position of the Western Electric recording system.

From about 1898 until 1922 the Columbia company had been managed in the UK as a branch of an American company called the Columbia Graphophone Manufacturing Company. In 1917 the Columbia Graphophone Company Ltd was registered as an English company, its shares being owned by the American company. Five years later the whole of the shareholding of the English company was acquired by British nationals and, from 1922, it carried on its business in the UK and in many other countries as a British company with British capital. Three years later, in 1925, the British company by a curious turn of the wheel of fortune secured the whole of the issued capital of the American company.

The General Manager of the Columbia Graphophone Company was Mr Isaac Shoenberg. By the beginning of 1929 he felt that Columbia's recording developments were not sufficiently successful and decided to recruit a first-class telephone engineer to work on a new recording system which would be free from Western Electric patents. In England, in 1929, the premier telephone company, with excellent R&D facilities, was Standard Telephone and Cables, which had previously been the International Standard Electric Corporation. It was likely that STC would be a suitable company from which to recruit such a person.

Now I L Tumbull, who briefly had been Blumlein's laboratory partner at the City and Guilds Engineering College, had shared a room at ISEC with Blumlein, Dr J Collard and Mr E K Sandemann, before transferring to the Columbia Company. Turnbull[11] was asked by Shoenberg if he had 'any information about anybody who would be capable of developing a new recorder' which would be free of payments to Western Electric. Some (unknown) person had given Shoenberg three names—E K Sandemann, A D Blumlein, and Carter (of Collard's group)—and he (Shoenberg) was keen to have opinions of their abilities.

Sandemann appeared to be a suitable candidate. When electric recording commenced, he went to the United States and was given training at Bell Telephone Laboratories on the Western Electric system. On his return to ISEC Sandemann assumed responsibility for servicing the Western Electric recording equipment which both Columbia and HMV were using. Turnbull was asked to assist with this work and so he was introduced to the activities of the Columbia Company. He was offered a position with the company in 1928. Sandemann was senior to Blumlein and Carter and he was offered the post at Columbia. However, he was not interested, and Tumbull was requested to say who was the best engineer of the remaining two possible candidates. He said Blumlein was undoubtedly a brilliant man and was told by Shoenberg to write to Blumlein and suggest that he should meet W S Purser, who was the recording manager at Columbia's headquarters in Petty France, London.

Blumlein, who knew of Dutton's work on electrical recording when he was at the City and Guilds Engineering College, subsequently applied and was successful. E A Nind[12], who also was with the Columbia Company at this time, has said that two applications, from Carter and Blumlein, were received. The applications were passed to him for comment and he, recollecting Blumlein's ability from his college days, recommended Blumlein.

In his application Blumlein stated[13]. 'I believe my ability lies more in a sound understanding of physical and engineering principals [*sic*] than in the knowledge of telephone engineering obtained by meticulous reading.' Shoenberg offered the job to Blumlein and within a week he was working in the old Columbia recording studios situated in the Westminster district of London. Later, Shoenberg used to tell how Blumlein had said he might not still be wanted, in view of the salary he was asking, but he (Shoenberg) had riposted by offering him even more.

Blumlein's change of employment in March 1929 was timely. First, there was the exciting challenge of working in a new field of activity which seemed to have a great future. Secondly, although the 1929 Wall Street crash did not have an immediate effect on STC's staff, the no par stock fell from $149 in 1929 to a low of just over $7 in 1931, and the company's profits plummeted in that year to £36 163 from a 1930 high of £576 224. By the middle of 1931 all 420 employees at the Hendon laboratory had gone and the vacated building became the police college[14].

From 1929 until Blumlein was killed in 1942 Shoenberg was the person to whom Blumlein effectively reported. Shoenberg must be given great credit both for recognising Blumlein's genius and for providing the resources, manpower and material to enable Blumlein to realise his potential. Thus a few words of introduction to Shoenberg are apposite[15].

'Isaac Shoenberg was born of a Jewish family in 1880 in Pinsk, a small city in north-west Russia. His great ambition in life was to be a mathematician, and it was this ambition that brought him to England. But because of the difficulties facing a young Jew in Russia at that time, he had to settle for a degree in engineering; reading mathematics, mechanical engineering, and electricity at the Polytechnical Institute of Kiev University. However, he was to retain his interest in mathematics throughout his life and, about 1911 he was awarded a gold medal by his old university for mathematical work.

'After leaving university, he worked for a year or two in a chemical engineering company, but very soon he joined the Russian Marconi company and so, in the first decade of this century, became involved in wireless telegraphy. From 1905 to 1914 he was chief engineer of the Russian Wireless and Telegraph company of St Petersburg. He was responsible for the research, design and installation of the earliest radio stations in Russia.

'In 1914 he resigned his good job and emigrated to England where, in the autumn of the same year, he was admitted to the Royal College of Science, Imperial College, to work under either Whitehead or Forsythe for a higher degree in mathematics.'

The outbreak of war brought this plan to a premature end and Shoenberg had to look for a job. He found one with Marconi Wireless and Telegraph Company at the princely sum of two pounds per week. His abilities and potential were soon noted and he became Joint General Manager and Head of the Patent Department.

Professor J D Magee, in his 1970 Memorial Lecture on 'The life and work of Sir Isaac Shoenberg, 1880–1963', has told how Shoenberg joined the Columbia Company[15].

'Shoenberg had no formal training as a musician, but he had an intense natural love of music; this made him a keen connoisseur of recorded music and the technique of recording. This in turn resulted in a friendship with Sir Louis Sterling, [the chairman of Columbia] and an invitation to join the Columbia Graphophone Company and to put into practice his ideas on recording. He was clearly successful in this and soon the Columbia company was competing effectively with the formerly almost unchallenged, prestigious firm The Gramophone Company or HMV. So began the association of Sterling with Shoenberg which was, to my mind, crucial in the field of television in the following decade. The shrewd businessman and financier, Sterling, completely trusted Shoenberg, an engineer, scientist or applied physicist with a large dash of the visionary.'

Soon after Blumlein moved to Columbia he invited Turnbull[16] to have dinner with him at a restaurant in Kew Gardens. The occasion must have been a very memorable one, for Turnbull could recall, more than 40 years later, how Blumlein had discussed with great enthusiasm his proposed scheme for a new recorder. It would be based on the use of hydraulic pulsations, as in the operation of the Constantinescu machine gun which had been used at the end of the First World War. Blumlein envisaged having a static sound transmitter – diaphragm arrangement which would couple sound waves into a fluid where they would be propagated and used to control the motion of a lightweight cutter. The notion was soon abandoned as being too limited in its scope. Tumbull vividly recalled the apprehension he felt during the journey home when Blumlein 'drew' diagrams with his fingers on the windscreen of the car while he was driving — 'He put the breeze up me!', Turnbull said.

When Blumlein joined the Columbia Graphophone Company the head of the company's laboratory was P W Willans. Among his forebears was one who had been a noted mechanical engineer who had designed a high-speed steam engine known as the Willans engine. P W Willans, himself, had previously been with the Marconi Wireless Telegraph Company, where he had developed an 'ideal' transformer to give a good bandpass frequency response.

Turnbull has opined that Willans was 'brilliant', but his various schemes rarely reached the stage where they could be used in production. He had an idea for an elliptical recording studio, in which the sound source would be at one of the foci and the recording microphone at the other focus. Another idea which was developed was intended to improve the recording of piano music. During the early days of recording, the piano was particularly difficult to record. In attempting to improve the situation, Willans had a 'honeycomb contraption made up of broomsticks' which could be put on top of the piano. The idea was to break up the sound waves and give a more uniform response: however, it was not effective. Turnbull's recollections of Willans and Blumlein were that Willans's schemes were plausible but not always practical, whereas Blumlein's schemes were always eminently practical. Willans and Blumlein seem to have enjoyed a good relationship. On one occasion, a weekend, Willans invited Turnbull and Blumlein to his cottage at the Ditchling Beacon, Sussex, where Willans, being 'almost a professional pianist', played to Turnbull and Blumlein. There was 'no question of [there being] any differences between them'. Later Willans joined EMI and worked with Blumlein on the 405-line television system. Subsequently, following a disagreement with Shoenberg, he left EMI went to Bairds and then joined Phillips as a technical adviser.

Staff at the Columbia Company were engaged in furthering two major projects, namely, the moving iron recording cutter, and the high-frequency microphone. The latter comprised two concentric coils, one of which was excited from a high-frequency Colpitts oscillator, and a mica diaphragm, carrying a thin metal annulus, positioned close to the coils. Vibration of the diaphragm altered the coupling between the two coils and enabled a modulated h.f. signal to be obtained from the unexcited coil. Anode bend rectification of the induced voltage provided the audio output.

Blumlein became associated with this microphone on 1st May 1929. He designed the coils and diaphragm, and estimated the stray capacitances of the system, and another engineer calculated the mutual inductance between the coils and the effect of the diaphragm's shorted turn. Frequency response tests were conducted in August using a Western Electric Wente type capacitance microphone, calibrated by the National Physical Laboratory as a standard, and were found to be non-uniform: there was a pronounced peak at 100 Hz and a dip at 1000 Hz. Blumlein redesigned the cavity into which the coils fitted and on 5th December the h.f. microphone was again set up and further tests carried out. It had been expected that a balanced input circuit would enable the noise from the oscillator to be balanced out, but when the wanted signal was also balanced out the h.f. microphone project came to an end. The inventor of the microphone is not known.

Columbia's moving iron recording cutter was similar to that of Western Electric, with the exception that it incorporated two stacks of thin pieces of rubber for damping instead of the long lossy rubber line of the Western Electric cutter. The Columbia cutter was supplied to a Japanese recording company, called Nipponophone. It was not wholly satisfactory and Columbia's engineer in Tokyo, E A Nind, reported that the cutter tended to overload and did not give a cut proportional to the amplitude of the signal. Nind complained that low frequency signals became peaky, and that Japanese music appeared particularly difficult to record. These problems tended to be compounded by the Japanese practice of recording at a very high level[17].

Blumlein and his colleagues confirmed by tests that the cutter was very non-linear. Indeed, at 375 Hz they determined that, for a sine wave input, the cutter was recording 150% second harmonic and 100% third harmonic. Comparable figures for the Western Electric cutter were 25% second harmonic and 5% third harmonic. In Blumlein's opinion the non-linearity arose from at least two sources. First, as the moving iron armature moved from its rest position the reluctance of the magnetic circuit decreased, causing an increase in the sensitivity; secondly, the stack of thin rubber pieces used for damping was a major source of non-linearity. Various solutions were proposed and tried, but without success.

One of the difficulties in the 1920s in engineering sound recording and reproducing apparatus stemmed from the paucity of instruments available for audio frequency measurements. Suitable spectrum analysers, wave-form analysers, harmonic analysers and the like were unknown: an analysis of the waveform of a recorded sound had to be undertaken by measuring the waveform of the cut by a microscope and then performing, by hand, a Fourier analysis — altogether a very tedious task. The audio engineer's ear was the main test instrument.

None of this was satisfactory to Blumlein. His engineering philosophy was founded on good design. Trial-and-error methods had no place in Blumlein's working life. He argued that engineering systems should be designable and that the experimentally determined responses and tolerances of a piece of equipment should agree with those theoretically evaluated. Accordingly he set about improving the quality and range of the measuring equipment at Columbia. He designed a Π-line attenuator and a gain set to permit of amplitude measurements, to a fraction of a decibel; he improved the laboratory's General Radio oscillator; and he designed an inductance bridge. He initiated, in June 1930, a rigorous testing procedure for the transformers and chokes which were commonly used in inter-stage amplifier couplings; and he investigated whether the differing sound qualities from the Western Electric and Columbia cutters arose from the different phase responses by adding phase changing networks to make the two phase responses identical.

The extant laboratory files of this period illustrate Blumlein's approach to a problem. He would undertake the necessary engineering analysis, usually from first principles, and then prepare a testing procedure. He would next draw up a large sheet bearing columns for the insertion of the measured results and would pass it to a junior engineer for completion.

The difficulty of designing suitable experiments with the instruments at hand placed a substantial premium on an engineer's sound theoretical understanding of the principles of engineering. With his ability to work from first principles Blumlein was able to circumvent many problems with skill.

Turnbull, who was in charge of the acoustic arrangements in the recording studios of Columbia, experienced this facility at first hand[18]. He was having trouble recording dance music satisfactorily in the heavily damped studio at Petty France, London, because the lack of reverberation made the recording 'too dry'. In an attempt to obtain some objective rather than subjective assessment, he devised a galvanometer technique, but without success. Now Blumlein and Turnbull often went to the Albert Hall for lunch. One day during lunch Turnbull was downcast because he had not obtained the results which he had anticipated. Blumlein asked Turnbull if he was using the correct formula and Turnbull said he was employing the Sabine formula. Whereupon 'Blumlein brought out some paper and worked out (between mouthfuls of roast beef and Yorkshire pudding) a formula, pushed it across [to Turnbull] and said: "That is what it should be from first principles." Blumlein's formula was original and more complete than Sabine's and took account of damping.'

Later, according to Turnbull, staff at Bell Telephone Laboratories produced a similar formula. As Turnbull pointed out in relating this anecdote, Blumlein had no special interest in studio acoustics, but was able to solve a problem from his knowledge of the fundamentals of engineering science.

Sometimes Blumlein's gift for tackling quite difficult problems outside his own work, did not always endear him to others, and 'in some ways it led him into a little bit of trouble. People [were] not always grateful, after having worked on a problem for a couple of years, to have Blumlein say: "You are doing it all wrong and should do it like this." What made it worse was he was [usually] right.' Turnbull has recollected that: 'Blumlein got on very well with [his] colleagues and Shoenberg, although he was unpopular at times because he didn't suffer fools gladly and could see problems more easily from the outside. [He] was not always tactful in telling people they were working on the wrong lines. [He] told them direct and did not mince words.'

Blumlein's experience of the Western Electric and Columbia moving iron recorders showed him some of the basic problems which had to be overcome in making a reasonably satisfactory cutter. In particular he was aware that the good performance of the Western Electric instrument was due primarily to the effectiveness of the rubber line damping system even though it was not wholly linear in its action. Clearly Blumlein had to base his electromagnetic recording cutter on a different operating principle, to circumvent Western Electric's patents, and it had to incorporate means which would, if possible, provide linear damping. Indeed it was preferable that all the principles of operation used should have linear characteristics.

In the 1920s voltmeters, ammeters, and various other laboratory electrical measuring instruments were predominantly of the moving iron or moving coil type. Both types had been much developed from about 1880 and both types utilised some form of damping to make the motion of the indicating needle 'dead-beat'. It was a matter of general observation that moving iron instruments had non-linear scales, whereas those of moving coil meters were exceedingly linear in calibration, provided the coils moved in uniform magnetic fields.

When a current is passed through a rotatable coil immersed in a uniform magnetic field a 'back' electromotive force is induced in the coil. Since the back e.m.f is linearly proportional to the angular velocity of the coil (i.e. the rate of change of its angular displacement), and acts in opposition to the applied e.m.f., a damping torque is exerted on the coil which, if the parameters of the system are carefully chosen, causes the coil to move in a non-oscillatory way. This electromagnetic damping torque can be easily calculated, unlike the damping torque provided by the 'lossy' rubber line of the Western Electric cutter.

With these considerations in mind, Blumlein decided to design and engineer Columbia s new recording cutter according to moving coil principles. Such a cutter, it was envisaged, would be linear in its function and, most important for Blumlein, would be the product of engineering calculations.

The earliest extant file in the EMI archives on the new cutter is dated 16th October 1929[19]. In this, Blumlein gives a drawing, which might be a preliminary sketch, of the first version of his cutter design, and shows a general cross-section arrangement and five details. There are analyses of the resistance and moment of inertia of the coil; the inductance and resistance of the signal field system; the angular velocity of the coil for a 'normal recording level'; the electromechanical conversion factors; and the equivalent circuit of the recorder for a magnetic field of one tesla.

As an aside: during his analyses Blumlein was concerned with the compliance of certain moving elements. Compliance is the inverse of stiffness and was measured in dynes per centimetre. This unit did not meet with Blumlein's approval, so a new unit, the 'flab', was invented to quantify the 'flabbiness' (compliance) of the various mechanical parts.

The moving coil of a meter such as an ammeter might respond to an excitation of a few hertz. In recording music the cutter has to respond to frequencies as high as 15 kHz so the moving part must have a very small moment of inertia about its axis and therefore must be exceedingly light. Blumlein ingeniously solved this difficulty by, first, making the coil a simple closed turn machined from solid aluminium, thus producing a winding with the lowest possible resistance for a given moment of inertia about its rotational axis, and secondly, by providing a magnetic circuit (comprising a laminated nickel–iron core) to couple it to a stationary multi-turn signal coil, the current in which would induce the signal into the moving coil.

In Blumlein's first version of his cutter, probably known later as the MC1, the axis of the moving coil was horizontal and the field was provided by a single coil. Calculations, some of which were undertaken by H A M Clark, who joined the Columbia Company after Blumlein, showed that the power dissipated in the field system would be unacceptable. Several alternative arrangements were considered, and eventually a double coil system with two coils on a U-shaped core tilted upwards at 22.5° was adopted. At approximately the same time, Blumlein realised that there would be an appreciable reduction in the moment of inertia of the moving

coil assembly by changing to a vertical rotation axis. The moving coil
was pivoted upon a pair of vertical knife edges, resting on hardened steel
plates, to reduce friction and stiction to a minimum, and carried a stylus
arm of magnesium–aluminium alloy, at one end of which the cutting
stylus was attached. Steel wires in torsion held the knife edges onto the
plates and provided the necessary restoring couple; they also absorbed the
assembly's end thrust. The stylus arm was held in a slot in the lower end of
the axis rod, and to minimise subsidiary resonances of the arm it was
balanced with a small lead counterbalance. Since the high-frequency cut-
off of the cutter was primarily determined by resonance of this arm, its
resonant frequency could be adjusted by the screws which clamped the
arm, via 0.003 inch thick cork or rubber shims, to the fork at the end of the
vertical stylus shaft. The mechanical design, to which H E Holman contributed significantly, is illustrated in Figure 3.3.

Electromagnetic damping, to control the resonance of the moving
system, was implemented by terminating, via a transformer, the moving

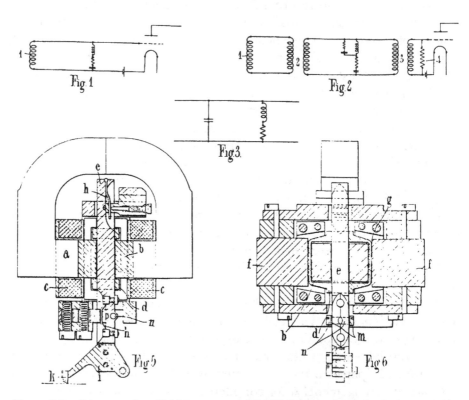

Figure 3.3   *Diagrams from British patent no. 350 998 showing the mechanical construction*
of the moving coil cutter, and the electrical equalising circuits

Source: British patent no. 350 998

coil with an external circuit, the impedance of which was such as to make the impedance of the moving coil circuit a minimum at the frequencies at which damping was required. Because this impedance usually approximated to a pure resistance, the optimum impedance for the external circuit was a very low impedance compared with that of the moving coil. However, this requirement for good damping was in conflict with the need for good power efficiency.

Blumlein resolved this matter by considering equaliser circuits which had a small impedance at frequencies approximating to resonance and a large impedance at frequencies far removed from resonance. His patent[20] (no. 350 998) with Holman on electromagnetic damping gives several circuit arrangements. Figure 3.3 shows the circuit which was incorporated in the recorder. The frequency response was from 50 Hz to 6 kHz and the damping provided was such that the mechanical impedance at the cutter stylus was much greater than that of the wax, so variations in wax temperature, groove velocity and depth of cut caused little change in performance.

To ensure the effectiveness of this form of damping it is essential that the electromagnetic coupling between the moving coil and the external shunt circuit should be as large as possible and that the polarised magnetic field should be as strong as possible. The first requirement was met by the use of the laminated nickel−iron core mentioned earlier.

'For the [second] requirement a DC excited polarised system was naturally employed and in order to keep its mass reasonable so that it could be "floated" over the wax disc with a sufficiently small moment of inertia, refinements such as the use of sectionalised windings, each having the optimum gauge for its radius, were utilised. Even so, using pure electrolytic iron, the maximum flux density obtainable of about 1.7 T, limited by saturation, proved inadequate. In a paper by an American named Yensen, reference had been made to the use of a cobalt−iron alloy with 33% cobalt, and having a saturation density of [2.4 T], but such an alloy required careful annealing in a vacuum in order to achieve usable mechanical properties. Such a material had never been made in England and the services of Messrs Darwin Limited in Sheffield were enlisted to produce such an alloy. This was eventually achieved and pole pieces working at [2.2 T] finally gave an acceptably high working flux in the gap.' (See Figure 3.4.)

To complete his recording system, Blumlein required a high-quality microphone. Because Columbia's high frequency microphone project had been abandoned and the use of Western Electric's capacitor microphones would have required the payment of royalties, it was necessary for Blumlein to design a microphone which embraced some novel and patentable features.

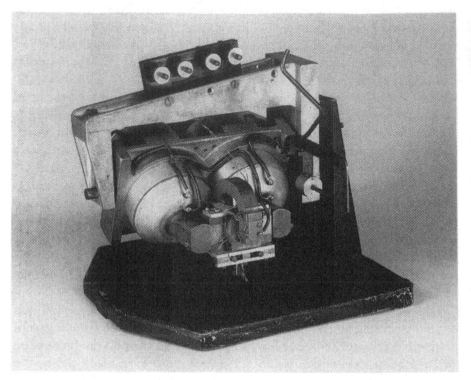

Figure 3.4   Developmental model of the Blumlein moving coil wax cutter, 1931

Source: EMI Archives

In the late 1920s various carbon, capacitor, moving coil, and crystal microphones had been developed and patented. The standard BBC studio microphone from 1926 to 1930 was the Reisz transverse current carbon microphone, which consisted of a marble block having in its front face a rectangular recess, 3 inches by 2.5 inches in area by 0.125 inches deep. The recess was filled with powdered coal, and was covered by a thin sheet of rubber. A transverse current, provided by a 6 V battery, flowed through the powder and was modulated by the sound pressure variations on the rubber diaphragm.

It would seem that the Reisz microphone was tried in the mid-1920s for gramophone recording by the Columbia Graphophone Company, but the instrument was never widely used for this purpose because of the presence of background hiss, at about 1 kHz, which was about 15 dB higher than from the noisiest capacitor microphone. Furthermore, the operating characteristic was non-linear and therefore unsuitable for studio recording. Western Electric, in 1919, endeavoured to produce a satisfactory microphone for broadcasting purposes by employing two carbon button assemblies, connected in push–pull, mounted on an air-damped steel disc having a resonant frequency of about 8 kHz. Amplitude distortion was reduced, but at the cost of a considerable reduction in sensitivity. The instruments do not appear to have been used for general studio recording.

The Western Electric Company also carried out investigations on crystal microphones, in which surface electric charges are generated on the crystals when they are subjected to stress along specific crystal axes. The pioneer of the application of this effect—the piezoelectric effect—was P Langevin. During the Great War he applied the piezoelectric effect to the design and construction of underwater sound transmitters and receivers for the detection of U-boats.

Later, in 1919, A M Nicholson, of Western Electric, developed several microphones, loudspeakers, and gramophone pick-ups which were based on piezoelectricity. Western Electric investigated whether crystal microphones and crystal headphones would be satisfactory for telephony problems, but the problems associated with the large-scale fabrication of the artificially grown Rochelle salt crystal inserts proved to be formidable.

A crystal pick-up, due to E W C Russell and A F R Cotton, was demonstrated in July 1925 at the London Gramophone Congress. They sold the rights in their invention to an American, C F Brush, whose company (Brush Laboratories) then developed, by a team led by C B Sawyer, the important 'bimorph' method of construction. By 1930, crystal microphones, pick-ups, and loudspeakers of acceptable performance could be manufactured in quantity. All the methods and apparatuses were, of course, patented.

Moving coil microphones date from 1919 when A F Sykes and H J Round developed a sensitive, if cumbrous, microphone for sound recording and broadcasting. In the Round–Sykes instrument the coil, wound as a flat helix, formed the diaphragm. This was stuck with petroleum jelly onto a wad of cotton wool, which served to dampen the diaphragm's motion, in the 6 mm gap of a cylindrical magnetic circuit. Some art, at which the early BBC technicians became especially adept, had to be acquired in deciding how much grease and cotton wool had to be used when different types of programmes were broadcast, but, remarkably, the Round–Sykes microphone was capable of a good performance.

A stretched diaphragm type of moving coil microphone was developed in 1929–30 by E C Wente and A L Thuras, of Western Electric. They used an acoustic network of cavities and slots positioned behind the diaphragm to attenuate its resonance.

Thus, when Blumlein in 1929–30 was required to devise a high-quality microphone for recording purposes, much work on carbon, capacitor, crystal and moving coil microphones had been undertaken. In particular, Western Electric, whose patents the Columbia Graphophone Company wished to avoid, had experimentally investigated all four types of sound transducer.

Blumlein based his microphone on the moving coil principle; that is, sound waves acting on a compliantly mounted diaphragm–coil assembly (the coil of which is positioned in a uniform magnetic field), lead to the generation of an e.m.f in the coil which is proportional to the velocity of the assembly. His diaphragm, the central part of which moved as a stiff piston, consisted of two circular aluminium sheets separated by a disc of balsa wood, the whole being riveted and waxed together to form a rigid structure of low mass. An extension of one of the aluminium sheets formed the surround to the diaphragm.

At first the flat extension, which had been thinned by etching, was merely clamped between the diaphragm clamping ring and the diaphragm mounting ring, but the surround's compliance was inadequate and the extension tended to be wrinkled. Two circular ridges which had being designed by Clark were next formed in it to ameliorate the difficulty. Blumlein decided that these ridges were too deep, probably because they affected the acoustic impedance of the air passages behind the diaphragm, and in the redesign four were provided. He took great care in the design of the coil and the surrounding cavity—about twenty different surrounds were made before he was satisfied.

The coil was of aluminium wire, anodised and bakelised or waxed onto the aluminium former (which had a slit to eliminate eddy currents): the former was riveted to the diaphragm. The diaphragm could be stretched by an adjustable ring so that the principal resonant frequency could be adjusted to 500 Hz. Evidence that the diaphragm–coil assembly moved as a stiff piston up to the highest operating frequency was obtained when the first modal resonance above the 500 Hz resonance was measured to be about 15 kHz. The mass of the diaphragm–coil system was just 0.7 g (Figure 3.5).

Blumlein designed the grille and the air cavity between it and the diaphragm, and the cavity behind the diaphragm, using an equivalent circuit. Since the constraint on the use of the Western Electric patent meant that the air in this cavity could not be used for damping, he included a calculated amount of cotton wool, to prevent cavity resonance, based on measurements made by Turnbull.

As with all his work, Blumlein paid great attention to points of detail. He realised that the supporting surround of the moving structure was also liable to resonate due to high-frequency sound waves incident on it, and he took steps to prevent the resonance so that the response of the microphone

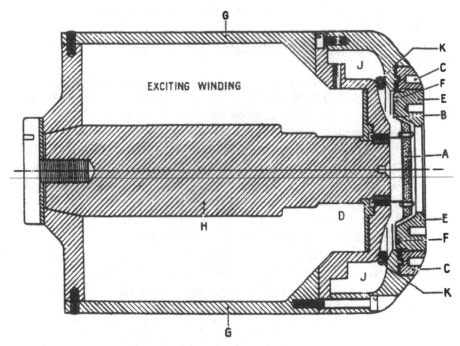

EXCITING WINDING

Figure 3.5   Schematic diagram of the Blumlein – Holman moving coil microphone

Source: *Engineering Science and Education Journal*, **2**, (3), 1993

was controlled and regular at high frequencies. Heavy air damping on the thin surround was provided by ensuring that the air gap between the back of the stretching ring and the surround was very small to prevent it from vibrating in unwanted modes.

The principal resonance at 500 Hz was damped electromagnetically, as in the cutter, because the residual air damping and eddy current damping in the coil former were insufficient for the purpose. Blumlein shunted the moving coil by a circuit which in its simplest form consisted of an inductor, a resistor and a capacitor connected in series. The circuit was designed to have a low impedance at frequencies close to the 500 Hz resonance but a high impedance at frequencies remote from resonance.

Blumlein decided that all manufactured microphones of the above construction should be utilised with only one type of equaliser circuit. This was accomplished by tuning the microphones precisely to 500 Hz and by adjusting the magnetic flux in the air gap of the magnetic circuit to a magnitude which made the damping of the standard equaliser appropriate for any individual microphone.

On 4th November 1930 the first completed microphone was ready for testing. It was eventually called the HB1A, the HB referring to Holman and Blumlein. Sometimes it was mentioned as the 'Hell's Bell's', or 'Hot and Bothered' microphone.

The first recording session with the new microphone and cutter was held on 8th December 1930 when a Mr Sparks played the piano (disc RWTT 535/1) and then spoke (disc RWTT53 5/2). Next day the van Phillips band was recorded on RWTT 536. No comments survive, but a frequency run dated 19th December showed a very good, smooth, even characteristic. Further piano and voice tests on the 19th with both the HB and the Western Electric capacitor microphones led to recordings RWTT 549/1 and RWTT 549/2; they were repeated on 23rd December, RWTT 554/1 and RWTT 554/2. Another band, the Greenings Band, was recorded on 6th January 1931; the numbers of the masters were RWTT 565/1 and RWTT 565/2.

Testing continued throughout February and March and during this period the HB1A, the HB1B (which incorporated some modifications) and the Western Electric microphones were compared. On 29th March three records were cut using the HB1B microphone and for each recording three different settings of 'brightening' were employed. The disks were coded RWTT 679/1, RWTT 679/2, and RWTT 679/3. HB1B microphones, after some additional minor developments, were used for many years both in EMI's recording studios and later in the Alexandra Palace, London, television station. Blumlein's design led to the general excellence of the piano recordings made by EMI during and after the 1930s. A later mark of the microphone, the HB1E which was in use by EMI to at least 1955, had a frequency response which extended from about 40 Hz to above 17 kHz[21].

Blumlein's achievement, with his collaborators Clark and Holman, was remarkable and outstanding. Here was a young engineer, not yet 30 years of age, with no prior experience of sound recording or sound reproduction, solving in just one year the problem which previously had engaged the attention of research staff at both the Columbia Graphophone Company and the Gramophone Company. Blumlein's success is even more note-worthy when set against that of Harrison and Maxfield of the Western Electric Company, a company which had vast resources and which employed in the early 1920s more than 3000 personnel on R&D. Blumlein not only engineered a better system, since it was based on linear operating principles, than the WE system, which was inherently non-linear, but he succeeded in his task in a shorter time. According to Nind[22] the wax cutter was extensively used before the commencement of the Second World War (Figure 3.6). It was finally replaced sometime in the 1960s by a Neumann cutter.

In addition to the work on the cutter, an excellent microphone had been developed, and everything else associated with the recording process

Figure 3.6   1928 Columbia gravity-driven wax-cutting lathe with 1932 production model Blumlein moving coil wax cutter

Source: EMI Archives

had been conceived and built in-house, from the microphone amplifiers, level controls, intermediate amplifiers and equalisers to the power amplifiers that operated the recorder. Test apparatus had been constructed and during the progression of the work the well known constant impedance volume control (the Π-line attenuator), which became a commonplace unit in professional equipment of all types, had been invented.

In the 1930s automatic frequency response curve plotters for testing amplifiers, microphones and large speakers did not exist. The testing of these systems was laborious and involved determining the output of the responses of the system at numerous 'spot' frequencies. From these responses a graph of output versus frequency could be plotted by hand. This procedure did not suit Blumlein and he decided to design and construct an automatic response plotter. For his purpose, he required a sum of £500, an appreciable portion of the weekly wages bill for the Central Research Laboratories. 'After a struggle he won the day.'[23]

'Full details of the plotter are not known, but the basis was a [General Radio] signal generator with its drive coupled to a paper tape recorder. There was an amplifier with a logarithmic output over a range of 10 dB, and a bank of relays which stepped the output of the amplifier to the next one after each 10 dB. The operating current of the relay coil was used to reset the amplifier and maintain a continuous curve on the record. All [of] this [was] in the days when log amplifiers and dB attenuators had hardly been thought of, [but Blumlein] had generated an AGC [automatic gain control] system with a range of 60 dB or more.'

The plotter was also used with a National Physical Laboratory calibrated HB microphone to measure the frequency response of some new loudspeakers which were being developed, by Dr Dutton, in the Works Design Department. He was experimenting with various cone materials and cone shapes (both circular and elliptical) to improve and extend the frequency response and performance of domestic loudspeakers. These tests were undertaken in an anechoic chamber, and typically a response test could be completed in a few minutes. Good loudspeakers were, of necessity, required for domestic record players which used electrically recorded 78 r.p.m. discs. 'As far as I can remember [M G Harker], the best compromise turned out to he an elliptical cone with an aluminium centre of about one-third the cone area, the rest being made of some kind of fabric.'[24]

Naturally, the Directors of the Columbia Graphophone Company were delighted with Blumlein's endeavours and in September 1931 they awarded him a bonus of £200, an appreciable sum for the depressed times in which it was made. (In 1931 the average weekly wage for a skilled male was £5.00.) Blumlein's letter of thanks on receiving the award, records his pleasure on learning that Holman and Clark had also been given bonuses. Since they played important roles in the recording project, a few words on their backgrounds seem appropriate.

Henry Arthur Maish Clark was born on 29th August 1906 in London and received his engineering education at Northampton Engineering College from 1920 to 1928. He graduated with a BSc (Eng), second-class honours degree in 1928 and was recruited as a research engineer by the Columbia Graphophone Company in the same year. From 1931 he was with Electric and Musical Industries and was responsible for the company's recording studio activities from about 1938; in 1953 he was appointed Technical Manager, Record Division, EMI Ltd. He became Technical Director, Records and International Division, EMI Ltd in 1957. On his early days at Columbia, Clark has recalled[25]:

'[In 1929] 1 had the good fortune to serve as AD Blumlein's assistant. . . . [Our] experiments spread into the field of moving coil microphones and cutters, the main object of which was the reduction of non-linear distortion present in all the then known moving iron devices. H E Holman was a tower of strength in all these projects as he was responsible for most of the mechanical design. In those early days I learned that the overall performance of even complicated devices could be calculated in advance before the device was made. Blumlein's wrath when the measured resonant frequency of a new model did not match the pre-calculated one is still a vivid memory.'

During his four-year degree course Clark received industrial training at Columbia under the general guidance of P W Willans and the direct supervision of Holman.

Herbert Edward Holman,[26] born on 4th April 1892, was educated at Redhill Technical School. He served a mechanical engineering apprenticeship and subsequently worked in various branches of precision engineering, toolmaking, and electrical and mechanical instruments before joining Columbia in 1924. He has recalled some of the projects which were being developed prior to Blumlein's appointment in 1929. Holman's first task was to develop a simple form of moving coil microphone. This followed contemporary practice by having a glass diaphragm to the centre of which was attached a coil having several layers of 47 SWG enamel wire, the layers being cemented together with a thin film of Bakelite varnish. According to Holman: 'It worked moderately well, but due to insufficient damping, the inherent resonances of the diaphragm were all too apparent'.

'In another part of the laboratory experiments were proceeding with a different type of moving coil microphone; this had a coil of flat spiral form attached to a diaphragm and arranged so that half of the coil moved in its own plane between the poles of an electromagnet.

'Both of these microphones, although inferior in output and frequency characteristic to the best carbon resistance types, yet possessed one great advantage in that they did not hiss. This was important to the surface-noise-conscious recording company. The recorder or cutter was similar to a watch telephone receiver, and its pivoted stylus bar was attached to the centre of its diaphragm.'

Work on these devices ceased when in 1925 Columbia began to use the Western Electric recording apparatus. Holman's labours were then concerned with the calibration of this apparatus and the adjustment and maintenance of WE's capacitor microphones. In this work he was 'most ably' assisted by Clark. Holman also engaged in the development of a moving coil recorder which 'was capable of limited use'. It utilised 'a transverse current carbon microphone in which the diaphragm was triangular in shape, in order to minimise the stiffness of the diaphragm and to reduce the tendency of the diaphragm to vibrate in localised zones'. This was patented by Purser, Holman and Clark. On Blumlein, Holman has said:

'In 1929 we were joined by A D Blumlein, a young engineer of whom I cannot speak too highly. I was closely associated with him in many projects, and his capacity for original thought, coupled with a thorough knowledge and appreciation of basic engineering principles considerably impressed us. His first important project, with Mr H A M Clark and myself, was the design and development of the Columbia moving coil microphone and recorder together with the associated amplifiers and equalising systems. The results obtained with this apparatus were extremely good. In fact, records made with this equipment were far superior to any previously made; this was largely due to the introduction of moving coil rather than moving iron devices, and, also, because the careful design of each item afforded a high degree of electromagnetic coupling which enabled equalising circuits to provide effective damping of the moving elements. Blumlein's ability as a circuit designer here came to the fore; equalisers not only had to produce an ideal response curve, but had also to provide for wide deviations by simple strapping of terminals if such were required. This was a characteristic of ADB, nothing must be left if an improvement could be made. He spared no effort himself and inspired all those who worked with him'.

Although Columbia commenced using the Blumlein recording system in 1930, it was not until July 1931, after the merger of the Columbia Graphophone Company and the Gramophone Company in April 1931 to form Electric and Musical Industries Ltd, that the system began to be employed extensively. At the time of the formation the Gramophone Company had still not fully developed any recording method which was independent of Western Electric's patents and the company perforce was obliged to pay considerable royalties to WE. Nind who, as mentioned previously, worked for Columbia and then EMI, had 'not the slightest doubt' that Blumlein's non-infringing recording method and apparatus made the amalgamation possible.

Soon after EMI was inaugurated, with Mr Alfred Clark as Chairman, and Sir Louis Sterling as Managing Director, the new recording studios in Abbey Road were opened by Sir Edward Elgar; and the principal factory for the manufacture of gramophones and radio sets was at the Hayes site. The number of employees increased to more than 15 000.

In May 1931 EMI decided to standardise, as far as possible, the recording activities of the two former companies, while allowing at the same time some flexibility so that the two constituents of EMI could retain some of their idiosyncratic recording characteristics. Accordingly, F A Dart and J White, of the Recording Division and the Product Engineering Department respectively at HMV, visited the studios at Petty France to inspect the Blumlein recording process. It received 'guarded if somewhat grudging praise' from the two representatives, with the consequence that one of the recording units was transferred to Hayes so that the HMV engineers could 'become properly familiar with it'. Subsequently the Blumlein equipment was operated daily and found to be 'in every way satisfactory'[27]. One former employee of HMV, Mr D Bicknell, reminiscing in about 1960, well remembered Blumlein's recording apparatus:

> 'Right from the start Studio 3 was pretty satisfactory. The old Western Electric system had been pretty good, except for piano recording, but Blumlein's was much smoother altogether and, with this improvement, piano recording in Abbey Road was placed far ahead of any other recording studio in the world.'

In July six sets of recorders were ordered and by the end of November 1931 the Blumlein moving coil system had been installed in the newly opened St John's Wood studio. From this time HMV commenced recording some of their discs with this system and slowly all EMI records were eventually produced by the Blumlein process. HB1B microphones were also supplied to the BBC.

The establishment of EMI led to large savings. Under the contract, dated 1st May 1925, which the Gramophone Company had negotiated with the Western Electric Company, the UK company agreed to pay royalties on all recordings made with WE's cutters. These royalties were[28]:

on the first 2 500 000 records sold	1 cent per record
on the next 2 500 000 records sold	0.75 cent per record sold
on the next 5 000 000 records sold	0.50 cent per record sold
in excess of 10 000 000 records sold	0.25 cent per record sold

The agreement was to be effective until 1st January 1936 and thereafter until terminated by either party on six months' notice.

A similar but financially more onerous agreement had been signed just eight days previously, on 22nd April 1925, by the Columbia Graphophone and Western Electric companies. Under the terms of the contract the royalties would be determined as follows:

on the first 5 000 000 records sold	1 cent per record
on the next 5 000 000 records sold	0.75 cent per record
on the next 10 000 000 records sold	0.50 cent per record
in excess of 20 000 000 records sold	0.25 cent per record

The agreement was to continue until 22nd October 1935 and thereafter until terminated by six months' notice.

In addition to these terms, both contracts stipulated that a minimum annual royalty of $25 000, free of British income tax, would be payable to Western Electric.

Table 3.1   Record sales

Sales	Records sold	Royalties payable to Western Electric
Columbia Jan. To Dec. 1931	20 170 000	£28 350
HMV Aug. 1930 to July 1931	22 600 599	£20 004
Totals	42 770 599	£48 354

The record sales of the Columbia and HMV companies when EMI was formed in 1931 are shown in Table 3.1. British income tax amounted to £2250, so the total cost of the royalty payments from the two record companies for 12 months' use of the Western Electric recording process was £50 604.

In 1932 Blumlein designed an improved cutter, the MC4. It was lighter than the MC1 which had a high power consumption and was somewhat unsuitable for location recording. The MC4 had short rubber blocks for suspension instead of the steel shaft and springs: the mass of the moving parts was reduced thereby and this led to a decrease in the power required for recording. The MC4A had an excellent frequency response — up to 11 Hz, mainly due to the lower moving mass — but the poor signal-to-noise ratio of the shellac 78 r.p.m. disc was inadequate to justify commercial exploitation.

Another design, known as the MC4B, had an even lower power rating. Though the upper frequency limit was just 7 kHz, this was considered suitable for location recording. New amplifiers based on PX25 audio output triodes instead of the DEM3, were designed by Blumlein: he used negative feedback to give the correct output impedance to match and load the new cutter.

Blumlein's achievements in monophonic recording and reproduction were recognised in 1958 by the Audio Engineering Society when it presented a posthumous citation to him. (Figure 3.7)

Technical note

Blumlein's patents on monophonic recording and reproduction comprise the following British patents:

1 350 954, March 1930 (with H E Holman), on a cutting head for gramophone recording — the mechanical arrangement of moving parts;

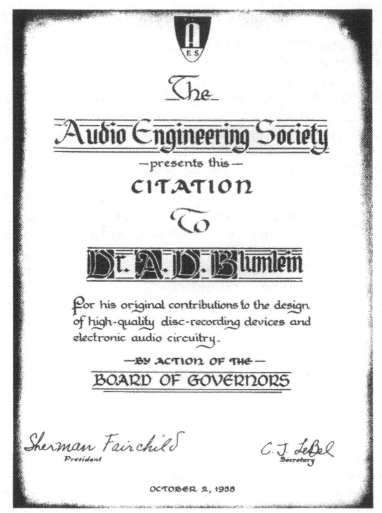

Figure 3.7   AES certificate

Source: S J L Blumlein

2 350 998, March 1930 (with H E Holman), on a cutting head for gramophone recording—the electromagnetic damping arrangement;

3 362 472, July 1930, on a constant impedance, variable attenuation network;

4 361 468, September 1930 (with P W Willans), on a gramophone pick-up—the mechanical arrangement of the parts;

5 363 627, September 1930, on a cutting head for gramophone recording—the electromagnetic damping of the main resonance combined with the mechanical damping of the minor resonances;

6 368 336, December 1930 (with H E Holman), on a gramophone pick-up—the mechanical arrangement of the parts;

7 369 063, May 1931 (with H E Holman), on a moving coil microphone with electromagnetic damping of the main resonance;

8 417 718, March 1933, on a gramophone pick-up or cutting head for recording—the mechanical arrangement of the parts;

9 425 553, September 1933 (with H A M Clark), on a negative feedback power amplifier with both current and voltage feedback;

10 496 883, June 1937, on an ultra linear amplifier circuit in which the screen grid of the valve is tapped on the output transformer to improve linearity.

Blumlein's monophonic recording and reproduction system was comprehensive. Everything from the microphone onwards was designed and constructed in-house, including the microphone amplifiers, level controls, intermediate amplifiers and equalisers to the power amplifiers that drove the recorder and, of course, the recorder itself. Blumlein's insistence on the need for good correlation between theory and practice demanded the ready availability in the R&D laboratories of appropriate apparatus for determining gains, frequency responses and the amplitudes of the harmonic distortion components. Where suitable equipment could not be obtained commercially, it was designed and built in the laboratories.

References

1 SCOTT, E. L.: French brevet no. 31 470, 25th March 1857
2 EDISON, T. A.: British patent no. 2909, 1877
3 GELAT, R.: 'The fabulous phonograph 1877–1977' (Cassell, London, 1977), p. 34
4 Ref. 3, p. 60
5 Ref. 3, p. 46
6 Ref. 3, p. 84
7 FAGEN, M. D.: 'A history of engineering and science in the Bell system' (Bell Telephone Laboratories, 1975), chapter 4
8 MAXFIELD, J. P., and HARRISON, H. C.: 'Methods of high quality recording and reproducing of music and speech based on telephone research', *Journal AIEE*, March 1926, pp. 242–253
9 History project manuscript, vol. 3, EMI Music archives
10 CLARK, H. A. M.: 'Some highlights in the history of sound recording', 14th October 1961, EMI Music archives
11 TURNBULL, I. L.: taped interview with the author, personal collection
12 NIND, E.A.: taped interview with the author, personal collection
13 Quoted in BENZIMRA, B. J.: 'A D Blumlein —an electronics genius', *Electronics and Power*, 1967, **13**, pp. 218–224

14 YOUNG, P.: 'Power of speech. A history of Standard Telephones and Cables' (George Allen and Unwin, London, 1983), p. 65
15 MCGEE, J. D.: 'The life and work of Sir Isaac Shoenberg, 1880–1963', *Royal Television Society Journal*, May/June 1971, **13**, (9), pp. 204–216
16 Ref. 11
17 BLUMLEIN, A. D.: Report No. 33, 'Columbia recorder in Japan', 25th October 1929, EMI Central Research Laboratories' archives
18 Ref. 11
19 LODGE, J. A.: 'Blumlein and audio, disc recording and microphones', unpublished paper, 1993, 12pp, personal collection
20 BLUMLEIN, A. D., and HOLMAN, H. E.: 'Improvements in apparatus for recording sounds upon wax or other like discs or blanks', British patent no. 350 998, 10th March 1930
21 Ref. 10
22 Ref. 12
23 CHARMAN, F.: letter to the author, 20th August 1991, personal collection
24 HARKER, M. G.: letter to the author, 26th July 1991, personal collection
25 BRIGGS, G. A.: 'Audio personalities' (Wharfedale Wireless Works, Bradford), pp. 72–75
26 Ref. 21, pp. 194–199
27 Ref. 9
28 Ref. 9

Chapter 4

Stereophonic recording and reproduction

The five-year period 1875–1880 was a remarkable one in the history of society. The practical inventions of the telephone (1876), the phonograph (1877), the incandescent lamp (1878) and the nascent ideas concerning the means to achieve cinematography (1876) and 'seeing by electricity' (1879) were eventually to transform the living and entertainment standards of the world's developing countries.

On 9th November 1876 Wordsworth Donisthorpe, an English barrister, filed a patent[1] for a moving picture camera. Donisthorpe's patent had for its object:

'to facilitate the taking of a succession of photographic pictures at equal intervals of time, in order to record changes taking place in or the movement of the object being photographed, and also by means of a succession of pictures so taken of any moving object to give to the eye a presentation of the object in continuous movement as it appeared when being photographed.'

At that time magazine dry-plate cameras were being introduced and Donisthorpe described a version in which plates were to be changed rapidly, the exposures taking place while they were stationary. Positives were to be printed on a long roll of paper and viewed in rapid succession. No clear idea as to how the necessary intermittent viewing was to happen was given in the patent.

Subsequently in a letter[2] to *Nature* dated 24th January 1878, following Edison's 1877 invention of the phonograph, Donisthorpe wrote:

'By combining the phonograph with the kinesigraph [Donisthorpe's invention] . . . the life-size photograph shall itself move and gesticulate . . . the words and gestures as in real life. . . .

'I think it will be admitted that by this means the drama acted by daylight or magnesium light may be recorded and re-acted on the screen or sheet of a magic lantern, and with the assistance of the phonograph the dialogues may be repeated in the very voices of the actors. . .'

Donisthorpe did not achieve success at this early date, but in the late 1880s he collaborated with W C Crofts in designing another motion picture camera and projector which were patented on 15th August 1889. Only a few frames, taken at 8 to 10 frames per second in Trafalgar Square, London, in 1890, survive but they indicate that Donisthorpe's and Croft's approach seemed to work well.

In the meantime E J Muybridge, an English photographer, had acquired some considerable fame in analysing, by sequence photography, the locomotion of animals[3]. He developed projection apparatus (initially called the zoogyroscope, and then the zoopraxiscope) which the *Photographic News* in 1882 described as 'a magic lantern run mad (with method in the madness) [by which] the animals walked, cantered, ambled, galloped and leaped over hurdles in a perfectly natural and lifelike manner'.

The publication of Muybridge's work stimulated others, including E J Marey and O Anschutz, to investigate the possibilities of sequence photography. However, it was the 'wizard of Menlo Park', Thomas Alvar Edison, who provided the ideas and the means which made cinematography practical. Muybridge had lectured in New Jersey on 25th February 1888 and on the 27th had visited Edison to discuss with him the possibility of combining the zoopraxiscope and the phonograph. Edison was intrigued with the notion and set his assistant W K L Dickson, a Scotsman who had emigrated to the USA in 1879, to work on the project. This led to the invention of the kinetophone, a machine in which the moving images were synchronised with the sounds of the phonograph. Dickson later described his task as a 'harrowing experience', but he succeeded in October 1889 in making a film of 12 seconds duration. It showed Dickson himself raising his hat and greeting Edison: it was the first snippet of a talking film. The kinetophone did not lead to great success for Edison and Dickson—it was only briefly marketed in the 1890s—but it encouraged inventors to attempt to devise apparatus which would record and reproduce both visual scenes and sound simultaneously.

In 1896 the Frenchman Auguste Baron patented a synchronised sound-film system and, in 1898, with the support of F Mesguich, he set up a studio in which he made short films, just four minutes in duration, of people singing. A single motor drove the camera and the phonograph. Further demonstrations of talkies were given at the World Fair of 1900, where a visitor to a small booth of the Phono-Cinema-Theatre could see and hear Sarah Bernhardt declaim Hamlet. Two other 'talking cinemas', the Theatroscope and the Phonorama, were opened at the turn of the century, and in 1902 the French film producer L Gaumont introduced his 'chromophone'.[4]

This was followed by Gaumont's 'chromegaphone' in 1906, the Walterdaw 'cinematophone' and Hepworth's 'vivaphone' in 1907; Jaepes's 'cinephone' and Thomassin's 'cinematophone' in 1909; and Donisthorpe's stentorphone' in 1910. In Germany, the film producer-director, O Messter, began to release all his films with recorded musical scores from 1908.

These early talkies suffered from several major defects. The synchronising of the sound and vision left much to be desired; the sound quality was poor; the actors had to position themselves very close to the horn of the phonograph because the low sensitivity of the sound recording apparatus militated against excessive movement; sometimes the horn was visible in a scene; and the films were necessarily short. Nevertheless many hundreds of such films were made between 1902 and the commencement of hostilities in 1914 and the general public felt the film had considerable entertainment value.

Contemporaneously with the various attempts to link synchronously cameras and phonographs some inventors were endeavouring to record sound directly on film. In 1902 W du B Duddell obtained a patent[5] for an instrument in which a spot of light of constant intensity was deflected across a film (running at a speed of three feet per second), by a mirror attached to an oscillograph to which the sound signals were applied. The result was a variable area sound track record. After film processing, sound reproduction could be obtained by scanning the track with a light beam which passed through the film onto a photoconductive selenium cell. Five years later E Lauste, using an electromagnetic shutter and aperture arrangement, produced a sound-on-film system in which the sound signals varied the optical density of the film.

Advancement of all the above methods was severely handicapped by the lack of valve amplifiers. However, as noted in Chapter 3, during the First World War radio communications had developed rapidly and the ready availability of valves and components after the war engendered a renewal of interest in talking films.

In 1919 de Forest, the inventor of the audion (triode valve) patented an optical sound-on-film process[6] which he called 'phonofilm', and between 1923 and 1927 he produced more than 1000 synchronised sound-on-film shorts for specially wired theatres. The films had a popular appeal, but the major Hollywood producers to whom de Forest endeavoured to sell his system were unimpressed. Talking pictures were viewed as an expensive novelty and the prospect of a good financial return seemed slight.

Soon afterwards, in 1924, Western Electric, which had perfected, as an offshoot of its electrical recording activity, a sophisticated sound-on-disc system called Vitaphone, attempted to persuade the US motion picture producers of the merits of simultaneous sound and vision. Western Electric's representatives were rebuffed, like de Forest, and all offers of licensing, with one exception, were declined. Sam Warner and his brothers, who ran a small film studio, took a gamble and bought both the Vitaphone system and the right to sublease it to other producers. Experiments began in their Brooklyn, New York, studio in July 1925 and after just one year, on 6th August 1926, the first full-length picture film— *Don Juan* — was given its debut. The lavish costume drama starred John Barrymore, was directed by A Crosland, and was provided with a musical background recorded in the New York Opera House by the New York Philharmonic Orchestra conducted by H Hadley. Each 1000 ft reel of film was provided with an accompaniment on a 16 inch, $33\frac{1}{3}$ r.p.m., lateral cut, low-noise shellac disc. The groove pitch was adjustable between 93 and 75 per inch, the groove width and depth were 0.007 in and 0.0025 in respectively, and the included angle was about 90°. Each pressing had a mass of 24 oz and was reproduced by an oil-damped pick-up using standard steel needles. The frequency response was from 50 Hz to 6.5 kHz[7].

Such was the enthusiastic response to *Don Juan* that Warner Brothers announced that all its films for 1927 would be released with synchronised musical accompaniment. Their next film *The Jazz Singer*, also directed by Crosland, with Al Jolson, in the star role, was a phenomenal success. Its popular songs, orchestral score and incidental dialogue virtually ensured the demise of the silent flicks. When the *Jazz Singer* was released in the early part of 1927 there were two 'sound stages' in Hollywood. By the end of the year there were six, and by 1929 there were 116.

Apart from Western Electric's sound-on-disc system, there were in early 1927 several sound-on-film systems which were technically superior to the Vitaphone. The rights to these were mostly owned by William Fox, President of the Fox Film Corporation. He had seen that sound offered an extra dimension in the making of films and had set about trying to corner the market. In the summer of 1926 he acquired the rights to the Case–Sponable sound-on-film system and formed the Fox–Case Corporation to make shorts under the trade name Fox–Movietone. Six months later he clandestinely purchased the US rights to the German Tri-Ergon process. In addition, to cover himself completely, Fox negotiated an agreement with Western Electric whereby each licensed the other to use its sound systems, apparatus and personnel[8].

The sound-on-film system eventually prevailed over the sound-on-disc system because it allowed both the image and the sound to be recorded simultaneously on the same medium, thereby disposing of the difficulty associated with synchronisation. The decline of the sound-on-disc system might have temporarily marked the end of Western Electric's participation in the cinema industry, but fortunately for them E C Wente, of Western Electric, in 1922 had devised and patented a light valve modulator which could easily be adapted for use as a variable density modulator for sound-on-film recording[9].

Although Warner Brothers had purchased the right to sublease the Western Electric system to other film-makers, the latter were not too keen to lease sound equipment from a direct competitor. They combined together and forced Warner Brothers to relinquish its rights to the Vitaphone system in exchange for a share in any new royalties earned. By May 1928 almost all studios, major and minor, in Hollywood were licensed by Western Electric's newly established marketing subsidiary, Electric Research Products, Inc., to utilize Western Electric equipment with the Movietone sound-on-film recording system.

When Blumlein, in 1929, joined the Columbia Graphophone Company, the Western Electric Company was well placed to benefit from its new electrical sound recording process for gramophone discs and its variable density sound-on-film recording apparatus. Their successful ventures highlight the importance of the patent system and the need for companies to engage in research and development, of the type exemplified by Columbia's innovative sound recording system, to avoid royalty payments and licence fees.

With the rapid growth of the 'talkies' from 1927, the late 1920s and early 1930s seemed opportune years to advance further the sensory experiences of cinema goers. If added realism, based on a new method of vision or sound recording and reproduction, could be given to cinema audiences, there was a prospect of financial success and stability for the successful company.

Sometime in 1930–31 Blumlein gave some thought to this matter and decided that, since the monaural variable area, and variable density sound-on-film methods were well patented, the way forward was to experiment with binaural (now called stereophonic) recording and reproduction. In his master patent, No. 394 325, dated 14th December 1931, he explicitly stated[10]:

'This invention relates to the transmission, recording and reproduction of sound and is more particularly directed to systems for recording and reproducing speech, music and other sound effects, *especially when associated with picture effects as in talking motion pictures*'. (author's italics)

Blumlein mentioned that the fundamental object of the invention was to provide a system 'whereby a true directional impression may be conveyed to a listener thus improving the illusion that the sound [was] coming, and [was] only coming, from the artist or other sound source presented to the eye'.

Prior to Blumlein's initiative, HMV in 1928–30 had developed an independent monosound-on-film method but had withdrawn it, possibly in the face of competition from RCA and Western Electric, after producing just two films. At the end of the decade both HMV and Columbia were still making many 16 inch pressings, for use in movie theatres, and it may be that these companies' associations with the film trade inspired Blumlein to consider how to improve the effectiveness of cinema sound.

The earliest patent on stereophonic reproduction was filed in 1881 by a French engineer named C Ader. His patent, titled 'Improvements of a telephone equipment in theatres', unambiguously describes his concept:

> 'The transmitters (i.e. the telephone mouthpieces) are distributed in two groups on the stage — a left and a right one. The subscriber has likewise two receivers [head phones], one of them connected to the left group the other to the right one. . . . This double listening to sound, received and transmitted by two different sets of apparatus, produces the same effect on the ear that the stereoscope produces to the eye'.

Ader's notions were tried out during the Paris Exposition of 1881, when sound signals were transmitted from the stage of the Paris Opera to the homes of subscribers[11]. However, the poor quality of the received sound led to the suspension of the service and soon afterwards the idea of stereophony became dormant. Some mention of Ader's musical telephone was made in the Paris press, not because of its stereophonic characteristic, but because the telephone permitted home listeners to 'attend' the opera dressed in their bathrobes and pantoufles. The press editorialised at length about the impropriety of such attire.

Stereophony was revived in the First World War when it appeared that the directional location of sound sources, such as gunfire and aircraft, could assist observers to determine the position of hostile targets, but it was not until the early 1930s that the principles of stereophonic recording and reproduction were examined. Blumlein, now of Electric and Musical Industries, and A Keller and his co-workers, of Bell Telephone Laboratories, were the principal investigators. Both Keller and Blumlein filed patents[12] but the *modus operandi* of their solutions to the binaural problem were entirely different.

If a vertical, two-dimensional array of many microphones (and associated apparatus), placed between a sound source and a listener A, were to be connected to an identical array of loudspeakers, the original radiated wave would be reproduced unaltered and a listener B would experience the same binaural effect as listener A. Linear, horizontal, arrays of microphones and loudspeakers would similarly give perfect binaural location, but in the horizontal plane only. For most purposes such an arrangement would be adequate, since location in a vertical plane seems to be impossible except by an inclination of a listener's head.

During the early 1930s engineers at Bell Telephone Laboratories investigated this method of binaural reproduction. Figure 4.1 shows a practical arrangement of the microphones and loudspeakers[13]. Tests were made with a sound source positioned at one of nine locations in the transmitting studio. The apparent locations established by observers in the listening studio corresponded 'fairly accurately in breadth, and to a reasonable degree in depth', when the observers were sited close to the centre line of the set-up.

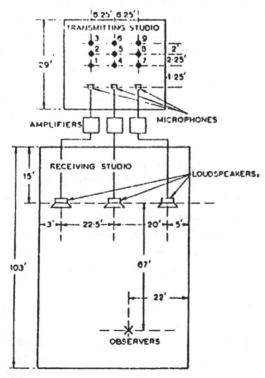

Figure 4.1   *Arrangement used by Bell Laboratories, in 1934, during their work on stereophony*

Source: *IRE Transactions on Audio*, July – August 1957

The use of the three independent amplifier and loudspeaker channels, although appropriate for public use, was uneconomical for general domestic application. For such purposes, two channels would be ideal. Investigations demonstrated that if two microphones were placed, say, 10 ft apart in front of a sound source, and the outputs were amplified and fed to two loudspeakers, also 10 ft apart, then the original sound field was not accurately reproduced: there was a predominant gap between the loudspeakers in which the sound was weak. The gap was later referred to as the 'hole in the middle'.

Blumlein's approach to stereophony was to consider the local conditions at an observer's ears. He believed that the main factors used by a listener's ear–brain system in determining the direction of a sound source were the phase differences and the intensity differences of the sounds that the observer hears, the former predominating at frequencies up to about 700 Hz and the latter prevailing at frequencies above 700 Hz. Blumlein[14] explained his reasoning as follows:

'At low frequencies the head is relatively small, so that there is little difference of pressure at the two ears, therefore phase difference constitutes the only method of determining direction. Similarly with high frequencies where the wavelength is short there would be ambiguity in the apparent direction were phase difference used; therefore, as the head forms an effective baffle for these short wavelengths, the relative intensities are the criterion of direction. For transient sounds of a high frequency the difference of arrival time of the transient is observed very similarly to the low frequency observation.'

Assuming this hypothesis, Blumlein endeavoured to devise a means to recreate the spatial distribution of sound heard by an observer sitting at some distance from a cinema screen, which essentially would be the same as that experienced by the cameraman on the film set. Thus, the electrical outputs from two microphones suitably spaced to represent human ears and positioned close to the film camera would have to be processed so that the acoustic outputs from two loudspeakers spaced on either side of a cinema screen would establish the sound field which existed at the time of film shooting.

Blumlein proved mathematically that if a listener's ears are replaced by two pressure microphones and the outputs of these microphones are amplified separately and fed to two spaced loudspeakers, which an observer faces, the resultant acoustic pressures at the observer's ears differ only slightly in magnitude but not in phase. Since phase differences at the observer's ears are necessary for determining the direction of a sound source, the above situation must be modified if a true binaural effect is to be produced.

According to Blumlein[15]:

'One method of obtaining a binaural illusion is to convert the low frequency phase differences of the pressure microphone outputs into amplitude differences. Thus an oblique low frequency sound would produce phase differences in the microphone outputs, which phase differences would be electrically converted to include amplitude differences, thus producing differences in output intensity of the two speakers. Difference of intensity of the two speaker outputs produces, for a central observer phase differences at his two ears at low frequencies, which phase differences gave the correct directional illusion. At the high frequencies it is necessary to amplify the difference of microphone outputs before conveying these to the loud speakers. The modification of microphone output described above may be called "shuffling".'

Blumlein's shuffling method (Figure 4.2) was quite brilliant in conception and ingenious in execution. First, the acoustic pressures on the two pressure microphones are converted to electric currents, which are then summed and differenced electrically (by a suitable connection of transformers) to form two new currents. These currents are in phase quadrature if the microphone currents are equal in magnitude but differ in phase. Secondly, the difference current is processed by an electrical operator which rotates the differenced current phasor by 90° so that it is in phase with the summed current phasor. This operation is in effect an integration with respect to time. Additionally, the summed current may be attenuated. Thirdly, the new summed and differenced currents are now added and subtracted (by a similar arrangement of transformers) and, after suitable amplification, are used to excite two spaced loudspeakers — the loudspeakers being positioned, for example, at the sides of a cinema screen.

It can be proven mathematically that the loudspeaker currents differ in amplitude but not in phase. Further mathematical analysis shows that the sound pressures at the left and right ears of an observer sited centrally with respect to the loudspeakers are equal in amplitude but differ in phase. Blumlein deduced the mathematical relationship between the various parameters in order to create the correct apparent position of a sound source with any given loudspeaker arrangement. Thus, as he noted, 'it is possible to convert low frequency [sound pressure] phase differences into [loudspeaker sound pressure] amplitude differences, so that the relative intensities of the loudspeaker and outputs may be made to vary in accordance with the direction of a sound arrival at the two pressure microphones'.

Blumlein extended his work to include velocity (i.e. ribbon) microphones. Such microphones have polar responses which vary as the cosine of the angle between the incident sound direction and the normal to the

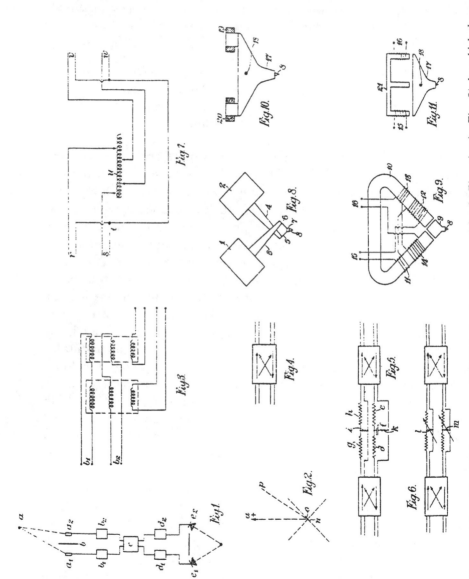

Figure 4.2   Diagrams, from Blumlein's patent no. 394 325, showing the means (the shuffling circuit, Fig. 3) by which the sum and difference signals were obtained; together with various suggestions for stereo pick-ups

ribbon strip. Consequently, intensity differences in the outputs of two velocity microphones, mounted close together with their axes pointing in different directions, can be obtained without the need to convert phase differences into amplitude differences. Shuffling is not necessary, though Blumlein thought it might be desirable to adjust the stereo image width. Blumlein's method has the advantage of being a 'field-free' system, the observer being able to move his head without producing a corresponding movement of the apparent source[16].

On 1st November 1931 Blumlein and his R&D colleagues from Columbia were transferred from the Columbia Laboratories to the recently built (1928) research building of Electric and Musical Industries Ltd at Hayes. Six weeks later, on 14th December, the provisional specification of Blumlein's famous British patent, the complete specification for which is no. 394 325, was filed. It is one of the great patents of the recording literature.

Genius, as defined by Ezra Pound, is 'the capacity to see ten things where the ordinary man sees one, and where the man of talents sees two or three, plus the ability to register that multiple perception in the material of his art'. This definition is exemplified by Blumlein's work on stereophony. His thoughts in this field were comprehensive in their scope. He considered not only the use of pressure type and velocity type microphones, as mentioned earlier, but also he dealt with the recording of stereophonic signals on cine film; he described means for recording such signals in a single groove cut in a gramophone record; he proposed the use of a material of the nature of cellulose acetate, rather than wax, for recording purposes; he outlined various designs of stereophonic pick-ups; and he considered stereophonic radio broadcasting. Such was the profusion of Blumlein's ideas that the 24-page patent document contains no fewer than 70 claims.

In recording the signals associated with the two channels of Blumlein's system, it is essential that the recording medium should permit the two recorded sound tracks to be permanently linked together so that the correct phase relationship between the two channels is preserved throughout any subsequent copying or replay operation. The media available in 1931 for this purpose were photographic film and wax discs. For disc recording Blumlein proposed that the two-channel signals should be recorded in a single groove. Such a method would minimise any differential amplitude or frequency distortion between the two recorded tracks.

In the United States, Keller[17] of Bell Telephone Laboratories, in September 1935, patented a system in which the two-channel signals were separately recorded in two spiral tracks on a given disc. Apart from the disadvantage of requiring two recording cutters and two pick-ups, and the difficulty of locating two pick-up styli in the correct grooves, the system was electrically unsatisfactory. The frequency response and distortion characteristics of the two tracks differ markedly when there is an appreciable difference between the track diameters. Furthermore, the playing time of such a disc is just half that of a normally recorded disc.

Hill-and-dale cut, and lateral cut, V-shaped record grooves were features of Edison's phonograph and Berliner's gramophone respectively, and by 1930 many thousands of cylindrical and disc recordings had been pressed and sold to the general public. With this knowledge Blumlein proposed a method of stereophonic recording in which the sum and difference signals of his two channels would be applied to a complex recording cutter in which the stylus would be capable of movement in two orthogonal directions. The sum and difference channel information would be represented by lateral and hill-and-dale modulations respectively of the record groove. Such a recording process would be free from the limitations of the two-track system. Additionally, a record produced by this process could be played on a monophonic gramophone to give the equivalent reproduction of a single-channel recording.

Blumlein was well aware that another recording possibility existed. In his patent he observed[18]:

'If the two channels being recorded are directly picked up from two microphones, or are intended to work unmodified into two speakers, that is with intensities and qualities similar to those of the original sounds received, it is preferred not to cut one track as lateral cut and the other as hill-and-dale, but to cut them as two tracks whose movement axes lie at 45° to the wax surface, or at some other convenient angle dependent on the relative available intensities from lateral cut and hill-and-dale respectively.'

This feature formed an important claim in the patent. The $+45°/-45°$ cut was reinvented by the US company Westrex in the 1950s. It appears that Westrex, which had originally been part of the Western Electric Company, was, with others, in ignorance of Blumlein's 1931 patent and had considered that $+45°/-45°$ recording was a new idea[19].

While at Columbia and during the early days of EMI, Blumlein had wide-ranging responsibilities for sound recording and sound reproduction. He continued to make improvements to the wax cutter (Figure 4.3), microphones and amplifiers of his monophonic recording system, and it may be that these activities delayed the onset of experimental work on his binaural sound recording and replay system.

On 21st July 1932 Blumlein wrote a memorandum[20], for Shoenberg, in which he described in non-mathematical language his various ideas on binaural reproduction. The tenor of the memo suggests that he was

Figure 4.3   The Blumlein stereo wax cutter, 1933, before restoration

Source: EMI Archives

seeking Shoenberg's formal permission to commence experimental work. In his conclusions Blumlein noted[21]:

> 'It would appear that if a true directional illusion can be obtained new possibilities are added to sound recording, and there would be a large increase of realism.
>
> 'It is by no means certain that the systems described above will work, but it is believed that previous investigators have not got beyond the stage of discovering that whereas pressure microphones gave a good binaural effect on head receivers, the effect was lost on loudspeakers. We at least understand why this effect is lost, so that we may be on the road to discovering how to produce it effectively in loudspeakers. It is therefore proposed to attempt to make the systems described above operate, and so to produce some binaural transmission in order to determine whether there is any likely field for development in this direction.'

It seems that Shoenberg was not particularly eager for R&D effort to be allocated to binaural recording and reproduction, since experimental work in this field did not commence until about the beginning of 1933, that is approximately one year after the date of the patent. When he received Blumlein's July 1932 memorandum, the number of staff engaged in research and development at EMI was quite small. Shoenberg had to allocate priorities and resources to the projects which were being tackled and which possibly could be tackled. Monophonic recording and reproduction, using the Blumlein system, was being pursued with vigour because the recording, pressing and sale of records was a source of income for the company. In addition, the directors of EMI had agreed to extend the television activities which had been initiated, before the formation of EMI, by the Gramophone Company. With just a few staff the development of television was being pushed vigorously and on 30th November 1932 N Ashbridge, the Chief Engineer of the British Broadcasting Corporation (BBC), had been invited to a demonstration of medium-definition television.

EMI was very anxious that some form of television service should be started soon on ultra short waves and had asked the BBC to take up its system—on an experimental basis—for about seven or eight months and then later for regular use on their programmes.

The developments of monophonic recording and television were major endeavours of EMI's 1932 R&D team. Shoenberg had to ensure that progress was not jeopardised by a spread of interests. And so, possibly with these considerations in mind, Shoenberg delayed the commencement of the binaural recording experimental investigation until January 1933.

There was much to be done before actual listening tests could be carried out. Shufflers, equalisers, pre-amplifiers and power amplifiers, stereophonic reproducing pick-ups, stereophonic recording cutters and microphone assemblies all had to be designed and constructed in the research laboratories. Fortunately the Western Electric monophonic recording system in EMI's studios was being replaced by the Blumlein moving coil system and the surplus Western Electric apparatus which became available could be adapted for use in the new experimental programme.

The stereo wax cutter, for example, comprised two modified Western Electric moving iron armature units coupled to a single stylus by a light-weight lever arrangement so that the stylus could move both vertically and horizontally. It was available for tests from about July 1933.

Some reminiscences of the early days of EMI's binaural group have been given by M G Harker[22].

'My first meeting with Alan Blumlein was on the 3rd of October 1932 when he interviewed me for the job of research assistant in the Recording Research Department [RRD] of the EMI Research Laboratories at Hayes. Much to

my surprise he offered me the job, as I felt I had not done very well at the interview, having cut my teeth on megawatt, rather than milliwatt, systems and
devices.

'The RRD was headed by Blumlein as co-ordinator and chief systems engineer, supported by H A M Clark as chief circuit engineer and H E Holman as
chief mechanical engineer. I was allocated to Clark, who already had A L
Westlake for an assistant; and Holman had L F Bury to help him. It was a
very small, dedicated team. Its brief was to produce the highest quality disc
recording and reproduction system that the state of the art allowed. . . .

'[After several months] the department became involved in building an experimental binaural [stereo] sound system and Clark's team was increased by
the addition of F R Trott, and later, by P Vanderlyn and Terry. I L Turnbull was
also transferred from the Abbey Road recording studios to the department.'

Practical testing of Blumlein's binaural ideas commenced in about
August 1933, in the Research Laboratories' main drawing office/
auditorium; this had been built in 1928 when the Gramophone Company
began its work on sound-on-film. The drawing office/auditorium
measured 100 ft by 50 ft by 30 ft in height and was provided with a projection room, a stage and a proscenium arch: the polished maple floor had no
rake and there was no fixed seating.

'The sound pick-up transducer consisted of a cluster of two HB microphones
mounted closely side-by-side, together with a pair of single crystal microphones similarly mounted and positioned at the front of and immediately
above the other pair. The four signal channels so produced were reduced to
two channels by subsequent frequency cross-over networks, the HB microphones contributing the signals below the cross-over frequency and the crystal
microphones, likewise above. A phase sensitive network then transformed
phase differences in the signals at every frequency into amplitude differences
which constituted the left and right hand binaural signal channels . . .'

Listening took place in a room, situated on the first floor of the
research building, which contained two 12 in moving coil paper cone loudspeakers mounted on large balsa wood baffle boards.

'The initial trials of this system had to be carried out after normal working
hours as the main drawing office [was in use during the day]. The microphone
cluster was set up at the front and we used to walk backwards and forwards
across the width of the room, threading our way between the rows of drawing
boards, endlessly reciting the days of the week, the months of the year, or even
nursery rhymes while the results were assessed [in the listening room].'

Appendix 1 describes the experimental apparatus (Figure 4.4) and
testing programme in detail. The programme was assisted by the HMV

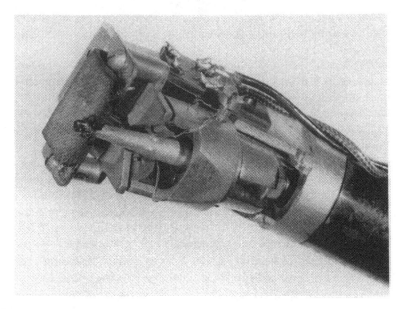

Figure 4.4   The Blumlein – Holman moving coil stereo pick-up, 1933

Source: EMI Archives

Dramatic Society, which was requested to perform short dramatic sketches.

By December 1933 all was ready for wax disc stereophonic recording. In this month ten 10 in discs were cut, each consisting of the sounds produced by persons – often Blumlein himself – talking and walking before the various microphone arrangements. Early in 1934 further recordings were made in EMI's Abbey Road studios, first with a dance band, and then with a trio of pianos. On 19th January nine now famous sides were cut with Sir Thomas Beecham's London Philharmonic Orchestra playing Mozart's *Jupiter* Symphony[23]. Details of all these recordings are given in Technical note 1 at the end of this chapter.

Recorded with a hill-and-dale/lateral disc cutter with the lateral modulation of the groove containing the sum information and the hill-and-dale modulation the difference information, the experimental discs proved conclusively that the methods and apparatuses described by Blumlein in his patent could indeed record and reproduce sounds binaurally.

On the quality of these discs one observer[23] has written, in 1991:

'Heard today, these pioneering recordings still sound surprisingly solid and musical, even though Blumlein's omni-directional EMI HB-1B microphones, spaced 20 cm apart, could not produce the strong directional effects that bi-polar designs were to make possible, ideas which Blumlein himself had predicted in his patent and which he developed in later experiments.'

A system of coincident bipolar (i.e. velocity) microphones (Figure 4.5) has the important property that the root mean square sum of the outputs of the loudspeakers is constant for a sound source at a constant distance from the microphones, regardless of its direction. It is this property which permits a uniform sound field to be reproduced between two spaced loud-speakers without the 'hole in the middle' effect.

The next phase of the testing programme was to record binaural sound tracks onto a film. Suitable subjects/objects were moved across the field of view of the camera and the sounds and visual images recorded.

'For the first film the camera and microphone were set up in a field to record the company's fire engine dashing from side-to-side with someone furiously ringing the bell! For the next film the camera was set up on the roof of a tall building to film [the] trains [which passed] below — the microphones [being]

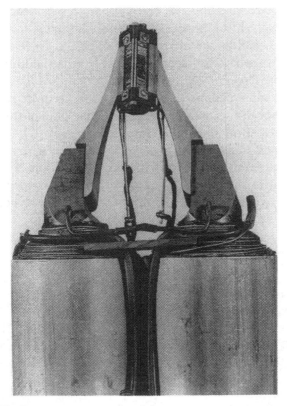

Figure 4.5   The Blumlein stereo microphone, 1934

Source: EMI Archives

set up at ground level beside the track to pick-up the sound. Finally, the famous restaurant scenario was filmed in which a couple, seated at a table too close to the orchestra to hear themselves speak, asked the head waiter to move it across to the other side of the room, which was conveniently done electronically.... Having proved experimentally the viability of this binaural system it was then demonstrated to a number of VIPs, which included the Prince of Wales.'[24]

EMI's work on stereophony stopped in 1935 and remained dormant until 1953. Shoenberg knew when to stop research: 'There is no future in that, drop it. Never continue when there is no future in something', he would say[25].

According to Nind[26], EMI had had talks with some cinema companies to determine if stereophonic sound reproduction in cinemas had a viable future. The company wanted to install its system into movie theatres, to obtain a commercial return from its endeavours but, though some experimental motion-picture films were made incorporating twin-track recordings, and these were reproduced in 1935, nothing came from EMI's venture. Turnbull's recollection[27] of this episode was that Sterling, EMI's Managing Director, felt the benefit of stereo sound in cinemas would not be worthwhile: it would be the introduction of colour which would be the next major step forward. A list of the early EMI binaural films is given in Table 4.1

In 1935 the commercial prospects for stereo records and domestic stereo gramophones also seemed bleak. The shellac discs of the 1930s were noisy and did not permit the subtle sounds associated with present-day stereophonic LP and CD recordings to be reproduced with sharp clarity. Blumlein's process required a record material which had a much lower surface noise than the shellac–slate mixture then in current use. Working with a chemist, who prepared a number of different compositions,

Table 4.1   Early EMI binaural films

Subject	Date	Length (ft)	Running time
Fluxite—early telecine recording		85	54 s
Men walking outside building		103	1 m 06 s
Trains		487	5 m 11 s
Men on stage with stick	12.07.1935	116	1 m 15 s
Men on stage counting	16.07.1935	115	1 m 14 s
Men on stage walking	16.07.1935	35	22 s
Fire engine		117	1 m 16 s
Playlet	26.07.1935	396	4 m 13 s
Playlet		336	3 m 33 s

Blumlein measured the noise levels of records stamped from these compositions but no suitable matrix was discovered.

Moreover, in 1935 EMI's engineers and scientists were heavily engaged
in the evolution of high-definition television systems. The Television
Committee, constituted in April/May 1934 under the chairmanship of
Lord Selsdon to consider the development of television, had worked
rapidly and had submitted its report to the Postmaster General on 14th
January 1935. Among the recommendations were:

1 'High definition television has reached such a standard of development as to justify the first steps being taken towards the early
   establishment of a public television service of this type.
2 'Baird Television Ltd and Marconi—EMI Television Company
   Ltd should be given an opportunity to supply, subject to conditions, the necessary apparatus for the operation of their respective
   systems at the London station.'

Blumlein became involved in EMI's television project in April 1933
and made quite outstanding contributions to the world's first, public, all-
electronic, high-definition system. Shoenberg had by now a brilliant television R&D team, but it was Blumlein who was *primus inter pares* in the team.
It was Blumlein who accompanied Shoenberg when in June 1934 he gave
evidence to the Selsdon Committee. Consequently it was inevitable that
Blumlein's active association with monophonic and stereophonic
recording and reproduction should cease sometime in 1933/34.

From this time until the commencement of hostilities in September
1939, EMI's Central Research Laboratories were primarily concerned
with work associated with high-definition television. This included the
equipping of the London television station; the provision of mobile outside
television broadcasting units; the development of television apparatus for
use in aircraft; and the evolution of the Super Emitron television camera.
In 1938–39 the company began to give some consideration to the defence
needs of the country. A new sound locator, based on Blumlein's binaural
concepts, was designed and subsequently manufactured in quantity; and
as a private venture, the company investigated a radar system due to
Blumlein and White.

On 2nd September 1939 all television broadcasting in the United
Kingdom ceased and for the whole of the war period EMI's efforts were
controlled by Government Ministries. Among the contracts awarded to
EMI were three in which Blumlein played a major role and which had
profound effects on the well-being of the UK, namely, the radar systems
known as AI Mark IV, AI Mark VI, and H_2S Mark II/ASV Mark III.

After the war, EMI returned to its record-making business. In October 1948 the company introduced its new BTR-1 tape machine at classical recording sessions, and by the end of the year new RS (Recording Studio)-1 disc cutting units had begun to replace Blumlein's original moving coil models. The recording studios were also re-equipped with several new microphones (Figure 4.6); and, with an eye to the future, the rights to Blumlein's stereo patent, no. 394 325, were extended from its 1949 expiry date to December 1952.

By 1955, EMI's record manufacturing facility was larger than any of its British competitors. The company's principal British plant was at Hayes, but overseas EMI had plants scattered from France to India and Brazil. Much of EMI's recording was undertaken within the Abbey Road complex, which housed not only the three main recording rooms but also the cutting, transfer and editing suites.

With its record business booming, EMI in the early 1950s again considered the outlook for stereophonic recording and reproduction. P B

Figure 4.6   An experimental stereo ribbon microphone

Source: EMI Archives

Vanderlyn, who joined EMI's Research Laboratories in 1935 and was in charge of the acoustics group of the Laboratories, wrote a report dated November 1953 in which he summarised 27 technical papers on stereophony. He concluded that for reasons of efficiency and economy the Blumlein method was 'likely to prove at least as cheap as any other system considered', including those based on the work of Bell Telephone Laboratories in the 1930s[28]. Subsequently, H A M Clark, G F Dutton, and P B Vanderlyn engineered EMI's 'stereosonic' recording and reproducing system, and described the principles on which their work was based in an important paper[29] published in July–August 1957. In this the authors referred to some post-war demonstrations, which had been given by Philips at Eindhoven, of a system using two microphones placed in an artificial head, the outputs of which supplied two widely spaced loudspeakers. Unlike the Blumlein system, directional effects at frequencies below 700 Hz were neglected and only 'a measure of spatial location', due to the difference of sound intensities at the two microphones, could be achieved.

On recent work in the United States, Clark, Dutton and Vanderlyn reported:

'... several American companies have made recordings using two widely-spaced microphones driving loud speakers via the medium of twin-track magnetic tapes, and one has issued some disks carrying the two channels, one on the outer half and the other on the inner. This has the serious disadvantage of halving the playing time'.

The ready availability of magnetic tape onto which two independent but synchronous sound channels could be recorded made possible the commercial exploitation of Blumlein's work, resulting in the 'stereosonic' system. This was demonstrated to members of the recording profession and to representatives of the press in April 1955. In April 1956 a full-scale public demonstration at the Royal Festival Hall, London was given to an audience of 1800.

EMI did not have a monopoly on stereophonic recording and reproduction in either the UK or overseas. After the rights of Blumlein's patent no. 394 325 expired in December 1952, the way was open for other companies to utilise either the hill-and-dale/lateral or the $+45°/-45°$ recording methods. In Britain both Pye Records Limited and Decca Records Limited were keen to record, press and market stereo records, but by 1957 no general agreement existed which defined the characteristics of the disc recording process. There was a danger that a multiplicity of disc standards would have an adverse effect on trade, so it was desirable for recording companies to reach an agreement.

On 28th November 1957 the first European meeting of the major recording companies was held in Zurich to discuss the standards of stereo disc recording. The meeting, which had been initiated by A Haddy, the Technical Director of Decca Records Limited, was attended by all of the leading recording companies and accord was reached that the $+45°/-45°$ principle should be adopted. Shortly afterwards Haddy went to the United States to persuade the American recording companies to follow the European consensus.

He has left a delightful anecdotal account of his experiences in the US and of the impact of Blumlein's stereo patent on the American companies.

'I went over to the States when [Decca] were developing stereo because [there] were three systems, [namely] the carrier system, the hill-and-dale/lateral system and the $+45°/-45°$ [system]. . . . And so I could see utter chaos coming if we had hill-and-dale/lateral and someone else had $+45°/-45°$. So I said to the old man:"I'm going over to the States to see if I can't pull all the boys together."

'I took a chap called Bob Goodman with me and all our reproducing equipment. We got into New York on Saturday night and on Sunday morning we went down to the London Record building to unpack all our equipment. Bob and I promptly got arrested in New York for working in New York on a Sunday without a permit — which cost the company about 50 dollars each.

'We gave a demonstration, that afternoon, of stereo to RCA and then the old man packed us off on a plane that night. It was a charter plane — the food in a cardboard box — [and we flew] up to Los Angeles. I had an appointment with Dr Frayn [of Bell Telephone Laboratories], the world big noise [in recording]. I was met by all the boys from the big record companies. . . . They were marvellous and couldn't believe we had brought all the equipment [with us].

'We arrived up at the Bell Telephone Laboratories and [met] old Dr Frayn. All our equipment was 220 V and we wanted an auto transformer."Oh", he said, "no such thing in this place". The boys took him out to lunch and treated him very well, and when he came back in the afternoon he went to sleep. Someone produced an auto transformer and we put all our equipment on it. We gave a demonstration of hill-and-dale/lateral — basically the record we put out later, "The journey into stereo sound" — and eventually they produced a 16 inch record recorded at 33 and a third [r.p.m.] which the Western Electric Company had recorded. . . . I know now that what we heard was their recording of "My Fair Lady" and it was b. . . .y good. So they went and fetched old Dr Frayn. Well, he [had] never heard the record played [so] well . . . and he became quite amiable.

'We managed to tell him that we didn't want to sell the equipment. All we wanted to do was to get the industry off on the same foot. Then he turned round and said: "Well that's impossible, because you see we've got patents on this".

'I then laid Blumlein's patent on the table and there was dead silence for about five minutes. They didn't even know it existed and we ended up with an agreement that we wouldn't do anything for three months and they wouldn't do anything for three months to allow them to evaluate the systems.

'Well, they did not keep to their word. About three weeks afterwards they were down in New York at the Radio Fair demonstrating this, so the old man said: "Well, we pull no punches", and we went.

'That was the start to all the companies getting together to standardize.'

In March 1958 the Recording Industry Association of America (RIAA) approved the Westrex $+45°/-45°$ system for stereo recording.

When the Audio Engineering Society, in 1958, devoted an entire issue of its journal to stereophony, it published Blumlein's 1931 stereo patent in its entirety, an almost unique event. In the editorial accompanying the various papers the writer noted:

'It [the patent] is of historic importance in the development of stereophony.... When it is realized that many of the ideas, pyscho-acoustic, mechanical and electrical set forth in this document of 1931, are only now gaining wide popular currency, one may reflect on the magnitude of the economic forces which control the viability of inventions.'

Subsequently on 2nd March 1978 the Audio Engineering Society certified its recognition of Alan Dower Blumlein (Figure 4.7)

'Because of his fundamental inventions in stereophony to be an
Audio Pioneer
Worthy of ranking with Alexander Graham Bell,
Emile Berliner, Thomas A Edison and other Noted
Founders of the Arts and Practices of Audio Engineering.'

Figure 4.7   *A D Blumlein, c. 1932–33*

Source: Mr S J L Blumlein

Technical note 1

List of records made with binaural recorder in auditorium

Test/Record no.	Test no.	Description of tests in auditorium	Channel control settings	
			A	B
5756 19 Dec. 1933	1	Blumlein, Westlake, Trott and Harker talking in the auditorium (Blumlein and Westlake in front	18	21
	2	Blumlein, Westlake, Trott and Harker talking in the auditorium (Blumlein back)	"	"
	3	All walking and talking in 'muddle' (changing position)	"	"
	4	Talking in turn (two pairs) more slowly (muddled changeover)	"	"
	5	Same as no. 4. Heavy shuffle – difference channel reduced by 3 dB	"	"
	6	Blumlein walking in auditorium. Heavy shuffle; difference channel reduced by 3 dB	19	22
5768 19 Dec. 1933	1	Talking and walking in auditorium, Blumlein only. Heavy shuffle	18	21
	2	Same as no. 1 Light shuffle	"	"
	3	Westlake, Trott, Harker and Turnbull talking in turn. Various shuffles	"	"
	4	Same as no. 3	19	22

Records made at Abbey Road with binaural gear

TT.1557-2 11/12 Jan. 1934		Dance band with microphones *c.* 45 ft distant. Heavy shuffling	18	18
TT.1557-1		Microphones as above. Heavy shuffling	24	24

WT. 5769		Three pianos converging towards standard HB microphone. Outer pianos make angle of *c.* 60° with each other. Heavy shuffling		
	1	Binaural microphones *c.* 12 ft from tip of each piano. 'Recording too heavy' (Hungarian Rhapsody, Brahms)	26	22
	2	As no. 1. (Hungarian Rhapsody, Brahms)	22	18
	3	Repeat of no. 2. (Hungarian Rhapsody, Brahms)	22	20
	4	Binaural microphones advanced 6 ft nearer piano (now *c.* 18 inches behind HB microphone) (Ride of the Valkeries)	18	16
	5	Microphones restored to original position, as for nos. 1, 2, and 3 (Ride of the Valkeries)	22	20
	6	Microphones withdrawn to *c.* 25 ft from ends of piano (Snippets of rehearsal after session had finished)	26	24
WT.5771 19 Jan. 1934		Sir Thomas Beecham and the London Philharmonic Orchestra playing Mozart's Jupiter Symphony		
	1	Microphones *c.* 13 ft high and *c.* 12 ft from nearest 1st violin. Heavy shuffling	24	22
	2	As no 1. Part 2 of symphony	"	"
	3	Microphones moved *c.* 6 ft further back and 1 ft higher	"	"
	4	As no. 3	26	24
	5	Microphones *c.* 7 ft further back, i.e. *c.* 25 ft from nearest 1st violin	"	"
		Part 3 of symphony	28	26
	6	As no. 5	30	28
	7	As no. 5	26	24
	8	As no. 5. Part 4 of symphony	28	26
	9	As no. 8.		

Technical note 2

Blumlein's monumental patent, no. 394 325, was so wide-ranging in its treatment of stereophony, and in its claims, that few additional patent applications were necessary to safeguard his ideas and methods.

When Blumlein commenced his study of binaural recording, only pressure type microphones were readily available to him. Later, when velocity type microphones, having figure-of-eight polar responses, were obtained, Blumlein experimentally investigated their use for stereophonic recording. His work on microphone types and microphone spacings led to three patents. The relevant patents are:

1 394 325, December 1931, on stereophonic recording and reproduction (the system description and details of techniques);
2 429 022, October 1933, on stereophonic sound (use of differently spaced microphones for different frequency bands);
3 429 054, February 1934, on stereophonic sound (two channels obtained from sum and difference of outputs of pressure and velocity microphones);
4 456 444, February 1935, on arrays of microphones with outputs mixed to give various polar diagrams. These ideas derived from the techniques used in stereophonic sound.

In addition to these, no. 505 079 of October 1937, on: 'Vertically and horizontally polarised transmitting stations in adjacent areas, to reduce co-channel interference', describes an application whereby binaural sound signals may be broadcast by radio. Each of the two binaural sounds modulates a radio frequency carrier wave. One modulated carrier is applied to a horizontal dipole antenna and the other is applied to a vertical dipole antenna. Since orthogonal dipoles do not interact, the two binaural sound channels can be broadcast without any crosstalk occurring.

Also no. 581 920 of July 1939 (with E L C White), on 'Direction finding on modulated signals, or reflections of modulated signals, by comparing the time of arrival of the modulation at separated antennas', is based partly on the principles enunciated in patent no. 394 325.

After 1935 Blumlein's only other involvement with stereophony was with aircraft location by sound. This work is considered in Chapter 11.

References

1 DONISTHORPE,   W.:   'Apparatus   for   taking   and   exhibiting photographs', British patent no. 4344, 9th November 1876

2 DONISTHORPE, W.: 'Talking photographs', a letter, *Nature*, 24th January 1878, **18** p. 242

3 BURNS, R. W.: 'Television, an international history of the formative years' (Peter Peregrinus Ltd, London, 1997), chapter 4 on 'Persistence of vision and moving images', pp. 63–77

4 TOULET, E.: 'Cinema is 100 years old' (Thames and Hudson, London, 1995), p. 48

5 DUDDELL, W. du B.: British patent no. 24 546, 1902

6 'Encyclopaedia Britannica': article on 'Motion pictures', 1972, p. 392

7 FORD, P.: 'Audio in retrospect', part 24, sound films (1), *Hi Fi News*, February 1962, p. 585

8 Ref. 6

9 Ref. 7

10 BLUMLEIN, A. D.: 'Improvements in or relating to sound transmission, sound recording and sound reproducing systems', British patent no. 394 325, 14th December 1931

11 HOSPITALIER, E.: 'The telephone at the Paris Opera, *Scientific American*, 31st December 1881, **45**, p. 422

12 KELLER, A. C.: 'Sound recording', US patent no. 1 910 254, 23rd May 1933. See also Ref. 10

13 FLETCHER, H.: 'Auditory perspective', *Bell System Technical Journal*, April 1934, **13**, Bell pp. 239–244

14 BLUMLEIN, A. D.: 'Binaural reproduction', 21st July 1932, EMI Central Research Laboratories' archives

15 Ref. 14, p. 5

16 CLARK, H. A. M., DUTTON, G. F., and VANDERLYN, P. B.: 'The "stereosonic" recording and reproducing system', *IRE Transactions on Audio*, July–August 1957, pp. 96–111

17 Ref. 16, p. 107

18 Ref. 10

19 DAVIS, C. C., and FRAYNE, J. G.: 'The Westrex stereo disc system', *Proc. IRE*, October 1958, **46**, pp. 1686–1693

20 Ref. 14

21 Ref. 14, pp. 7–8

22 HARKER, M. G.: letter to the author, 26th July 1991, personal collection

23 GRAY, M.: untitled paper, *TAS Journal*, 23rd September 1991

24 Ref. 22

25 LOCKWOOD, Sir Joseph.: taped interview, National Sound Archives, UK

26 NIND, E. A.: taped interview with the author, personal collection

27 TURNBULL, I. L.: taped interview with the author, personal collection

28 Ref. 23

29 Ref. 16

30 HADDY, A.: taped interview, National Sound Archives, UK

Chapter 5
Pre-EMI television history

On 2nd November 1936 the world's first, public, regular, high-definition television service was inaugurated at Alexandra Palace, London[1]. The decision to establish the service was made by Parliament following the submission to it of the report of the Television Committee. This Committee was constituted in May 1934 'to consider the development of television and to advise the Postmaster General on the relative merits of the several systems and on the conditions — technical, financial, and general — under which any public service of television should be provided'.

The Committee, chaired by Lord Selsdon, a former Postmaster General, comprised Colonel A S Angwin and Mr F W Phillips of the General Post Office, Vice Admiral Sir Charles Carpendale of the BBC, Mr O F Brown of the Department of Scientific and Industrial Research, and Sir John Cadman of the Anglo-Persian Oil Company.

Lord Selsdon and his colleagues worked with commendable speed and tendered their recommendations, a total of 17, to the Postmaster General, the Right Honourable Sir Kingsley Wood, on 14th January 1935. During their work the committee had examined 38 witnesses, had received numerous written statements from various sources regarding television and had visited Germany and the USA to investigate television developments in those countries.

The principal conclusion and recommendation of the committee was that 'high definition television had reached such a standard of development as to justify the first steps being taken towards the early establishment of a public television service of this type'.

Marconi – EMI Television Company Ltd and Baird Television Company were the two companies who were invited to submit tenders for studio and transmitting equipment, and for a short period from 2nd November 1936 to 13th February 1937 both companies transmitted television programmes, on an alternate basis, from the London station. Subsequently, until the commencement of hostilities in September 1939, only the Marconi–EMI equipment was in operation.

The founding of the high-definition service was a remarkable achievement for, until 1931, just low-definition systems, operating on a 30-line or 60-line standard, predominated for public use, and the first demonstration of rudimentary television had been given by Baird just five years previously on 26th January 1926.

The early history of television from about 1877 to 1926 has been described and considered in copious detail in the author's book *Television, an international history of the formative years* (Institution of Electrical Engineers, London, 1998) and is not repeated here, suffice it to say that by the mid-1920s the time was opportune for low-definition television experiments to commence. Several factors can be adduced to account for the growth of television from about 1925. Much thought had been given by inventors and scientists to the problem of 'seeing by electricity', following Willoughby Smith's important discovery of the photoconductive property of selenium in 1873, and by the 1920s the principles of television, and particularly of scanning, had been broadly enunciated. Contemporaneously, during this period, considerable work on the physics of photoelectric emission, of thermionic emission, and of cathode rays had been undertaken. Again, the 1914–1918 war had given an impetus to the utilisation of valves in signalling systems, and had so stimulated developments in circuit and radio equipment techniques that by 1918 triode valves could be manufactured to cover a wide power and frequency range and were suitable for both receiving and transmitting purposes. Consequently by 1920 the time was ripe for the establishment of sound broadcasting: the radio systems were available and public demand was growing. In the USA the Westinghouse Electric and Manufacturing Company, which owned an experimental station KDKA, noting in 1920 that its broadcasts were popular, established a regular broadcasting service. Sound broadcasting on a permanent basis commenced in the UK during November 1922.

The growth of commercial radio telephony and domestic broadcasting influenced the progress of television. By the early 1920s all the basic components of a rudimentary television broadcasting system appeared to be available. Whereas before 1920 only a few isolated attempts had been made to investigate, on an experimental basis, the subject of 'distant vision', from about the start of the 1920s determined efforts to advance television were being made in the UK, the USA, France and Germany. Initially these endeavours were mainly those of individuals working in isolation from others. J L Baird of the UK, C F Jenkins of the USA, E Belin of France, and D von Mihaly, a Hungarian working in Germany, were four of the principal investigators in this period. For a short time in 1923, Zworykin pursued some personal work on an all-electronic television camera while at the Westinghouse Electric and Manufacturing Company, but the only determined effort by a public or a private organisation appears to be that which was initiated at the Admiralty Research Laboratory, Teddington, in 1923[2].

During the second half of the decade 1920–30 other large manufacturing organisations took an interest in television. In the USA, the General Electric Company, Westinghouse Electric and Manufacturing Company, Bell Telephone Laboratories (of the American Telephone and Telegraph Company) (Figure 5.1), and the Radio Corporation of America, in addition to a number of smaller companies, began to be associated with television projects; in Germany both Fernseh AG, and Telefunken AG became active in this field by the end of the decade. Leading companies in the UK adopted a rather reserved position on television matters until 1930, and before then only the Baird companies (Television Ltd, Baird Television Development Company Ltd, and Baird International Television Ltd) had vigorously engaged in the pursuit of television

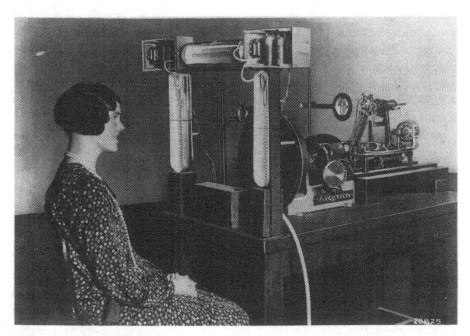

Figure 5.1   Low-definition transmitter scanner of Bell Telephone Laboratories for producing 'head and shoulders' type images, c. 1927

Source: AT&T Bell Laboratories

research and development. The Marconi Wireless Telegraph Company, and the Gramophone Company initiated their television activities in 1930[3].

J L Baird's approach to the television problem was circumscribed by his financial position[4]. He had little money, no laboratory facilities for the construction and repair of equipment, no access to specialist expertise, and no experience of research and development work in electrical engineering. He had to carry out his experiments in the unsuitable surroundings of private lodgings. His scanners were based on variants of the 1884 Nipkow disc. Nevertheless, a crude form of television was first successfully demonstrated by Baird on 25th October 1925, and subsequently was shown by him to about 40 members of the Royal Institution on 26th January 1926. The demonstration consisted, partly, of the reproduction of an image of a person's face.

Baird had to devote a great deal of time and labour to acquire a patent holding which would place his companies in favourable positions commercially, and until about 1930 he engaged in this task almost single-handedly. From the start of his work in 1923 to the end of 1930 Baird applied for 88 patents; the number of patents which originated from other members of the Baird companies in the same period total four.

Not surprisingly, Baird had little time for writing scientific papers and engaging in extensive field trials. He tried to anticipate every likely development and application of the new art. Daylight television, noctovision, colour television, news by television, stereoscopic television, long-distance television, phonovision, two-way television, zone television, and large-screen television were all demonstrated in a basic way by Baird during the four-year period following his 1926 demonstration.

Baird's plans for television were ambitious and extensive and he hoped to establish his system in many countries including, of course, the UK. Baird wished to create a television monopoly in this country, but the rather aggressive methods employed by his business associates caused antagonism with the BBC. The lack of enthusiasm shown by the Corporation towards Baird's low-definition system was the source of much concern and frustration to him and his supporters and resulted in delays in the execution of their objectives.

Essentially, the BBC was not interested in participating in the advancement of television on the basis of a system which could not reproduce images of, say, a test match at Lord's or tennis at Wimbledon; the BBC felt that low-definition television was inappropriate to its services. As a consequence, the BBC's policy towards Baird's work was necessarily negative in outlook, and was not conducive to the rapid advancement of Baird's aims.

Patronage and encouragement are important factors in the early progress of an invention. In America and elsewhere, facilities for television broadcasting were given by broadcasting stations in the 1920s, but in Britain the chief engineer of the BBC opposed the use of the BBC's stations for this purpose. This opposition led to Baird's business associates pursuing a vigorous policy to establish a low-definition television service. They were successful and on 30th September 1929 an experimental 30-line service was inaugurated, followed on 22nd August 1932 by a 30-line public service. The televised images were of the 'head and shoulders' type.

In April 1927 Bell Telephone Laboratories demonstrated well engineered apparatus for the transmission and reception of television images by landline, and by radio links (Figure 5.2). The Laboratories were formed in 1925, when the engineering department of Western Electric was reorganised and became Bell Laboratories with a total staff of approximately 3600. With its vast resources, the Laboratories' impressive television demonstration represented the best which could be shown anywhere. Later, Bell Laboratories demonstrated colour television, two-way television, and zone television[5]. All were engineered with an excellence consonant with the prestige of the American Telephone and Telegraph Company (the company which included Western Electric and Bell Laboratories). From 1925 to 1930 the company approved the expenditure of $308 100 on low-definition television developments, and a further $592 400 on other aspects of television from 1931 to 1935 (inclusive). Baird's financial resources from 1923 to 1927 were minimal, so inevitably his equipment and demonstrations could not emulate those of the Bell Laboratories.

The director of television research at Bell Laboratories was Dr H E Ives. In an important appraisal, published in 1931, of the progress which had been made by his group, he outlined the difficulties which faced television workers in the late 1920s. His prognosis for the future of television was gloomy in outlook. For Ives, the statement[6] of the problem that had to be solved was simple:

'An electrically transmitted photograph five inches by seven inches in size, having 100 scanning strips per inch, has a field of view and a degree of definition of detail, which, experience shows, are adequate (although with little margin) for the majority of news events and pictures. It is undoubtedly a picture of this sort that the television enthusiast has in the back of his mind when he predicts carrying the stage and the motion picture screen into the house over electrical communication channels.'

The difficulty of achieving this desirable result was readily apparent. In the photograph the number of picture elements is 350 000, and, at a

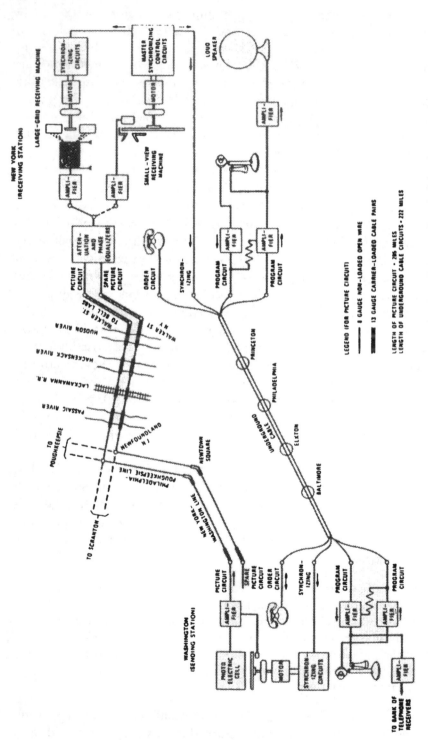

Figure 5.2   Schematic diagram of the line and radio circuits used in the 1927 Bell Telephone Laboratories television demonstration

Source: AT&T Bell Laboratories

repetition speed of 20 per second (24 per second had now become standard with sound films), 7 000 000 picture elements per second would have to be transmitted. The bandwidth required would be 3.5 MHz on a single side band basis. Ives compared the criteria for high-definition television and the results which had been obtained in the USA, and observed: 'All parts of the television system are already having serious difficulty in handling a 4000 element image'. (This was the number of image elements used in the 72-line pictures of the Labs.' 1930 two-way television link.) (See Figure 5.3.)

The obstacles which had to be overcome before a high-definition system could be implemented were the limitations of the scanning discs at the transmitter and the receiver, the photoelectric cells, the amplifier systems, the transmission channels and the receiving lamps. Ives noted that the Nipkow disc was entirely impractical when really large numbers of image elements were required, and opined: 'As yet however, no practical substitute for the disc of essentially different character has appeared'. After surveying the problems he concluded: 'The existing situation is that if a many-element [i.e. high-definition] television image is called for today, it is not available, and one of the chief obstacles is the difficulty of generating, transmitting, and receiving signals extending over wide frequency bands'.

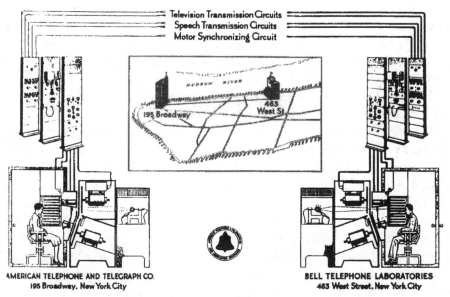

Figure 5.3   *Schematic diagram of the line circuits and equipment used in the 1931 Bell Telephone Laboratories demonstration of two-way television*

Source: AT&T Bell Laboratories

A partial solution to the problem of medium/high-definition television was to employ multiple-scanning and multiple-channel transmission. Interestingly, in 1931, the Gramophone Company (HMV), Baird Television Ltd, and Bell Telephone Laboratories all demonstrated multi-channel television: independent advancements in television engineering were now following convergent paths.

HMV's interest in television effectively dates from October 1929 when a 'case opening' report on the subject was prepared[7]. Mechanical scanning was regarded as the more practical method of analysing and synthesising images, but the advantages of cathode ray scanning systems in both transmission and reception were noted. A few months later, in January 1930, a Television Section was set up in the Research Department and an allotment of £800 was given for work, which was on mechanical scanning lines, for the period January to June 1930.

In April 1930 A Whitaker, of the Advanced Development Department, visited RCA which had recently recruited Zworykin's television section from Westinghouse. Demonstrations of 60-line television using cathode ray tube displays were given to Whitaker and though their definition was still not satisfactory for the commercial market, nevertheless the brightness of the cathode ray tubes used was 'so remarkably in advance of anything previously known', that he returned with the belief that this method of reception was worthy of progression. Indeed, 'the prospects of this line of attack seemed so bright' that in the second half of the year, July to December 1930, a further allotment of £800 was requested. An additional sum of £1200 for work on the transmission system was also sought. This followed an HMV report which recommended the utilisation of the 3 m to 5 m short-wave band for the system. Prior to this report, the General Post Office had been approached and asked to waive some of the restrictions which had been applied to experimental transmissions.

During the second half of 1930 some investigations were carried out by HMV on cathode ray tube reception. The correctness of this course of action was confirmed when G E Condliffe, of HMV, visited RCA's Camden works in November 1930. These visits by Whitaker and Condliffe to RCA entirely changed their views regarding television receiving apparatus. Before April 1930 HMV had adjudged cathode ray tube reception as 'an interesting theoretical scheme but probably impracticable owing to the instability and low luminosity of the tubes available at that time'. After their visits and the demonstrations of the new tubes, which had enhanced brilliance, focus and efficiency, they were converted to the opinion that the RCA way was the right one. As Whitaker noted: 'Our decision to change our line of attack was caused by these RCA demonstrations ... and as we had no wish to start repeating all the work which [had] been so excellently carried out in America, we decided to investigate the possibility of using a very considerably greater number of picture elements than [had] been possible' (Whitaker to Dr Goldsmith, Vice-President of RCA, 22nd August 1930).

Such a plan necessitated a wide bandwidth and therefore a very high radio carrier frequency. Whitaker in his letter of 22nd August informed Goldsmith that HMV were 'at present' working with wavelengths from 2 m to 5 m although the radio section had recently been experimenting as far down as 30 cm. Goldsmith replied on 4th September 1930: 'I am indeed glad to learn that your research group is working along somewhat different lines from those which are being carried forward in Camden. The field is so new that duplication of effort would be uneconomical.... I note with some astonishment that you propose to use 250 [kHz] side bands.'

The Gramophone Company's attitude to the television problem was cautious and sensible. After a watching brief from 1926, the company naturally was keen to avoid some of the pitfalls which had been encountered by others. Apart from the visits to RCA, representatives of the company had been to the Baird Television Company's studios in Long Acre, London, and had had discussions with O G Hutchinson, J L Baird's business manager, about current and future trends. Moreover, HMV had subjected each Baird patent specification to a careful examination for claims which might subsequently 'be awkward in operating a television system'.

The company also had sought confirmation, from both the General Post Office and the BBC, of statements made by the Baird companies concerning future licences and wavelength allocations.

On the practical side HMV wished to advance from a position where some knowledge of the properties of the essential components of the television system had been accumulated by carefully conducted experimental studies. The photoelectric cell and the light-valve were two of these components and hence laboratory work was initiated on investigations of photoemissivity and the Kerr effect. These were pursued by W F Tedham and W D Wright respectively and reports were prepared by the researchers in September and November 1930.

Meanwhile, C O Browne, a colleague of Tedham, was giving some thought to a system of five-channel television. Browne's report[8], dated 26th September 1930, indicates that the Gramophone Company was anxious to become a competitor in the infant but growing television industry, for he wrote: 'In the proposed system of television as far as possible known results and data are utilised with a view to producing a workable system in a short time with a reasonable chance of success. It is for this reason largely, that a number of transmission channels are to be used'.

The advantages of this procedure were later described by Browne in a paper[9] presented to the Institution of Electrical Engineers:

'The total frequency band necessary to transmit a given picture may, in the case of a multi-channel television system, be divided into a number of channels, each of which accommodates a frequency band given by the total frequency range divided by the number of channels. On this account the difficulties of design, not only of the apparatus situated at the transmitter and receiver but also of the transmission line between the two stations, are considerably reduced. Apart from this advantage of the multi-channel system, the amount of light available for illuminating the receiver screen is increased in proportion to the number of channels used. Further the velocity with which the scanning spots travel over the surface of the picture to be transmitted is decreased so that the accuracy necessary for synchronising is reduced as the number of channels is increased.'

Browne succeeded in his task and his plan was engineered in the remarkably short time of approximately four months. The equipment was demonstrated, in January 1931, at the Physical and Optical Society's Exhibition, London. Five channels were chosen and each picture was scanned at the rate of 12.5 per second. The line standard was 150 lines per second.

HMV's television apparatus did not represent true television, in which reflected light is received from an object and allowed to fall onto a photoelectric cell, but was more in keeping with the very early silhouette system of Baird and others, in which a powerful light source was situated behind the object to be televised. There was an important difference, however: whereas the early workers had used opaque objects, Browne utilised cinematograph film. Half-tones were thus taken into account as in true television. Film was used mainly because it was plentiful and it enabled the conditions existing for any particular transmission to be repeated, with some accuracy, on any subsequent occasion. Browne, like the engineers of the Bell Telephone Laboratories who contemporaneously were examining the possibilities of multi-channel television, was particularly interested in investigating the problems associated with the transmitter and the receiver, and in examining the electrical and optical conditions necessary to secure good results. Repeatability was thus an important point to be considered and the use of film allowed this to be achieved.

The demonstrations at the Physical and Optical Society's Exhibition created great interest and queues of people had to wait outside the small theatre, which the Gramophone Company had set up, to view the new system. G A Atkinson[10], the film critic of the *Daily Telegraph*, referred to the

This peep into the future was drawn by FITZ for the *Daily Mirror* in June 1932. 'Will it prove to be a help or hindrance to life?' he asked.

Figure 5.4   '*The future of television . . . How the cartoonists saw it in the 'thirties'*

Source: *Daily Mirror*, June 1932

notable gain of the system and wrote: 'It marks a considerable technical advance on any system yet demonstrated, especially in the direction of bringing television rapidly into use for entertainment purposes'. The *Evening Standard*[11] noted that television now seemed to be within sight of realisation. (See Figure 5.4.)

During the exhibition the Gramophone Company showed televised images projected onto a screen measuring 24 inches by 20 inches. No longer were 'head and shoulders' shown, but instead the audience saw images of buildings, soldiers marching, cricketers walking on and off the field, and so on. 'Everything was easily recognisable. An LCC tram car showed up so clearly that its number on the front was decipherable without difficulty. The pictures were steady on the screen. They were in good focus at very short range. They would probably stand enlargement up to four or five times the size of the screen actually used. . . . The general effect was that of looking at a performance of miniature films lacking full illumination.'[12]

These somewhat enthusiastic newspaper reports did not quite match the more objective report of Dr H E Ives, of the Bell Telephone Laboratories, on his team's three-channel television system.

Ives[13] found that the multi-channel apparatus yielded results strictly in agreement with the theory underlying its conception and observed that the 13 000 element image was a marked advance over the single-channel 4000 element image.

'Even so, the experience of running a collection of motion picture films of all types is disappointing, in that the number of subjects rendered adequately by even this number of image elements is small, "Close-ups" and scenes showing a great deal of action are reproduced with considerable satisfaction, but scenes containing a number of full length figures, where the nature of the story is such that the facial expression should be watched are very far from satisfactory. On the whole the general opinion ... is that an enormously greater number of elements is required for a television image for general news or entertainment purposes.

In the 1920s several inventors had given some thought to the means, based on the use of cathode rays, which would allow an all-electronic television system to be implemented. Pre-eminent among them were V K Zworykin and P T Farnsworth, both of the USA. Their early schemes were to be much developed and were to be the only viable all-electronic television schemes, of those put forward in the 1920s, which were implemented in the 1930s[14].

References

1 BURNS, R. W.: 'British television, the formative years' (Peter Peregrinus Ltd, London, 1986)
2 BURNS, R. W.: 'Early Admiralty and Air Ministry interest in television', Conference papers of the 11th IEE Weekend Meeting on the 'History of electrical engineering', July 1983, B/1–B/17
3 BURNS, R. W.: 'Television, an international history of the formative years' (Institution of Electrical Engineers, London, 1998)
4 BURNS, R. W.: 'J. L. Baird: success and failure', *Proc. IEE*, September 1979, **126**, (9), pp. 921–928
5 BURNS, R. W.: 'The contribution of the Bell Telephone Laboratories to the early development of television' (Mansell, London, 1991), pp. 181–213
6 IVES, H. E.: 'A multi-channel television apparatus', *J. Opt. Soc. Am.*, 1931, **21**, pp. 8–19
7 Ref 3, chapter 14
8 BROWNE, C. O.: 'Proposed system for five channel service', Report GCI, 20th September 1930, EMI Central Research Laboratories' archives
9 BROWNE, C. O.: 'Multi-channel television', *J. IEE*, 1932, **70**, pp. 340–349
10 ATKINSON, G. A.: 'Great advance in television tests. Nearing practical success. Broadcasts of film and plays. When all may see the Derby', *Daily Telegraph*, 6th January 1931

11 ANON.: 'Success of new tests today. Film of everyday life projected. Ride on a bus. Everything clearly recognizable', *Evening Standard*, 6th January 1931
12 Ref. 10
13 Ref. 6
14 Ref. 3, chapters 15 and 16

Chapter 6
EMI and high-definition television

Three months after the Gramophone Company's demonstrations of television, EMI was formed to acquire the ownership of the Gramophone Company Limited and the Columbia Graphophone Company Limited. The research groups of the two companies merged. Shoenberg's Columbia team included A D Blumlein, P W Willans, E C Cork, H E Holman, H A M Clark and others, and they jointed the HMV team at Hayes. This team comprised the research manager G E Condliffe, W F Tedham, C O Browne and W D Wright.

EMI's directors agreed that HMV's television development effort should continue to be an item of the new company's R&D programme. One of the first questions which had to be tackled was whether this work should proceed on mechanical or electronic lines. Mechanical scanners had the advantage that they had been made successfully, whereas electronic scanners were still in the development phase. On the other hand, electronic scanning had many potential advantages for high-definition television, and HMV's exploratory work on television had been based on the strategy that the company should aim to develop an effective cathode ray tube picture receiver, but should leave any endeavour to fabricate a television camera to others.

EMI felt that the most promising attack on the problem of home television should be on the lines initiated by HMV. There were thought to be divers factors which made it desirable for EMI to proceed independently of RCA: these have been discussed at length in one of the author's books[1], and are not repeated here. It was agreed that television development should proceed on the following lines[2]:

1 '[that EMI should get the] 150-line scanning equipment, and 150-line cathode ray receiving equipment into operation by the end of [1931];

2 'that [EMI] should commence a small amount of experimental work on the production of cathode ray oscillographs suitable for television;

3 'that [EMI] should keep in mind for investigation and development, as soon as possible, the short wave transmitting and receiving gear which [would] be essential for the commercial utilization of television services.'

As an indication of the timescale of the programme it was suggested that: first, the experimental system should be completed by December 1931; secondly, the cathode ray tube work should be well advanced by June 1932; and thirdly, the problems of a short-wave radio link should be solved by June 1933. Such a timescale would enable EMI to make provision for the commercial utilisation of their television work by the middle of 1933.

Actually this programme was soon modified. While the view had been taken that the company's business was in the field of receiving, and not transmitting, apparatus it was essential to have, for the testing of the receivers, a source of video signals. Brown's 150-line equipment was unnecessarily complicated, with its multi-channel requirements, and was not really suitable for this aspect of the work. Consequently, four months after the merger, work started on the design of a single-picture channel system based on a 120 lines per picture, 24 pictures per second, mirror drum film scanner.

By the end of August 1931, Wright had constructed a cathode ray tube receiver, using sawtooth scanning, and in a report dated 18th August he had recorded[3]: 'Some good receptions of the Baird television broadcasts have been obtained, but not consistently from day to day, nor during any one broadcast'.

In September 1931 A G D West, of EMI's Research and Design Department, visited Zworykin's RCA laboratory in Camden, New Jersey. West was shown the current results of RCA's television work and reported[4]: '. . . television is on the verge of being a commercial proposition. They [RCA] intend to erect a transmitter on top of a New York skyscraper in the autumn of 1932'. West had seen a televised image about 6 inches by 6 inches in size and had observed that its quality was comparable to that of an ordinary cinema picture viewed from a position near the extreme back of a large theatre. The planned selling price of the receiver was to be about $470 (about £100) and all the receiving apparatus for sound and sight was to be contained in a single cabinet of the size of an ordinary radiogram. The vision and sound wavelengths were to be 6 m and 4 m respectively.

West's report had an immediate effect: EMI reviewed its position concerning television. As Whitaker noted: 'Television is apparently coming rather more quickly than even the most optimistic of us have considered likely during the last couple of years'. Consequently there had to be a reappraisal of EMI's television project. On the one hand, in the UK, the Baird companies had achieved a great deal of favourable publicity for their 30-line system; the BBC had reluctantly given the Baird companies some degree of official recognition by allowing them to use the BBC's transmitters for experimental broadcasts during non-programme hours; there seemed to be a prospect that the Baird companies would be granted permission for their equipment to be housed in the BBC on a permanent basis, and additionally it had been stated that short Baird transmissions would take place during normal BBC broadcasts. On the other hand, the depressed business conditions of 1931 had led to a curtailing of EMI's expenditure on television and because of this EMI could not give the demonstration at the end of the year, which it had originally planned; there was a possibility that, if Baird's system became firmly established in the BBC, future television standards — and 'there were at least a dozen points which [needed] to be considered in great detail with a view to securing standardisation' — could be influenced by the standards of the Baird system; and there was a 'most urgent necessity' for EMI to establish 'a real demonstrated competition to the Baird system, which at the same time would give EMI prestige in television'.

For Whitaker there appeared to be only one short-term solution to the problem of gaining a position of authority in television matters and that effectively was to buy in expertise. RCA was further advanced than EMI in its work on 'seeing by electricity', so Whitaker suggested that approval should be given for the expenditure of roughly $50 000 (about £13 000) for the purchase of transmitting equipment (including 4 m and 6 m transmitters for sound and vision, studio equipment and demonstration receiving sets), and for the expenditure of not more than £2000 for the installation of this equipment in England. Whitaker's figure of $50 000 was based on an approximate quotation he had received from RCA. A later quotation received on 2nd December 1931 gave a total figure of $83 732[5].

EMI's Executive Committee was persuaded[6] by Whitaker's arguments and decided to recommend to the Board of Directors the proposals outlined by him. However, this was adamantly opposed by Shoenberg, the Head of the Patent Department.

By October 1931 it had become clear that technical cooperation with RCA was going to be 'very difficult and unreliable'[7]. Fortunately, the RCA rights on the transmitting side belonged to the Marconi Company and a favourable offer had been received by EMI from the company for the hire of radio transmitting apparatus. Hence the idea of attempting to work in parallel with RCA faded out and Whitaker's proposition was not implemented. Instead, a programme of work on television began independently of RCA, except for the patents to which EMI was entitled[8].

The challenge which faced EMI, and which was being faced by Farnsworth and by Zworykin's group at RCA, was immense. Photoelectricity, vacuum techniques, electron optics, the physics of the solid state and of secondary electronic emission, and wide-band electronics and radio communications, were all in a rudimentary state of development. Many fundamental investigations would have to be undertaken before a high-definition television system could be engineered[9].

A report, written by Condliffe and dated 16th November 1931[10], on 'Programme of work for the Advanced Development Division' lists the cathode ray tube developments which were being progressed: construction of cathode ray tubes according to the present RCA Victor design; improved tube design for lower modulation voltages; investigation of fluorescent screens for rate of decay of fluorescence, and colour; development of commercial methods of manufacture; development of high-voltage tubes for image projection; development of a cathode ray scanning device, since this was considered to be an important line of development for the production of scanning devices for outdoor use; and construction of large vacuum photocells for studio work. The list illustrates the intentions and commitment of EMI regarding all-electronic television at the end of 1931.

On 21st October 1931 Condliffe and Whitaker had talks with Colonel Angwin and Mr F Gill of the Engineer-in-Chief's Department, General Post Office[11]. They sought permission to set up dual short-wave transmitters, operating on 4 m and 6 m, with a maximum power output in each antenna of 2 kW, the bandwidths of the radiated signals being 500 kHz and 25 kHz. Angwin did not foresee any objections provided EMI did not deliberately attempt to transmit material of high entertainment value and did not cause interference with the transmitter stations of the War Office and Admiralty situated in Westminster, London.

A few weeks later on 8th December 1931 a licence[12] was sent to EMI allowing them to establish, for experimental purposes, a wireless sending and receiving station at the Hayes premises of the company[13]. In compliance with this licence, EMI could operate within the following bands:

c.w. and telephony   62.01 – 61.99 MHz
television   44.25 – 43.75 MHz

and at a power of up to 2 kW into the antennas. (The licence details show that sometime between 21st October and 8th December 1931 EMI had abandoned its plan to use RCA transmitting equipment which worked on 4 m and 6 m.)

The practical association between EMI and the Marconi Wireless Telegraph Company Limited, which was to prove so beneficial to both companies, began late in 1931 when Shoenberg invited MWT to supply a low-power v.h.f. transmitter, complete with a modulator, for use with EMI's film scanner. N E Davis, MWT's transmitter expert, was instructed to prepare a transmitter and modulator for this purpose and subsequently the units were sent to EMI in January 1932. The transmitter had an output of 400 W at a frequency of 44 MHz and employed grid modulation of the final stage.

Progress was now swift and on 11th November 1932 Shoenberg invited[14] the BBC's chief engineer, Mr N Ashbridge, to a private demonstration both in the transmission and reception of television. 'In my humble opinion', wrote Shoenberg, 'they would be of quite considerable interest to you'.

Ashbridge visited the Hayes factory on 30th November and was shown apparatus for the transmission of films using four times as many lines per picture and twice as many pictures per second as Baird's equipment. He was impressed and thought the demonstrations represented by far the best wireless television he had ever seen and felt they were probably as good as or better than anything that had been produced anywhere else in the world. He wrote[15]: '... there is not the slightest doubt that a great deal of development, thought and expenditure [has] been expended on these developments. Whatever defects there may be they represent a really remarkable achievement. In order to give some idea of the cost of such work, I might mention that the number of people employed is only slightly less than that in the whole of our research department.'

The actual demonstration consisted of the transmission of a number of silent films, over a distance of approximately two miles, by means of an ultra short-wave transmitter using a wavelength of 6 m and a power of about 250 W. On the quality of the images Ashbridge reported:

'The quality of reproduction was good, that is to say one could easily distinguish what was happening in the street scenes and get a very fair impression of such incidents as the changing of the guard, the Prince of Wales laying a foundation stone and so on. A film showing excerpts from a play was in my opinion not so good although it was possible to follow what was going on all over the stage. On the other hand excerpts from a cartoon film were definitely good. I think they could have given a better demonstration had they been in possession of better films. The ones they showed had been in use for several years. The size of the screen is about 5 inches by 5 inches but they have a second machine which magnifies this by about four times in area. The quality of the reproduction can be compared with the home cinematograph but the screen is smaller.'

EMI was very keen that some form of television service should be started on ultra short waves, and, following up Ashbridge's visit to Hayes, Mr Alfred Clark, the Chairman of the firm, paid a visit to the BBC to have discussions with the Director General, J F W Reith. Clark was anxious to know what television standards would be adopted for television. He hoped Reith would say that the number of pictures per second and the number of lines per picture would be 25 to 30 and 120 to 180, respectively. EMI would then have been in a position to have started an experimental service, with equipment in the BBC's Broadcasting House, on ultra short waves early in 1933, and probably before Baird's companies were in a position to do so.

The emergence of a competitor in the form of EMI caused J L Baird and his associates much unease[16]. They could not acknowledge for some considerable time that EMI's television system was being engineered by British workers in a British factory using British resources. For them the Radio Trust of America, through its associated companies in London, was the mainspring of EMI's progress. A very noticeable bitterness is evident in the letters emanating from the Baird companies, during the early 1930s, on the progress, and support from the BBC, of the Hayes company. Baird was always ready to point out that his firm could match the steps towards high-definition television which were being made by EMI and that therefore the pioneer company, i.e. Baird's, should be preferred.

The controversy which is described and discussed at length in the author's book *British television, the formative years*, (Peter Peregrinus Ltd, London, 1986) was resolved when the Postmaster General opined[17] 'that it would be right to postpone a decision in regard to the institution of tests of the EMI apparatus at Broadcasting House, London until demonstrations of Baird's apparatus [had] taken place. The arrangement was that both EMI and Bairds would give demonstrations to be witnessed by the BBC and the Post Office and that a decision on the installation of the EMI apparatus at Broadcasting House should be postponed until the results of these demonstrations [had] been considered.' The demonstrations were scheduled for the 18th and 19th of April 1933.

The Baird apparatus was demonstrated using a wire link between neighbouring rooms in the companies' Long Acre premises. It was described by Simon, one of the GPO's officials, in the following terms[18]:

'The transmitting apparatus was of a makeshift type and, at the receiving end, pictures about 3 inches by 3 inches were produced in black and white on the broad end of a funnel-shaped cathode ray tube in two cases, and by a Nipkow disc in a third. Films were fed into the transmitters: but the received pictures were in all cases indistinct, jerky and erratic. It was stated that arrangements were being made for the presentation of a picture 9 inches by 5 inches. The best that could be said for the demonstration was that it was an interesting experiment in picture transmission with rather crude apparatus.'

At Hayes the EMI apparatus was demonstrated by wireless transmission, the transmitting apparatus being at the works and the cathode ray tube receiving set in a cottage two miles or so away. Simon observed that the complete receiving apparatus for sight and sound was complicated and involved the use of 25 valves; but it was claimed that the number of valves could be reduced and the apparatus simplified so as to reduce the cost of the television set to about £80 or £100. (Baird claimed that his receiving set could be manufactured in bulk for about £30 or £40.) The demonstration consisted of a reproduction of films on the screen giving an image size of 6.5 inches each way. In one case the receiving set gave a black picture on a white background and in the other on a green background.

'The action on both pictures could be followed clearly throughout, without the guidance afforded by the accompanying speech; but the detail on the green background was superior to that on the white. A very high degree of stability was achieved. The company are experimenting with various substances on the cathode ray tubes with the object of securing a black and white picture without loss of detail. They are also experimenting in the further magnification of the received pictures and a demonstration screen picture 9 inches square was shown. The Post Office engineers, who were present at the previous demonstration in February last, considered that marked improvements had been achieved.'

The use of film and associated equipment as a source of test signals for television purposes persisted until EMI's electronic camera tube was fully developed and able to generate 'live' pick-up signals. W Turk, a member of EMI's staff, has related an amusing anecdote[19] about one of the test films.

'One of these [BBC] test films was of Irene Prador (now [1991] appearing as the landlady in 'Dear John' on BBC as a re-run) performing a rather active dance routine. Her bra was a rather loose fit and during the sequence didn't do a very good job! The engineer-in-charge . . . was convinced that one day the garment would give way completely and so he never missed a showing. It was some time later on a trip to Lime Grove studios that I discovered that the film was a continuous loop so, at the appropriate frame, the dance sequence started again with [the] intriguing garment firmly fixed in position!'

Following the Baird and EMI demonstrations, a conference was held on 21st April 1933, at the Post Office, between the BBC and the Post Office representatives. It was agreed that[20]:

1 the EMI results were vastly superior to those achieved by the Baird Company;
2 the results were incomplete because of the different transmission methods (line and wireless) used in the two cases—also the effect of electrical interference and absorption could not be tested;
3 further tests by wireless in a town area were essential to determine the range of reception and the effect of interference;
4 whatever system of synchronisation was adopted in the first instance for a public service might be liable to standardise the type of receiving equipment;
5 a test of one system could not be a reliable judgement on the results achieved by the other.

During the discussion the BBC said it was anxious to start trials of the EMI system, but considered the inability of the system 'at the moment' to produce direct television a disadvantage, as the cost of film—about £30—might be prohibitive.

Meanwhile, from about May 1932 Tedham and J D McGee had been undertaking many investigations on the chemistry of the preparation of electronic camera mosaic signal plates and on the physics of the mechanisms operating at the surfaces of the plates, and both single-sided and double-sided signal plate camera tubes had been constructed[21-27] (see Burns, *op. cit*) (Figures 6.1 and 6.2).

In July 1933 Zworykin presented a paper[28] on his electronic camera tube (which he called an iconoscope) at a meeting, held in London, of the IEE. McGee attended the lecture, since he was especially interested to learn about the practical construction of the iconoscope, but he noted[29], when reviewing the situation at this time: 'it was quite clear that [we] could learn nothing from Zworykin's paper'. No practical information was given and the preparation of the mosaic by a 'special process' was not described. Apart from the knowledge that successful development was feasible, McGee's group had to develop all the necessary techniques themselves.

One effect of Zworykin's lecture was the further recruitment of staff into Shoenberg's R&D team. By June 1934 the Research Department comprised 32 university graduates, 32 laboratory assistants, 4 glassblowers, 4 girl vacuum pump operators, 1 coil winder, 3 mechanics, 25 instrument and tool makers, 5 girl assistants, 7 draughtsmen designers, and 1 designer draughtsman, a total of 114 persons. Of the 32 graduates, 9 had PhDs, despite the fact that PhDs were not particularly common in the early 1930s, and 10 had been recruited direct from Oxford and Cambridge Universities[30].

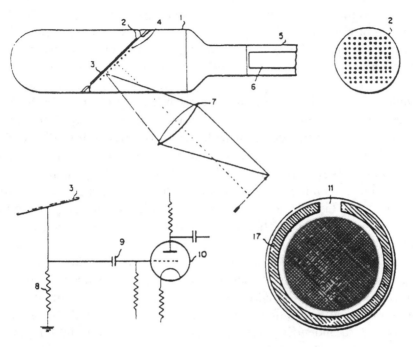

Figure 6.1   *Single-sided mosaic target plate camera tube*

Source: British patent no. 406 353

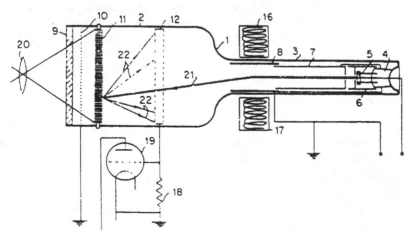

Figure 6.2   *Early McGee (EMI) double-sided mosaic signal generating tube*

Source: British patent no. 419 452

The two most senior members of Shoenberg's staff were G E Condliffe, formerly of HMV, and A D Blumlein. Condliffe was an efficient and able research manager: Blumlein had a roving commission. Prior to early 1933 Blumlein had been engaged on stereophonic recording and reproduction, but from about March 1933 he spent most of his time on the television problem. Effectively from this date he became the project leader. Such was his genius that he had Shoenberg's authority to enter any of the laboratories and engage in discussion with any of the research staff. He had a seminal effect on all work undertaken in the Research Department and contributed greatly himself to all aspects of EMI's television research project. Being so full of ideas and of a modest and a generous disposition, he went out of his way to give credit to others. This generosity engendered complete trust so that staff would discuss with him even the most 'half-baked' ideas[31].

Organisationally the research staff were split into groups, each having a group leader. McGee[32] has described how Shoenberg maintained control of the various research projects of these groups. Quite frequently the group leaders, individually, would be summoned by a 'Royal Command' to Shoenberg's office, which was in a different building, where 'elevenses' or afternoon tea would be served.

> '. . . then he'd start to grill us, asking questions about our programme, what we were doing and why . . . he would really put us through our paces. I suspected at the time and I have thought [so] ever since, that as often as not he was playing the devil's advocate: he was advocating a point of view that he didn't really believe in but wanted to provoke us into quite vehemently defending what we were doing. . . . He would spend an hour or so on this business and then he would finish. He would brood over [what had been discussed] and he would put every one of us [the group leaders] through the same kind of process. [Afterwards] he would make up his mind as to what was to be done. He kept a pretty close check on exactly what was [happening] and of course every now and again he came round the labs and saw what was going on.'

On Blumlein, McGee has recorded:

> '. . . it was only when you got to know him and talked to him and worked with him that you really appreciated the tremendous stature of the man, both intellectually, in the widest sense, and technically as an engineer and as a human being. He was a man of such tremendous integrity; I think that is the most outstanding characteristic of Blumlein that I can remember. He was the kind of person who I don't think would ever cheat anybody else of the smallest thing. This was of course particularly important in this question of credit for ideas for the work that was done. If he had been a 'go-getter', out to grab everything — every idea that he possibly could get the credit for and exclude his colleagues — he could have completely disrupted [the] group, by causing jealousy and friction, but there was never the slightest hint of that.'

Shoenberg's decision to allow Blumlein to have a roving commission was certainly wise and led to the resolution of many technical difficulties. In the early 1930s there was a distinction, almost a barrier, between the electrical engineers who applied electronic devices to the solution of engineering problems, and the physicists who conceived the devices and who were concerned with the physical processes operating within them. At EMI many experimental investigations had to be undertaken on photo-emission, secondary emission, the motion of electrons in electric and magnetic fields, and fluorescence, before satisfactory electronic camera and display tubes could be properly designed. This was the field of the physicist. The task of the engineers was to integrate these electron tubes into a completely engineered system of television. Fortunately Blumlein's questioning and fertile mind recognised no barriers and the problems of the physicists were of as much interest to him as were the difficulties of the circuit designers. McGee has said[32]:

'I vividly remember how he would push open the door of my office and say in a diffident tone of voice, "Do you mind if I make a silly suggestion?" Of course his suggestions were anything but silly and this was usually the prelude to a very interesting animated discussion of some idea or problem. These discussions were of immense help to me and demonstrated quite clearly that even if he had not been trained as a physicist he knew more physics than most physicists. I'm sure that my tube colleagues [W F Tedham, Dr L Klatzow, Dr L F Broadway, Dr G Lubszynski, Dr Bull and others] would all agree that these discussions contributed much by way of stimulation and clarification of ideas. Anyone who has tried to do research will know how valuable such discussions can be, especially if they are with a person whom you can trust not to snatch ideas, and it is in this respect that Blumlein's integrity was of paramount importance.'

An appreciation of Blumlein's prowess in solving problems has been given by W Turk[33], a member of Dr Lubszynski's group:

'Lubszynski and I had on test a complicated vacuum tube aimed at verifying some particular parameter and had accumulated around it the then usual array of accumulators and h.t. batteries to feed its many electrodes via wander plugs and multi-coloured wires. We had tried for several days to get the result we wanted and knew was possible but the meters just wouldn't behave. The day in question had been particularly frustrating. Lubszynski and I were still fighting the circuitry at 10.00 p.m. in the dark when [there was] a knock at the door. It was Blumlein, who apologised for the interruption, said he'd seen a glimmer of light under the door, was worried in case something had been 'left on', drew on his pipe, and asked what we were doing. Lubszynski explained briefly and Blumlein said: "Why not put that lead here instead of there?" Lubszynski did and everything clicked into place! We were mad! Here was a circuit engineer who knew nothing of tubes (in the heat of the moment we forgot his work with

Tedham on c.r.ts [cathode ray tubes]) coming along in the dead of night with no previous knowledge of the work we were doing, summing it up in a few seconds and solving the problem which had been taxing us for several days.

'Needless to say, what friendship there had been between Lubszynski and Blumlein was considerably strengthened!'

Funding for the television project was amply sufficient. Any equipment considered necessary to advance the research projects was agreed at once. Initiative was encouraged.

Moreover, working conditions for the staff were very agreeable. Shoenberg believed in giving his engineers time to relax and to reflect: the annual vacations varied up to eight weeks in duration depending on seniority and length of service. For most firms in the 1930s a holiday allowance of two to three weeks was the norm. Such munificence had its reward for EMI. The staff were highly dedicated and, for many, standard working hours meant nothing. 'There was terrific enthusiasm; it wasn't just a kind of slave driving. I don't think I have ever known a team of people work so enthusiastically and so willingly — because it was all so frightfully interesting' (McGee[34]).

Blumlein, as noted, would work until late in an evening if he was pursuing a particular idea, and if a thought occurred to him after hours on a Saturday or a Sunday he would return to the laboratory and conduct practical work to test the idea's validity. Overtime was not paid.

I L Turnbull[35], who first met Blumlein in the second year of the engineering degree course at the City and Guilds Engineering College (now Imperial College of Science Technology and Medicine), remembered an occasion when one Sunday he was playing tennis, at a club not far from Blumlein's house, with J Hardwick, a member of the television group. During the match Blumlein arrived, 'looked over the fence and said to Hardwick: "I think we had better go back to Hayes and sort out the amplifier we are working on." The result was Hardwick was yanked off to Hayes'. As Turnbull observed, it is seldom that 'anyone [is] such a hard worker and yet [is] so brilliant. Often [it is] one or the other.'

Sometimes Blumlein's interest in the problems of others would ruffle a few feathers. On one occasion a member of McGee's group, who was working on an experimental cathode ray tube, found, on returning to the labs after a particular weekend, that his experimental apparatus had been disturbed[36].

'Things were not as he had left them and he was very upset and came to me: "Someone has been in my lab. messing about with my apparatus over the weekend. This is not good enough." I reported this to Condliffe, and literally within half an hour I had a shame-faced Blumlein apologising. During the weekend, at home, he had had some thought ... got into his car, went to the labs. and started twiddling the knobs [of the experimental rig] trying to find out whether his ideas would work or not. Of course this upset [the researcher] quite a bit. But Blumlein was so genuine and so generous about it that one just could not hold any grievance for [more than] five minutes.'

According to Nind[37]: 'All [the staff] worked jolly hard under Blumlein's inspiration. It wasn't necessary for anyone to be in charge. If anyone was in charge of the television department it was Browne.' Nind recalled 'splendid relations' with Blumlein and the atmosphere of 'a community of friends'. 'There may have been some jealousies with his brilliance but there was no sense in competing with it. They [the staff] used what little bit of brilliance they had.'

To support his powerful research team Shoenberg succeeded in persuading the EMI Board to invest about £100 000 per year in EMI's R&D work on television. The expertise and funding of EMI's Research Department represented an ominous situation for Baird Television Ltd, which could not match EMI's staff and financial resources.

In the early 1930s the only efficient photocathode surface was the S1 which had been discovered by Koller in December 1930. McGee and Tedham knew from their 1932–33 experiments that such a surface could be formed on a silver layer by first oxidising the silver in a discharge in oxygen and then exposing it to caesium vapour at about 160°C. The difficulty in forming the mosaic lay in the preparation of the array of the many thousands of individually insulated silver islands on the substrate material. There were three possible ways of achieving this: first, by cross-ruling a silver layer which had been deposited by evaporation in vacuo; secondly, by evaporating silver onto a substrate through a stencil mesh; and thirdly, by aggregating a thin layer of silver by heating to about 600°C in air.

Of these methods McGee had had experience of the second, since this was the method which Tedham and he had adopted when fabricating their 1932 camera tube. The technique had worked well but in 1932–33 the meshes available did not have a small enough pitch to give a fine-grained mosaic for a high-definition picture.

McGee has recorded[38] that the first ten experimental tubes produced at EMI used ruled silver mosaics on mica. While at the Cavendish Laboratory, he had acquired a good deal of practice in splitting thin sheets of mica, for nuclear physics work, so it was not long before he had sheets of mica 3 inches by 3 inches in area and 20 µm in thickness. These were metallised on one surface by painting with platinum and baking, followed by evaporation of silver, in a vacuum, on the other surface. The surface was then ruled in the workshop using a fine sapphire point, in a machine adapted from EMI's record ruling equipment, at about 100 lines per inch in two orthogonal directions. This had the disadvantage that the silver surface was exposed in a not very clean atmosphere for many hours.

Many problems in the processing of the tube had to be overcome as a result of the rather incompatible processes that were necessary: the degassing of the tube and electron gun; activation of the thermionic cathode: oxidation of the silver mosaic; and last and most difficult the introduction of the caesium in just the right amount to activate the silver mosaic to give good photosensitivity without impairing the mosaic insulation by bridging over with caesium the gaps between the adjacent elements. By operating the tube while it was still connected to the vacuum pump the sensitivity and mosaic insulation could be monitored and adjusted either by adding caesium, or removing it by baking.

Of the first ten experimental tubes, no. 6 gave 'a very presentable picture' on 24th January 1934. McGee's group demonstrated, for the first time, their all-electronic television camera to the Company Chairman, Mr A Clark, and the Director of Research, Mr Isaac Shoenberg, on 29th January 1934.

Meanwhile Dr L Klatzow had been investigating the process of aggregating the silver films on mica to form an array of silver globules. These films soon appeared to be so much superior to those formed by ruling, from the point of view of uniformity, cleanliness and fineness of structure of the mosaic, that from tube no. 12 onward mosaics were prepared by this process[39].

EMI applied for patents for both the stencil and aggregate methods, but were not successful in obtaining a patent for the latter method. S Essig[40], of RCA, had also discovered the aggregation process and his patent application preceded that of McGee by only a short time.

Progress was now rapid. 'Reasonable' tubes were being made by February 1934. When tube no. 14 was made, it was so much better in picture quality and sensitivity that a very experimental camera directed through the window of the laboratory, on 5th April 1934, enabled a daylight outside broadcast picture to be obtained[41].

The iconoscope, or Emitron as EMI called their version of this type of camera tube, had several defects which had to be overcome before a practical camera could be marketed. Given a single-sided mosaic onto which both the scanning electron beam and an image of the scene to be televised were incident, it is obvious that either the optical image must be projected normally onto the mosaic target and the scanning beam projected obliquely, thereby giving rise to keystone distortion of the raster, or the converse. In either case a considerable depth or field is required, either of the electron lens or of the camera lens respectively.

In 1934–35 the target area of the mosaic was 5 inches by 4 inches. This called for a long (6.5 inches) focal length lens, to cover it. At full aperture, f/3, the optical depth of field was insufficient for the second of the above alternatives. However, by lengthening the electron gun and by stopping down, the electron lens could be designed to give a uniform beam focus even when the beam was incident obliquely onto the mosaic. For this reason the arrangement given in Figure 6.3 was chosen. Necessarily the line scanning signals had to be modulated to give a constant width scanning raster on the mosaic.

The adoption by RCA and EMI of the same constraints for the configurations of their camera tubes naturally led to similar shapes and the innuendo that EMI had copied Zworykin's iconoscope. This imputation probably gained further credence because of the known RCA association with EMI. But, McGee has stated[42] 'categorically that there was no exchange of know-how between the two companies in this field during the crucial period 1931 to 1936'. Dr G H Lubszynski, Dr L Broadway, Mr I Shoenberg, Mr S J Preston, Mr C O Browne, N E Davis, and Mr A D Blumlein have also confirmed the independence of McGee's team from any RCA influence (see Figure 6.4). This matter is considered in detail in the author's *Television, an international history of the formative years*.

After McGee's group had fabricated their first batch of Emitrons it was found that the signals obtained tended to become submerged in great waves of spurious signals associated with some secondary emission effects. Shoenberg has said that because of this, there was a temptation to continue

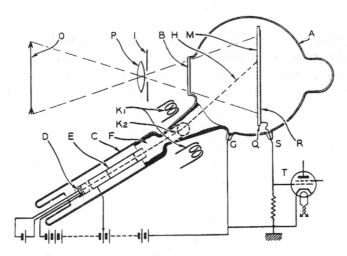

Figure 6.3   The Emitron television camera

Source: Professor McGee

Figure 6.4   *The photograph shows, from left to right: back row, Mr Blythen, Dr*
Allaway, Sir Joseph Lockwood, Dr Broadway, ?, Mr Birkinshaw and Mr
Bridgewater; front row, Dr Shoenberg and Mr Shoenberg (the sons of Sir
Isaac Shoenberg), Professor McGee, Mr Preston, Dr White, and Simon and
David Blumlein (the sons of A D Blumlein)

Source: Mr S J L Blumlein

their work using mechanical scanners. 'Instead we decided that the poten-
tialities of the electron scanning tube justified a great effort to overcome
the problems it [secondary emission] presented at that time.'

Figure 6.5A illustrates the spurious signals superimposed on the
wanted pictured signals and Figure 6.5E shows the signals required. To
effect the necessary transformation from A to E, correcting signals, which
became known as 'tilt' and 'bend', had to be electronically generated and
added to both the line and frame signals to annul the unwanted signals.
The large spurious pick-up signals that appeared during the line and
frame flyback periods (Figure 6.5C) were also suppressed (Figure 6.5D)
and appropriate synchronising signals inserted (Figure 6.5). 'Great credit',
recalled Shoenberg[43] in 1952, '[was] due to Blumlein, Browne and White
for the resourcefulness by which they devised tilt, bend and suppression
circuits to combat this evil'.

Strangely, Zworykin did not refer to the signals, or to keystone distortion in his 1933 IEE paper.

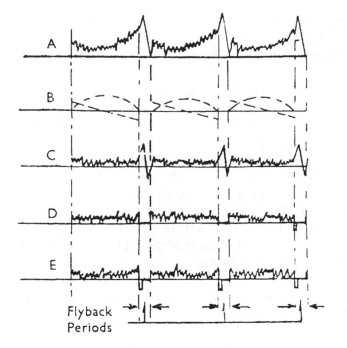

Figure 6.5   Spurious signals associated with the Emitron

Source: Professor McGee

At about this time, 1934, Blumlein and McGee invented[44] the ultimate solution—cathode potential stabilisation—to the problem of the spurious signals. In the conventional Emitron, the shading or spurious signals were due to the uncontrolled secondary electrons liberated by the scanning primary electron beam. Since the primary electrons struck the mosaic plate at a high velocity, thereby causing the secondary emission, Blumlein and McGee reasoned, independently, that if the primary beam approached the target in a decelerating electric field and struck the surface with substantially zero energy, no secondary emission would occur, and hence there would be no shading signals. Actually the primary electrons charge more and more negatively until no beam electrons are incident on the surface.

The method for cathode potential stabilisation (c.p.s.) of the target and a proposed apparatus were patented by Blumlein and McGee on 3rd July 1934. It seems that Blumlein formally informed Condliffe of his notions on c.p.s. and that Condliffe then asked McGee for his opinion.

"'Well', I said, "I have been toying with the same idea myself; here's my note-book to show you." He said: "It looks as though you've both got the same idea.

What I suggest is that you arrange to be co-inventors." Blumlein accepted this without the slightest hesitation. He was that kind of person; he wouldn't have thought of arguing that it was really his idea. Later on, it became the rule that the first person to actually write down a proposal in black and white and hand it to Condliffe was the [person] who got the credit for the patent.'

Cathode potential stabilisation has several important advantages:

1 the utilisation of the primary photoemission is almost 100% efficient, thus increasing the sensitivity by an order of magnitude;
2 the signal level generated during the scan return time, or 'picture black', in the picture;
3 the spurious signals are eliminated; and
4 the signal generated is closely proportional to the image brightness at all points; that is, it has a photographic gamma of unity.

These advantages became of great importance in the operation of signal generating tubes such as the c.p.s. Emitron, the image orthicon, the Vidicon, the Plumbicon, and every one of the Japanese photoconductive tubes up to the advent of the charge-coupled devices. Following the 1939–45 war, c.p.s. operation of camera tubes became the norm. The first public outside broadcasts with the then new c.p.s. Emitrons were from the Wembley Stadium and the Empire Pool during the 14th Olympiad held in London in 1948. The cameras were about fifty times more sensitive than the existing cameras and enabled a 'wealth of detail and remarkable depth of field' to be obtained (see Figure 6.6). Much enthusiasm for British television prevailed. Lord Trefgarne, Chairman of the Television Advisory Committee, felt that: 'In its television transmissions from the Olympic Games, the BBC has just given the world a striking visual demonstration of the technical excellence of the British system, using the latest British cameras, British transmitting equipment and British receivers. The results have been the admiration of our overseas guests, including the Americans...'.

However, when cathode potential stabilisation was put forward in July 1934, there were very considerable practical difficulties in implementing the principle. At that time the Television Committee, chaired by Lord Selsdon, was taking evidence from Baird Television Ltd, EMI, and others, and it appeared that the Committee would recommend the early establishment of a system of medium/high-definition television in the UK. Because of this, and the improvements in image quality which were being obtained from the experimental Emitron tubes, Shoenberg decided to concentrate McGee's group's efforts on these tubes.

Figure 6.6   CPS cameras at the Empire Pool, Wembley, for the 1948 Olympic Games

Source: EMI Archives

The early Emitron camera tubes had a low photosensitivity estimated at perhaps 25% of the ideal. In addition, the efficiency with which even this low photosensitivity was being utilised to produce picture signals was also very small — approximately only a few percent. Nonetheless, with all its faults the Emitron did have two major virtues: first, there was practically no visible lag in the picture, hence moving objects were reproduced clearly and second, it had a very acceptable gamma which resulted in quite pleasing images even when the scene illumination was excessively contrasty. The reason for both these observed facts did not become clear for some time.

Many visits to EMI were made by important persons during the formative period of EMI's system. These included Sarnoff, of RCA; the Prime Minister; Ashbridge and Kirk of the BBC; Vandeville of MWT; Reith, the Director General of the BBC; representatives of the General Post Office, and others. Of these private demonstrations, that given to Ashbridge in January 1934 is of some importance since his account of his visit is extant. He was very impressed[45].

'The important point about this demonstration is, however, that it was far and away a greater achievement than anything I have seen in connection with television. There is no getting away from the fact that EMI have made enormous strides.'

The demonstration had consisted of the transmission of films, with 150 lines per picture, from the EMI factory at Hayes to the recording studios in Abbey Road, a distance of approximately 12 miles.

'This was by means of an ultra short wave transmitter with a power of 2 kW on a wavelength of approximately 6.5 m. The results were extremely good and there was no question in my mind that programme value was considerable. The receivers used appeared to be in a practicable form and looked very much like large radiograms. On the other hand, it has to be said that the aerial arrangements were very elaborate, being directional in order to cut out interference.'

Shoenberg told Ashbridge that the policy of EMI was to develop television energetically since they believed there was a great commercial future for the firm which was first in the field with something practicable.

Two months later the Marconi–EMI Television Company Limited was formed 'to supply apparatus and transmitting stations'[46].

In the same month, March 1934, Baird Television Ltd showed their cathode ray tube receiver to a party comprising the Prime Minister, BBC representatives, Colonel Lee, and Colonel Angwin of the General Post Office, and because the cathode ray tube was of GEC manufacture, GEC personnel. Transmissions were given from the Crystal Palace using a wavelength of about 8.5 m for the vision signals and a longer wavelength for the sound signals. The pictures transmitted consisted of an introductory talk (showing the head and shoulders of the speaker), a violin solo by a lady violinist (also limited to head and shoulders), a talk on architecture illustrated by large-scale photographs, and short extracts from two new films.

Colonel Angwin[47] thought the standard of reception was approximately the same as that obtained in the demonstration given on 28th November 1933, with some limitations due to the radio link. 'Some interference was obvious from electrical sources, but as far as the radio link was concerned, the conditions were fairly good. The receiving aerial was at the top of the four-story building and fairly remote from motor car interference.'

The detail of both the head and shoulder subjects televised and close-ups of the films was reasonably good. For larger scenes the detail was much improved. It was not a demonstration to inspire Ashbridge, who wrote of his disappointment. The Prime Minister congratulated Mr Baird on the success he had obtained and the very great advance on his earlier attempts.

'The film transmission given by EMI is appreciably better than that shown by the Baird Company', Ashbridge observed. 'On the other hand, however, no opportunity has been available so far to compare a demonstration under absolutely strictly comparable conditions. Moreover, the EMI company have not so far attempted a demonstration with living objects.'[48]

In an attempt to settle the rival claims of the two companies, Reith[49] on 15th March 1934 wrote to Kingsley Wood, the Postmaster General, and proposed a conference 'between some of your people and some of ours to discuss the future arrangements for the handling of television'. Reith thought there were three aspect to discuss: the political, 'using the term in a policy sense and for want of a better one', the financial and the technical. He nominated Admiral Sir Charles Carpendale and Mr N Ashbridge. Kingsley Wood[50] agreed and put forward the names of Mr F W Philips and Colonel A S Angwin. The decision to ask for the conference had not been precipitated by the Baird demonstration alone, for Philips had noted four days before this that the BBC would probably be seeking an interview shortly to discuss the whole question of the future arrangements in regard to television[51].

The informal meeting, which was chaired by Philips, and to which Mr J W Wissenden of the GPO had also been invited, was held at the General Post Office on 5th April 1934. A number of general questions were examined by the BBC and GPO representatives, including:

1 the method of financing a public television service;
2 the use of such a service for news items and plays;
3 the relative merits of some of the systems available including those of the EMI, Baird, Cossor, and Scophony companies;
4 the arrangements necessary to prevent one group of manufacturers obtaining a monopoly on the supply of receiving sets;
5 the possible use of film television to serve a chain of cinemas.

The use of film for television purposes was of some importance at this time, because only the Baird company had shown direct television. EMI had refrained from exhibiting this form since they regarded its development in the early part of 1934 as unsuitable for commercial exploitation. They also felt that a film had a more lasting commercial value: their view of direct television was that by its very nature it was essentially transient.

With two rival companies campaigning for the creation of a television service—EMI for a new BBC station and Baird Television for a station of its own—it was agreed by the conference that a committee should be appointed to advise the Postmaster General (PMG) on questions concerning television. The BBC representatives were keen that this committee should be established as soon as possible, since difficult questions were arising, and would continue to arise, and they thought it would be helpful for the BBC and the GPO to have the weight of the authority of a committee behind them in any decision they might take. The PMG agreed and the Television Committee was constituted. The Chairman was Lord Selsdon.

The Television Committee worked with commendable speed and examined 38 witnesses, some of them on more than one occasion, and in addition sent delegates to the United States, and to Germany, to investigate and report upon progress in television research in those countries. By September 1934 the committee had so advanced its deliberations that it was able to commence the preparation of its report and, but for the absence of Sir John Reith (whom the committee wished to interview), towards the end of 1934, the committee's report would have been produced before January 1935 (the actual month of publication).

The first witnesses to be examined on 7th June were Major A G Church and Captain A G D West of Baird Television Ltd. EMI's and M–EMI Television Company's representatives were questioned on 8th and 27th June, the representatives being Messrs Shoenberg, Condliffe, Blumlein, Agate, Browne, and Davis; and Messrs Clarke and Shoenberg (accompanied by Blumlein and Preston) respectively on the two occasions. The presence of Blumlein at each examination highlights the regard Shoenberg had of him, and the position he held in the hierarchy of the Central Research Laboratories. Previously the committee had seen demonstrations of Baird Television Ltd's low- and high-definition television systems on 31st May and 2nd June, and demonstrations of EMI's system on 1st June.

Much of the committee's time was, of course, spent on EMI's proposed high-definition system. Shoenberg told Lord Selsdon that 'the reasons which guided [EMI] in the choice of the fundamental features of [its] system appear[ed] as convincing as the proof of a theorem in Euclid'. Lord Selsdon riposted by saying: 'I think the answer is that they [EMI] have picked the plums out of the pudding, in so far as they can find any plums'. Shoenberg challenged the Chairman to give him 'a more logical system'.

Any system of television is characterised not by the use of a particular type of camera, nor by the use of electrical or mechanical scanning, but by the specification of the system and the transmitted waveform. As Blumlein said[52]: 'The fixation of a waveform may almost certainly imply the use of particular types of apparatus at the transmitter, but such apparatus is largely subservient to the standard chosen'.

Of all the decisions which Shoenberg had to face, the most difficult concerned the specification of this standard. McGee has recalled[53]:

'There was, for example, a long debate as to whether d.c. or a.c. amplification should be used — and then, if a.c. how could the d.c. level (black level) be re-established? And as the picture definition was increased the bandwidth would be increased, so that the radio transmission frequency necessary also increased, limiting the range to approximately line of sight. Would this be acceptable? Then, should positive or negative modulation be used? And as the brightness of pictures increased, flicker became a serious problem so that the alternatives of sequential or interlaced scanning had to be decided. These problems landed squarely on Shoenberg's desk; to paraphrase a well known saying: "the buck stopped there".

'As the number of picture lines crept up from 120 to 180 and then to 240, the required picture-signal bandwidth increased; this increased progressively the problems of amplifiers, of transmitters, of tubes and of achieving adequate service area. It was clear that the quality of the picture increased very noticeably as the number of lines increased. Since it was possible to scan an electron beam at these much higher speeds, it was natural, indeed inevitable, that the advantages of the electronic system should be exploited to the maximum. But what was the practical maximum? It would clearly be difficult, if not impossible, to push the number of lines much further than 240. No one knew what disastrous snags we might meet if we attempted to reach still higher definition. Higher definitions were terra incognito, and not just the rather obvious line of development that it may now [1971] appear. . . .

'To us, then young men, it was a challenge and an adventure, but to Shoenberg it must have been a very worrying problem. On him fell responsibility to the company and to his staff to make the right decision. . . .'

And having made the decision, the task of delineating the television waveform which fulfilled the specification belonged to Blumlein. Among the research and development team of vacuum physicists and circuit engineers, Blumlein's role was that of overall system engineer.

He outlined the reasons for the choice of the 405-line waveform in an important paper[54] on 'The transmitted waveform' which he read at a meeting, held in April 1938, of the Institution of Electrical Engineers. The paper was one of three on the Marconi–EMI television system: the others were 'The vision input equipment' by C O Browne, and 'The radio transmitter' by N E Davis and E Green.

In any scene which is being televised by a television camera/scanner the camera/photocell tube generates signals having a value corresponding to the brightness of the various elemental areas which comprise the scene. The output from the camera/photocell tube is therefore unidirectional, albeit a fluctuating current above a certain zero datum line (corresponding to a completely black scene). This fluctuating, unidirectional current may be analysed into an alternating component which depends on the relative brightness values of the different parts of the picture and a direct component which conveys information on the brightness of the whole of the picture area. Thus the signals resulting from the televising of a completely black sheet of paper and a brightly illuminated white sheet of paper are the same except for the magnitude of the direct component in the two cases (assuming that the sheets of paper completely fill the field of view).

The EMI team thought it was essential to transmit the direct component so that the receiver would accurately reproduce the brightness values of the original picture, although the practice in the United States was to use the alternating component only.

Shoenberg has written[55]: 'A great controversy raged for some months among my staff as to whether we should use d.c. amplification or a.c. amplification with the d.c. component. In the early days Willans was the protagonist of the restoration method and later on Blumlein advocated it.'

Blumlein's first patent on d.c. restoration is dated July 1933. He subsequently, sometimes with others, produced during the period 1933 to 1938 a further six patents on this topic. However, it was Willans who, in 1933, proposed the accepted solution to the problem of d.c. reinsertion following Shoenberg's decision to adopt d.c. restoration. Willans had been with the Marconi Wireless Telegraph Company and was probably familiar with their work on the transmission of synthetic half-tone pictures by d.c. keying the radio transmitter.

In 1952 Shoenberg recalled[56]: 'It seems strange now that this matter should ever have been one for controversy, but at the time there were cogent arguments on both sides and in the end I made the decision in favour of restoration.' This decision meant that definite carrier values represented black, white and the synchronising signal. There was no mean carrier value, as this depended upon the picture brightness and therefore the system of transmission was analogous to telegraphy rather than telephony.

In addition to stressing, to the Television Committee, the advantage of transmitting the d.c. component in order to give a true picture, Shoenberg and Blumlein mentioned the gain in effective radiated power from the transmitter. Blumlein in particular gave the Television Committee an indication how this came about:

'If one is not employing d.c. working one requires synchronising signals which are approximately twice the picture signals and there is a further requirement of extra characteristic room to allow for the wander of the values of the synchronising pulse and the values representing full white. That gives a total possible amplitude of the wave which is 3.9 times the useful picture amplitude. By using d.c. working we can work with synchronising pulses which are of the order of half or possibly one-third of the picture signal, and there is no wander whatsoever. A fair estimate would put the total amplitude required as 1.5 times the useful picture amplitude. That is a comparison between 3.9 and 1.5. The power required to handle this amplitude range is proportional to the square of the amplitude range. But there is a further advantage in d.c. working; we can afford to work over the curves of the transmitter characteristics, which are normally unusable, with synchronising signals because they are square top waves, and a distorted square top wave is still a square top wave. We can therefore use our transmitter very efficiently without working to keep exactly on the straight line portion. We use the last dregs of the transmitter right down to zero.'

The result was, Blumlein stated, that the gain in effective radiated power was approximately eight by using d.c. working, although Shoenberg interposed 'ten or twelve times'.

On the question of the number of lines per picture necessary for entertainment purposes, several studies had been carried out during the period 1929–33.[57–60] The most important of these was that undertaken in 1933 by E W Engstrom[60], of the RCA Victor Company. For part of his investigation, Engstrom used an ingenious technique to make cine films having a detail structure equivalent to television images. His films included: (1) head and shoulders of girls modelling hats; (2) close-up, medium and distant shots of a baseball game; (3) medium and semi close-up shots of a scene in a zoo; (4) medium and distant shots of a football game; (5) animated cartoons; and (6) titles.

His findings were based on, first, the resolving properties of the eye; and, secondly, practical viewing tests of the cine films which were specially prepared to give pictures having 60-, 120-, 180-, and 240-line structures, and normal projection print quality.

In his work, Engstrom set as his standard the ability of the eye to see the elements of detail and picture structure. Another less exacting standard was the ability of images having various degrees of detail to tell the desired story. Taking as a standard the information and entertainment capabilities of the 16 mm home movie film and equipment, Engstrom estimated the entertainment value of the television images in comparison as:

60 scanning lines   entirely inadequate
120 scanning lines   hardly passable

180 scanning lines   the minimum acceptable
240 scanning lines   satisfactory
360 scanning lines   excellent
480 scanning lines   equivalent for practical conditions.

His conclusions agreed effectively with those of Wenstrom[61], particularly with regard to the number of lines per picture needed to gave excellent reproduction. Both investigators found that 360 to 400 lines per picture were required for this purpose. This finding was published generally in December 1933 and so was known to EMI.

When EMI was invited to give evidence to the Television Committee in July 1934, it submitted a memorandum outlining the characteristics which would be suitable for a commercial broadcast television system. Of the 11 points listed, the only one which was later altered concerned the number of scanning lines per picture, namely 243. By February 1935:

'. . . Shoenberg had made what was probably the biggest — and I[62] [McGee] consider the most courageous decision — in the whole of his career: to offer the authorities concerned a 50 frame/second, 405 line/picture television system. Remember that this meant a 65% increase in scanning rate and a corresponding decrease in scanning beam diameter in the c.r.t., a nearly three-fold increase in picture-signal bandwidth; and — worst of all — a five-fold decrease in the signal/noise ratio of the signal amplifiers and this lists only a few of the resulting problems.

'The cynic may say that this was a piece of gamesmanship planned to overwhelm our competitors. But no one who knew Shoenberg or who was aware of the real state of technical development at that time would give this idea a moment's credence. No — it was the decision of a man who, having taken the best advice he could find, and thinking not merely in terms of immediate success, but rather of lasting, long-term service, decides to take a calculated risk to provide a service that would last. . . .

'To us this decision was a stimulating challenge. To Shoenberg it must have been a heavy and worrying burden. In later years, he often recalled how colleagues in the higher management of the company had seriously questioned his decision, and had warned him that should he fail to fulfil his contract it would be disastrous for the company.

'Yet I cannot remember that he ever showed his worries to us at all obviously. The nearest perhaps was one day when things were particularly sticky. That day he finished up a rather depressing review of our progress with the comment "Well gentlemen, we are afloat on an uncharted ocean and God alone knows if we will ever reach port." '

Mr S J Preston[63], formerly Patents Manager of EMI, has observed that Shoenberg made his choice to adopt 405-line television 'knowing that receivers available at that time could not be expected to deal with the full bandwidth which 405 line scanning would require but his view was that it was better that the early pictures should be somewhat lacking in definition along the line so that later developments in receiver design which he was sure would take place could be usefully employed without any change of standards'.

EMI was led to consider interlacing when their cathode ray tube receivers began giving brighter pictures and it was obvious that 25 Hz flicker would be unacceptable in practice. Prior to the beginning of 1934 the company was working on a 180-line picture, with sequential scanning at 25 frames per second, using cathode ray tubes which were not at all bright. With a poorly illuminated picture at the receiver, flicker was not very objectionable, but as the brightness of the screen improved the flicker could not be tolerated

Flicker is of importance not only in television reproduction, but also in motion picture projection, and as the implementation of the latter art considerably predated the former one, some knowledge was available, on picture frame rates for flicker-free viewing, when television advanced to the stage of the public broadcast service.

In early cinematography, during the era of silent films, the picture rate was 16 frames per second. However, the use of this rate and a single-bladed shutter gave rise to an appreciable flicker effect and means had be found to maintain the picture rate with a higher viewing frequency. The solution was to adopt a three-bladed shutter which gave three light–dark changes during the projection of each film frame, and thus a viewing picture frequency of 48 frames per second. A reasonably flicker-free image was obtained, but, for many years after, the colloquial term 'the flicks' remained in popular usage as a name for motion pictures. With the introduction of the 'talkies' and the adoption of 24 pictures per second, only a two-bladed shutter was necessary to give the same flicker-free frequency.

In television the early workers were limited to the utilisation of a channel bandwidth of the order of 10 kHz, and as the bandwidth for television reproduction is proportional to the square of the number of lines employed and directly proportional to the picture scanning rate, a compromise had to be decided upon between the need to use a high picture rate for flicker-free viewing and a large number of lines for good definition. Baird adopted 30 lines per picture, 12.5 pictures per second as standard, but naturally this gave rise to some image flicker. The Gramophone Company also used the same picture rate in 1931, but when cathode ray tubes began to be made and used for television reproduction in the early 1930s, capable of giving bright pictures, the low picture rate was found to be unsatisfactory for prolonged viewing. Television engineers increased the rate to 25 pictures per second (or 30 pictures per second in the USA), and in 1934 Electric and Musical Industries Ltd proposed, to the Television Committee, the adoption of interlaced scanning. This suggestion enabled the frame rate to be increased to 50 frames per second, while the picture rate was maintained at 25 pictures per second, and paralleled the practice which had been adopted in the motion film industry. But there is an important difference between the projection of a cine film and the reproduction of a television image. In the former case the whole of the picture is shown at any instant of film projection, whereas in the latter the image is built up line by line. The employment of the technique used in the film industry to achieve a high frame rate clearly could not be used, so the principle of interlacing was adopted.

In his paper Blumlein expounded the reasons why 50 frames per second was chosen[64].

'Receivers built to operate from the 50-cycle mains always have some slight residual hum which either darkens or brightens parts of the picture, or produces a slight distortion of the frame. If the frame frequency is synchronous, or very nearly synchronous, with the supply frequency, these effects are stationary, or move very slowly across the picture, in which case they are not nearly so noticeable to the eye as when moving rapidly across the picture. Experiments on a cathode-ray tube receiver worked from 50-cycle mains showed that residual-hum effects which could not be detected when scanning took place at 50 frames per second were quite intolerable at 37.5 frames per second.'

The interlaced scanning system which EMI adopted was that devised by R C Ballard of RCA and patented by him in 1933.

Of the other factors which had be defined before a waveform specification could be prepared, the direction of line and frame scanning was arbitrarily chosen as left to right and top to bottom, to accord with normal writing practice. Both bilateral (zigzag) scanning and a type of scanning where the scanning velocity depends upon the picture brightness (velocity modulation) were considered by Blumlein, but were rejected for various reasons. The system which was adopted has all the lines scanned in the same direction at a uniform speed, each line being followed by a flyback: the method necessitates the transmission of distinct synchronising and vision signals.

In cinematographic practice, in the 1930s, it was usual to increase the contrast of the recorded scene by a factor, called gamma, having a value between 1.5 and 2.0, 'presumably to make up for lack of colour'. Since this same effect was required in television, the question arose as to whether this increase should be applied at the transmitter or the receiver. Blumlein concluded from his analysis of the problem that it was advantageous to transmit pictures with unity gamma and make any correction at the receiver. His arguments led to the decision that the transmitted signal should, as far as possible, be undistorted in frequency or amplitude, and should be representative of the brightness of successive elements of the picture scanned with a constant scanning velocity.

The ratio of picture width to picture height was defined as 5:4, as a compromise between 1:1 to give maximum picture area on the face of a circular cathode ray tube and 3:2 popular in photography. From the number of lines per picture, the number of pictures per second, and the aspect ratio, the bandwidth required was calculated to be 2.5 MHz.

The specified picture standards enabled the television waveform to be characterised, but there were several important issues which still had to be resolved:

1 whether the effect of interfering signals in producing white spots on a television picture (as in 'positive' modulation) would be preferable to the effect of such signals in greatly disturbing the picture scanning (as in 'negative' modulation); and
2 the forms of the line and frame synchronising signals had to be delineated.

Prior to EMI's decision to propose the 405-line standard, mechanical scanners had been constructed to produce 243 interlaced lines per picture, and synchronising pulses from these scanners had been used for the experimental Emitron camera. The method was inflexible; an electronic pulse generating system was required which could be set to different line standards. So, as White has written[65]:

'One Sunday in November 1934, Blumlein assembled four or five of us at his house to plan the circuits of such a system, which was constructed on bread boards in a few weeks.

'The electronic divider chain, which determined the number of lines in a frame, consisted of five stages. Initially each stage was set to divide by three giving $3^5 = 243$ lines. Various other odd numbers could be chosen, for example $7^3 = 343$, or $7^5 = 441$, but the simplest change was to make one of the threes into a 5: $3^4 \times 5 = 405$.

'This was tried and gave sufficiently promising results with the experimental emitron to be confirmed by Shoenberg as the standard [EMI] would offer to the BBC. It was good enough to last nearly 50 years.'

Shoenberg recalled, in 1952, that:

'Once the emitron's signals had become usable [by the use of tilt, bend and suppression signals], the way was open for a real high definition system. The choice as to the number of lines was no longer limited by mechanical considerations but by the bandwidth which could be dealt with at the time in the

transmitter and receivers. Great differences of opinion existed in the laboratory, but finally, early in 1935, I took my courage in both hands and chose 405
lines. This may seem a low standard now, but at the time I made the decision
there were many who thought I had taken a great risk.

'At the beginning of 1935, I was thus able to submit to the Television Committee a fully detailed specification for a high definition system operating on
405 lines. The specification was subsequently adopted, without any important
modification in the Alexandra Palace transmitter.'[66]

Figure 6.7 illustrates the transmitted waveform of the Marconi–EMI
television system. Some of the important features of the waveform are as
follows (they are quoted from Blumlein's 1938 IEE paper[67]).

1 d.c. modulation
'The picture brightness component (or the d.c. modulation component) is transmitted as an amplitude modulation so that a definite carrier
value is associated with a definite brightness. This has been called "d.c.
working" and results in there being no fixed value of average carrier,
since the average carrier varies with picture brightness.'

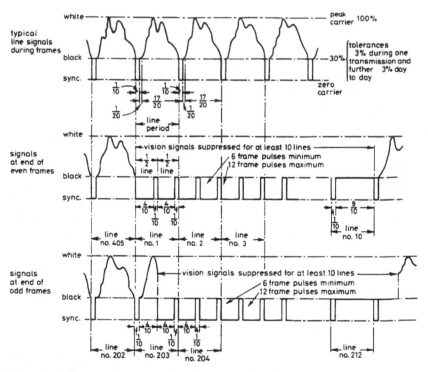

Figure 6.7   Marconi–EMI Television Ltd video waveform

Source; *J. IEE*, **83**, 1938

2 vision modulation

'The vision modulation is applied in such a direction that an increase in carrier represents an increase in the picture brightness. Vision signals occupy values between 30% and 100%. The amount by which the transmitter carrier exceeds 30% represents the brightness of the point being scanned.'

3 synchronising modulation

'Signals below 30% of carrier represent synchronising signals. All synchronising signals are rectangular in shape and extend downwards from 30% peak carrier to effective zero carrier.'

The three 1938 IEE papers on the Marconi–EMI television system attracted much interest and discussion. L H Bedford, of the A C Cossor Company Ltd, commented: 'Among the technical triumphs described in the papers three are outstanding, namely, the Blumlein waveform, the compensation [by Blumlein, Browne and White] of spurious signals by means of tilt and bend signals, and the choice of the 405-line standard. The last was a very bold step, in view of the circumstances existing at the time, but it has been well justified; at 405 lines the value of the entertainment receivable increases with increase of lines at a very much slower rate than does the cost. I expect 405 lines to remain a national standard for a considerable time if not indefinitely.' Actually the standard lasted until 1986.

Blumlein was subsequently awarded an IEE premium for his paper. In this, Blumlein acknowledged the valuable contributions of Browne and Willans. Later, after his death, the Electronics Division of the IEE established a premium associated with the names of Blumlein, Browne and Willans.

Shoenberg was very appreciative of Blumlein's role in the development of EMI's television system. In November 1934 Blumlein was presented with a gold pocket-watch inscribed:

AD BLUMLEIN

FROM

ELECTRIC & MUSICAL

INDUSTRIES LTD

CATHODE-RAY TRANSMITTING TUBE

& SCANNING EQUIPMENT

INSTALLED

HAYES. Nov. 1934

It is not known whether other members of Shoenberg's team received similar presentations. Nor is it known why Shoenberg chose to recognise Blumlein's contributions before the Television Committee had reported (on 14th January 1935) its findings to the Postmaster General. In November 1934 no public announcement had been made of the establishment of a high-definition television service in London. Presumably Shoenberg must have felt very confident that such a service would be inaugurated and that EMI's system would be selected.

As an interesting aside, it is pertinent to note the different attitudes of EMI, and of Bell Telephone Laboratories and the Radio Corporation of America to research publications. Both RCA and BTL were quite open with regard to their television interests, and both companies published important papers on their activities and findings from an early stage in their work. EMI's development plans, on the other hand, were kept strictly secret (except to members of the Selsdon Committee and some invited guests), and for some time it was Shoenberg's policy that no papers or publications should be prepared, or lectures given, by members of EMI's R&D team.

Some researchers were unhappy with this policy, but Shoenberg was quite adamant it should not be changed. 'Blumlein couldn't have cared less.'[68] Although there were many topics on which he could have written original papers, he was not an academic and was happiest when engaged in the practicalities of engineering systems. His publications list comprises just two learned society papers. The three 1938 Marconi–EMI papers mentioned earlier were the first to be published on EMI's television work from the time of the establishment of EMI in 1931.

One outcome of Shoenberg's policy was a general lack of knowledge about EMI's contributions to high-definition, all-electronic television. In 1952 he felt the need to comment on an IEE paper[69] on 'The history of television', as he thought the authors had 'perhaps not done full justice to the pioneer work carried out in Great Britain'. Another possible effect of the policy was the erroneously held view that the M–EMI system was primarily based on that of the RCA Company. The similarity of the iconoscope and of the Emitron and the use by the Hayes company of the Ballard patent on interlacing did nothing to dispel this opinion. This matter is discussed at length in Burns[70].

Shoenberg's decision to offer to the Television Advisory Committee a 405 lines per picture standard was made in February 1935, and the press notice which outlined the plans of Baird Television Ltd and Marconi–EMI Television Ltd was issued on the morning of 7th June 1935. M–EMI's ambitious proposals, while upsetting a number of companies which had developed television systems based on mechanical scanning, were quickly taken up by the Chairman of the TAC's Technical Sub-committee, Sir Frank Smith[71]. He thought it would be undesirable, if it could be avoided, that a standard should be fixed for the London Station which was inferior to the best that M – EMI were in a position to offer.

One of the most important issues which had to be considered by the TAC concerned the system of vision waveform generation to be used for film transmission, studio scenes, and outdoor broadcasts, and the lighting intensities required for each of these different modes of television.

In its tender Marconi – EMI proposed to utilise cathode ray scanning tube cameras (see Figure 6.8) for all three activities and considered that adequate illumination for studio working would be produced by 18 kW of roof lighting and 6 kW of directional lighting.[72]

Baird Television Ltd's reply[73] in the same section of the tender was much more extensive and highlighted the cumbrous nature of its equipment vis-à-vis the highly mobile Emitron camera. The company said that film transmission would be carried out first of all by means of disc scanning and then almost immediately by electron image scanning using an electron image camera (of the Farnsworth type). For studio scenes, the spotlight method with a scanning disc was recommended for the transmission of close-ups, while for studio scenes of all types the intermediate film apparatus was recommended. These systems needed substantial power

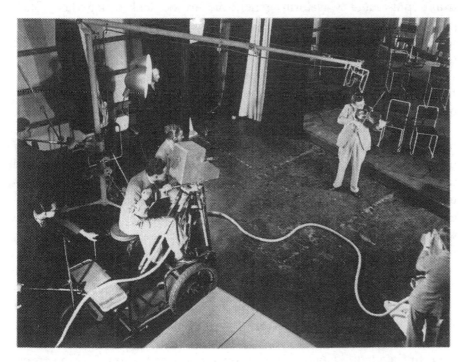

Figure 6.8   Studio 'A', Alexandra Palace, showing two Emitron cameras in use

Source: The Marconi Company

inputs, namely 28.5 kW for the supply arcs for the two telecine transmitters and the intermediate film transmitter; 31.5 kW for the supply arc of the spotlight transmitter; and 94.4 kW for the studio lighting for the electron camera.

An additional disadvantage of the intermediate film process was the cost of the 35 mm film stock and processing chemicals which amounted to £48 per hour or £12 per hour using split 35 mm film stock. Against this, the cost of servicing the Emitron cameras with tubes was £2.10 per transmission tube-hour. Subsequently, following some modifications[74], the tenders were accepted by the Television Advisory Committee[75].

The general layout of EMI's vision input equipment is illustrated in schematic form in Figure 6.9. In the system, the frequencies of both the line scanning and frame scanning circuits were synchronised to the mains power supply frequency (normally 50 Hz). This was achieved by means of a master electronic oscillator operating, under mains frequency control, at twice line scan frequency (i.e. 20 250 kHz), and a series of electronic frequency dividers. From these line scan signals (at $20\,250/2 = 10\,125$ kHz) and frame scan signals (at $20\,250/5/9/9 = 50$ Hz) were obtained.

The outputs of the dividers enabled the frequencies of the various electronic pulse and waveform generators to be locked together. Such generators were required to allow the electron beam, of the Emitron, to scan linearly over the signal mosaic plate; to black out the Emitron's output signal during the flyback period; to provide tilt, bend and suppression signals to annul the deleterious effects of the secondary emission from the mosaic plate; and to provide the necessary line and frame synchronising signals.

The extent of the studio and camera control apparatus was influenced markedly by the television production director's need to provide a variety of attractive programmes. Six camera channels were available, of which any two could be used in conjunction with two film scanners for the transmission of a continuous programme of film. All the channels were identical and therefore interchangeable, an essential advantage when 'live' studio scenes were being televised and reliability of presentation was of the utmost importance.

Furthermore all the cameras were synchronised, thereby enabling the producer to fade from one picture to another without resorting to a change-over of synchronising signals. An added advantage of this arrangement was the facility which allowed two or more pictures to be superimposed for special effects.

From the picture channel, the vision signal was fed to the vision transmitter. This comprised two parts. The radio frequency section consisted of a crystal-controlled oscillator and a number of multiplying stages for

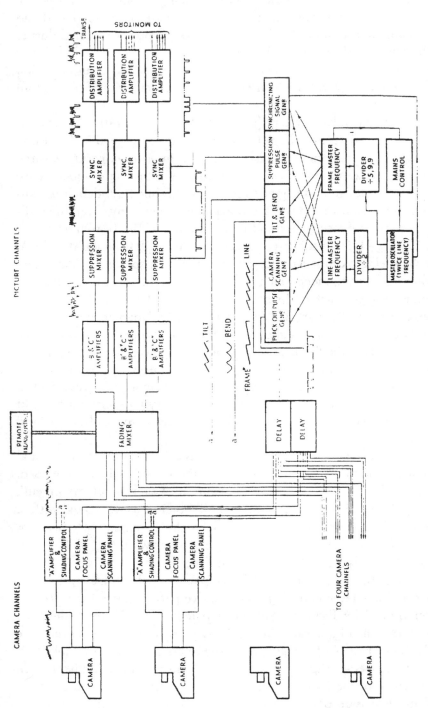

Figure 6.9 Schematic layout for the control equipment of the six-camera and three-picture channel television system

Source: *J. IEE*, **83**, 1938

generating a signal at the carrier wave frequency, followed by a cascade of six power amplifiers. The final power amplifier was grid modulated and gave a continuous output of 17 kW, which was the power radiated during the white parts of the picture. The modulator section, which included four vision-frequency power amplifier stages, provided a 2 kW output vision signal having a bandwidth which extended from zero to 3 MHz, and an amplitude of 2000 V.

More than 500 valves were utilised in the vision apparatus. Meticulous design ensured that the overall vision system was flexible in studio operation and reliable in transmission use. An important feature was the provision of means to monitor the performance of the many sub-systems. Altogether the design and engineering of the 405-line high-definition, all-electronic television system was a remarkable achievement given the rudimentary nature of electronics and camera tubes at the beginning of the decade 1930–40. In just four years, from January 1931, when the HMV multi-channel apparatus was demonstrated, to the spring of 1935, when full-scale studio equipment which answered the requirements for the 405-line service was completed, the bandwidths of the electronic circuitry had increased almost 30 times. EMI's R&D staff had had to engineer the system effectively *ab initio*. Amplifiers of various types (head amplifiers, video amplifiers, radio frequency amplifiers and power amplifiers), modulators, pulse generators, pulse forming circuits, equalisers, filters, attenuators and special power supplies were developed. A great deal of innovative activity was needed to implement the system and much credit must be accorded to Shoenberg and his brilliant team of physicists and engineers.

In this endeavour Blumlein played not only a leading and inspirational role, but in addition he evolved many of the circuits which were used. Browne, in the references to his paper[76] on the 'Vision input equipment', listed 17 relevant British patents; of these Blumlein was associated with 9. Other members whose names appeared on the 17 patent specifications were E L C White (6), C O Browne (5), J Hardwick (2), M Bowman Manifold (3), F Blythen (2), and E C Cork (1).

Of course, the total number of EMI patent applications, relating to high-definition television during its formative period, much exceeded the number quoted by Browne. Blumlein's patents total 52 from the time (early 1933) when he became actively involved in television to the date (2nd November 1936) of the opening of the London television station. This represents on average slightly more than one patent application per month for the whole of the four-year period—a most impressive contribution. Blumlein's television patents from 1933 to about 1939 range over many disparate fields of inquiry. A classification of the patents is given in Table

Table 6.1   Blumlein' television patents

Subject class	Number of patents
Cathode ray tubes (including camera tubes)	9
Cathode ray tube circuits	7
DC restoration	7
AGC circuits	6
Power supplies	10
Modulation	2
Miscellaneous television circuits	8
Antennas and cables	15
Miscellaneous electronic circuits	11

6.1. They total 75, or an average of 11 per year. Some of the patents are described and discussed in Chapter 9, which deals wholly with technical matters.

An assessment of Blumlein's work vis-à-vis other members of EMI's Central Research Laboratories has been made by the late Mr J A Lodge, who was responsible for the archives of CRL for several years. He has said (at an IEE conference held in 1992):

'There were some pretty brilliant men among that staff and a little while ago I asked myself: How does Blumlein stand up in that company?

'I took a list of his patents and, like everybody's patents, there are some good and some fairly good and some not so good. But there are a surprising number of very good ones, so I then said: Well there are these very good patents, how about the other 40 people — something like 40 people — of whom at least a quarter can be put in the top ten per cent, and how about the inventions they made. And there is no doubt if you take any one of Blumlein's inventions you can go along the other something like 600 or 700 patents done by the research laboratory and you can find a patent which is not far off being in the same calibre.

'So that seem[ed] to move Blumlein down in the scale a bit. But then I thought: Let's look at it the other way around. Let's look at the patents that the other people have done and see how far Blumlein on his own was able to match them. And, on a rough count, [of] something like 30% or 40% of the best patents made by all the other people Blumlein on his own had matched them. I think that means that, whatever a genius might be, he was an extraordinarily able and creative engineer. *He was as creative as half a dozen of the best.*' (author's italics).

Installation of Marconi–EMI's television equipment at Alexandra Palace was carried out during the period December 1935 to August 1936. Extensive building operations were necessary to accommodate the heavy machinery, and sound and vision transmitters, and various rooms had to be converted for use as studios, dressing rooms, control rooms and offices.

Major structural changes were effected in the southeast tower of the Palace to support the steel antenna mast. All the existing floors, and the windows on the south and east sides of the tower, were removed: fire-resistant floors and staircases were then constructed to provide five floors of offices, and bay windows were added to increase the lighting of the office spaces. The additional floors and staircases were supported by steel beams and the brickwork of the tower was tied horizontally by steel bars to provide a solid foundation on to which the mast could be fabricated.

The mast, shown in Figure 6.10, was square in section at its base and tapered up to height of 105 ft above the tower, the sides of the square being 30 ft at the base and 7 ft at the top of the tapered section. Above 105 ft the mast had an octagonal section, the distance from one face to the face opposite being 7 ft.

To maintain the structural integrity of the antenna mast during gale force winds, special means were used to transmit the wind loading (which could be as much as 100 tons weight of uplift) to the brick tower[77]. Four lattice-steel girders 30 ft long and 7.5 ft high were placed in the form of a square on top of the existing tower and the four legs of the mast were bolted to the corners of the square. Each corner was then embedded in 17 tons of concrete. Additionally, four substantial angle-shaped steel tie-bars, 50 ft long, were attached to the corners and taken down the inside of the brick tower where, after being tensioned to a force of 30 tons weight, they were firmly keyed into the brickwork of the tower.

The mast carried both sound and vision antennas. They were of similar design and each antenna array comprised eight, wide-band, push−pull, dipoles, evenly spaced around the octagonal section of the mast, with eight reflectors mounted between the dipoles and the mast. The antennas were connected to the transmitters via two special 5 in diameter concentric feeders.

Much attention was paid by EMI's engineers to the matching of the antennas to these feeders. If a mismatch existed a wave would be reflected back to the transmitter where, because of the high degree of mismatch at the transmitter end, the wave would again be reflected and arrive at the antenna end of the feeder scarcely diminished in amplitude, but delayed by twice the time of travel along the feeder. This time delay was about one microsecond, during which time the scanning spot on a receiver screen would have travelled one hundredth of the width of the screen, a distance equal to that occupied by about five lines in the vertical direction. Consequently, if a picture contained thin vertical lines, the reflections gave rise to a distortion consisting of 'echo' lines displaced from the original

Figure 6.10   *Alexandra Palace: Mr D C Birkinshaw with the Emitron camera transmitting the view from the London Station*

Source: BBC

image at intervals of 1/100 of the picture width. A sharp edge in the image would be followed by similar striae.

Blumlein was the first person to appreciate the effects of a mismatched feeder on the transmitted picture; he had a remarkable genius for analysing and resolving engineering problems. As a consequence of Blumlein's foresight, E C Cork and J L Pawsey, of EMI's R&D antenna section, sat at the top of the mast, day after working day, making measurements and introducing capacitances at suitable points in the feeder to annul mismatches. In their paper[78] on the Alexandra Palace antenna-feeder system Cork and Pawsey acknowledged the far-sightedness of Blumlein in this field.

According to Birkinshaw[79], Blumlein spent a 'lot of time' at Alexandra Palace during the installation of M−EMI's system and 'would not spare himself'. Birkinshaw, an ex-Cambridge science graduate who was appointed to the first BBC post in television, that of research engineer, has recalled that Blumlein was 'always there when the pace was hottest'. He worked very long hours and on one occasion passed out with fatigue and, in doing so, cut his head on the transmitter control desk.

On another occasion[80], while working at the top of the antenna tower, Blumlein dropped his pipe, which of course broke. Being unable to continue without it, he descended to ground level, walked to the nearest shops at Alexandra Park, purchased a pipe, walked back to the Palace, climbed to the top of the tower and then continued his work.

Birkinshaw, who saw much of Blumlein at Alexander Palace, has described him as 'a normal likeable uneccentric man' with a 'most active brain'. He 'never showed off' and 'would go into a pub, have lunch and then engage in a game of darts. In the middle of this he would say: "You know I think we can improve that suppression mixer circuit at Alexandra Palace. I have just been thinking about it", and proceed with the game.' Blumlein was 'not a bad darts player'.

What particularly impressed Birkinshaw was Blumlein's 'technical ingenuity and the speed at which he thought things out'. Birkinshaw was not ashamed to admit that he 'rather ran along behind'. Blumlein never kept 'things to himself' and was always willing to give an explanation. However, this was sometimes 'one that his brain considered [sufficient] when his disciples would have preferred something rather fuller'.

Apart from being an engineer of genius rating, Blumlein was also a very practical engineer. He could use a soldering iron or any other tool and get on with a task. He was very interested in the practicalities of how a device worked. Turnbull[81], who had responsibility for the sound system at Alexandra Palace (which used the same sound equipment that Blumlein had previously developed at Columbia), has related an episode when the clutch on his (Turnbull's) straight-8 Chrysler failed. He bought new parts for the clutch and repaired the car himself since it would have been too expensive to have had the work undertaken in a garage. 'The car was out of action for about six weeks. Blumlein was upset and said: "Look if you had only told me you were doing this I would have been only too keen to have assisted because I would have been very interested to have seen how [Chrysler engineered] it." '

A spirit of great goodwill seems to have pervaded the research and development department of EMI during the time the London television station at Alexandra Palace was being established. Turnbull remembered working on some occasions until 1.00 a.m., long after the last train of the day had run, and having to walk home from Hayes to Ealing. Although no overtime was paid, other members of the department did the same. There was a feeling that EMI was developing something very worthwhile and that consequently, with Blumlein providing genius, leadership and inspiration, there was only one course of action to be taken; that was to follow Blumlein's example.

After the passage of many years following a person's death, different highlights of a person's character are recalled. For H C Spencer[82], Blumlein's 'manageability of people' was an important factor in the success of Shoenberg's team. Spencer subsequently never met it in anybody else. He has told how Blumlein would visit once per fortnight everyone in the laboratories and engage in a discussion of their projects as if the previous discussion had taken place the previous day. He had had a remarkable memory and could recall details about the many different projects without prompting. All this was undertaken with good humour. Birkinshaw has referred to an 'atmosphere of hilarity' and has said that 'Blumlein, White, Browne and Willans used to get mischievous delight in twisting electronics to do ingenious things'.

On Browne, Birkinshaw thought he was 'terribly good at designing the production model while not originating original circuit design which was Blumlein's forte. The two together [were] most complementary'.

References

1 BURNS, R. W.: 'Television, an international history of the formative years' (Peter Peregrinus Ltd, London, 1997), chapter 19, pp. 431–478

2 WHITAKER, A.: 'Note on the development of television', 22nd April 1931, pp. 1–2, EMI Central Research Laboratories archives

3 WRIGHT, W. D.: 'Report of television progress to date', 31st August 1931, pp. 1–10, EMI Central Research Laboratories archives

4 WHITAKER, A.: 'A note on television', 1st October 1931, pp. 1–3, EMI Central Research Laboratories archives

5 Cable from Victor, Camden, received 2nd December 1931, EMI Central Research Laboratories archives

6 Executive Committee minute no. 20851, 1st October 1931, EMI Central Research Laboratories archives

7 MITTEL, B.: 'Television publicity'. 16th February 1936, p. 3, EMI Central Research Laboratories archives

8 WHITAKER, A.: 'Notes on a visit to Engineer-in-Chief's department General Post Office [on] 21st October 1931', 23rd October 1931, pp. 1–3 EMI Central Research Laboratories archives

9 SHOENBERG, I.: discussion on 'The history of television', *J. IEE*, 1952 **99**, Part IIIA, pp. 41–42

10 CONDLIFFE, G.: summary of 'Programme of work for advanced development division', 6th November 1931, pp. 1–3, EMI Central Research Laboratories archives
11 Ref. 1, pp. 252–253
12 WISSENDEN, J. W.: letter to EMI Ltd, 8th October 1931, BBC file T16/65
13 'Statement re—television', 10th October 1931, EMI Central Research Laboratories archives
14 SHOENBERG, I.: letter to N. Ashbridge, 11th November 1932, BBC file T16/65
15 ASHBRIDGE, N.: report on television demonstration at EMI, 6th December 1932, BBC file T16/65
16 BURNS, R. W.: 'British television, the formative years' (Peter Peregrinus Ltd, London, 1986)
17 PHILLIPS, F. W.: letter to Sir J. F. W. Reith, 13th March 1933, BBC file T16/42
18 SIMON, L.: memorandum on Baird and EMI demonstrations, 27th April 1933, Minute 4004/33, Post Office Records Office
19 TURK, W.: letter to the author, 30th April 1991, personal collection
20 Notes of a meeting held at the GPO, 21st April 1933, BBC file T16/42
21 TEDHAM, W. F.: 'Cathode ray tube for scanning', 12th May 1932, EMI Central Research Laboratories archives
22 CONDLIFFE, O.: 'Re—PO 382', 12th May 1932, memorandum to I. Shoenberg, EMI Central Research Laboratories archives
23 MCGEE, J. D.: 'The early development of the television camera', unpublished MS, p. 16, IEE Library
24 TEDHAM, W. F., and MCGEE, J. D.: 'Stencil mesh mosaic', British patent no. 406 353, 25th August 1932
25 MCGEE, J. D.: laboratory notebook, entry for 14th November 1932, EMI Central Research Laboratories archives
26 MCGEE, J. D.: 'A transmitting tube for television', EMI Central Research Laboratories archives
27 MCGEE, J. D.: 'Double-sided mosaic tube', British patent no. 419 452, 5th May 1933
28 ZWORYKIN, V. K.: 'Television with cathode ray tubes', *J. IEE*, 1933, **73**, pp. 437–451
29 Ref. 23
30 ANON.: Laboratory staff, Research Department, 12th June 1934, EMI Central Research Laboratories archives
31 LUBSZYNSKI, G.: 'Some early developments of television camera tubes at EMI Research Laboratories', IEE Conference Publication, 1986, (271), pp. 60–63
32 MCGEE, J. D.: taped interview with the author, personal collection
33 TURK, W.: letter to the author, 11th May 1991, personal collection
34 Ref. 32
35 TURNBULL, I. L.: taped interview with the author, personal collection
36 Ref. 32
37 NIND, E. A.: taped interview with the author, personal collection
38 Ref. 23

39 Ref 23, p. 35
40 ESSIG, S.: 'Aggregated silver mosaic', British patent no. 407 521, 24th February 1932
41 Ref. 23
42 Ref. 23
43 Ref. 9
44 BLUMLEIN, A. D., and MCGEE, J. D.: 'Improvements in or relating to television transmitting systems', British patent no. 446 661, 3rd August 1934
45 ASHBRIDGE, N.: Report on television, 17th January 1934, BBC file T16/65
46 Notes of a meeting of the Television Committee held on 27th June 1934. Evidence of Messrs Clark and Shoenberg on behalf of EMI and The Marconi–EMI Television Co. Ltd, Minute 33/4682, Post Office Records Office
47 ANGWIN, A. S.: memorandum, 12th March 1934, Minute 4004/33, Post Office Records Office
48 ASHBRIDGE, N.: report on demonstration to the Postmaster General and C(A), 12th March 1934, BBC file T16/42
49 REITH, J. F. W.: letter to Sir Kingsley Wood, 15th March 1934, BBC file T16/42
50 KINGSLEY WOOD, H.: letter to Sir J. F. W. Reith, 20th March 1934, BBC file T16/42
51 Notes on 'Conference at General Post Office, 5th April 1934', Minute Post 33/4682, Post Office Records Office
52 BLUMLEIN, A. D.: 'The transmitted waveform' *J. IEE*, 1938, **83**, pp. 758–766
53 MCGEE, J. D.: 'The life and work of Sir Isaac Shoenberg 1880–1963', *Royal Television Society Journal*, 1971, **13**, (9), May/June
54 Ref. 52
55 Ref. 9
56 Ref. 9
57 WEINBERGER, J., SMITH, T. A., and RODWIN, G.: 'The selection of standards for commercial radio television', *Proc. IRE*, **17**, (9), 1929, pp. 1584–1594
58 GANNET, D. K.: 'Quality of television images', *Bell Laboratory Record*, **8**, 1931, pp. 358–362
59 WENSTROM, W. H.: 'Notes on television definition', *Proc. IRE*, **21**, (9), 1933, pp. 1317–1327
60 ENGSTROM, E. W.: 'A study of television image characteristics', *Proc. IRE*, **21**, (12), 1933, pp. 1631–1651
61 Ref. 59
62 Ref. 53
63 PRESTON, S. J.: 'The birth of a high definition television system', *Television Soc. J.*, **7**, 1953, July/September
64 Ref. 52
65 WHITE, E. L. C. : 'Blumlein and television', *Eng. Sci. Educ. J.*, June 1993, **2**, (3), pp. 125–132
66 Ref. 9

67 Ref. 52
68 Ref. 35
69 GARRATT, G. R. M., and MUMFORD, A. H.: 'The history of television', *Proc. IEE*, 1952, **99**, Part IIIA, pp. 25–42
70 Ref. 1
71 Technical Sub-committee, minutes of 16th meeting, 5th June 1935, Minute Post 33/5533
72 Electric and Musical Industries, response to questionnaire of Technical Sub-committee on 'Proposed vision transmitter', Minute Post 33/5533
73 Baird Television Ltd: response to questionnaire of Technical Sub-committee on 'Proposed vision transmitter', Minute Post 33/5533
74 Television Advisory Committee: minutes of the 17th meeting, 14th June 1935, Post Office bundle 5536
75 Television Advisory Committee: minutes of the 18th meeting, 18th July 1935, Post Office bundle 5536
76 BROWNE, C. O.: 'Vision input equipment' *J. IEE*, 1938, **83**, pp. 767–782
77 ANON.: 'The London Television Station, Alexandra Palace' (BBC, London, c. 1937/38), 40pp.
78 CORK, E. C., and PAWSEY, J. L.: 'Long feeders for transmitting wide side-bands with reference to the Alexandra Palace aerial-feeder system', *J. IEE*, 1938, **83**, pp. 448–467
79 BIRKINSHAW, D. C.: taped interview with the author, personal collection
80 BBC radio programme: 'Move the orchestra' (B. Fox), 21st January 1990
81 SPENCER, H. C.: notes of telephone conversation with the author, 24th November 1996, personal collection

Chapter 7

The London Station

The opening ceremony of the London Television Station on 2nd November 1936 was a most modest affair[1]. Although the station provided the world's first, high-definition, regular, public television broadcasting service, the inaugural programme, arranged by the BBC and approved by the Television Advisory Committee (TAC), lasted hardly one quarter of an hour. Just a few short speeches lasting four minutes each, by Mr Norman, the Chairman of the BBC, Major Tryon, the Postmaster General, and Lord Selsdon, the Chairman of the Television Advisory Committee comprised the proceedings. The ceremony was witnessed by many invited guests from the BBC, the GPO, the Alexandra Palace trustees, Baird Television Ltd, Marconi–EMI Television Ltd and various other bodies. Among the M–EMI representatives, there were Messrs I Shoenberg, G E Condliffe, A D Blumlein, C O Browne and N E Davies.

The experiences gained by the British Broadcasting Corporation of the operation of both television systems under service conditions from 2nd November to 9th December were described in an important report[2] written by Gerald Cock, the BBC's Director of Television. It proved highly damaging to the Baird interests; indeed it meant the end of the company as a supplier of television studio and transmitting equipment for the Corporation's stations and studios. Cock stated that the Marconi–EMI equipment proved capable of transmitting both direct and film programmes with steadiness, and a high degree of fidelity.

'Its apparatus being standardised throughout, reproduces a picture of consistently similar quality and requires only one standard of lighting, make-up, and tone contrast in decor. Its studio control facilities are convenient and comparatively simple. It has proved reliable, and has already established a large measure of confidence in producers, artists and technicians. Outside broadcasts and [those] of multi-camera work have added considerably to the attractions

of programmes. With improved lighting, additional staff and studio accommodation, single system working by Marconi – EMI would make a service of
general entertainment interest immediately possible.'

Unfortunately, the Director of Television found it difficult to say
anything complimentary about the Baird Television system. The programmes were being transmitted under practically experimental
conditions and the prospect of anything approaching finality in the studio
stages of transmission seemed remote.

'Alterations in apparatus were constantly taking place. Breakdowns, with little
or no warning, and, even more serious, sudden, unexpected, and abnormal
distortions are a frequent experience. In such cases, it is difficult and embar-
rassing to make a decision to close down, since there is always the possibility
that faults may be corrected within a short time. This inevitably leads to criticism of television by those who may only have observed it in adverse conditions.'

In studio operations, the Baird system made use of the spotlight
scanner, the intermediate film method, the electronic camera and the
telecine scanner for 35 mm commercial film. The first three systems each
required a different technique in lighting and make-up to add to the difficulties of the producers and cameramen. Cock's views on these methods
were as follows:

1 Spotlight scanner
'This apparatus is limited to double portrait reproduction. Distortion
of picture tone and shape still appears to be intrinsic and unavoidable.
No reading is possible in a spotlight studio, nor could any artist
depending upon looks and personality be expected to televise by this
method. The result is a caricature of the image televised.'
2 Intermediate film method
'This is extremely intricate, and depends upon so many processes that
it causes continual anxiety. It is inflexible and rigid in operation, being
confined to "panning" in two planes. Changes of view can only be
effected at the cost of lens changes and "black outs"; otherwise the
picture is static. Its quality is variable; the delay action is extremely
inconvenient for timing and other production purposes. The maximum
continuous running time is at present limited to approximately 16
minutes, which adds to the difficulty of arranging programmes. The
cost for film alone is £12 per hour (rehearsal or performance). Sound,
recorded on 17.5 mm film and subject to development at high speed, is
invariably of bad quality, whole sentences having occasionally been
inaudible. It is consequently an unsuitable method of presenting any

programme item in which quality of music or speech is important. Mechanical scanning produces line bending and twisting in a variable degree. Black and white contrast is however generally good.'

3 Electron camera

'These cameras have quite recently been improved, but future progress is likely to be seriously handicapped by destruction of the Baird research plant and technical research records in the Crystal Palace fire. Bending and twisting of the lines are pronounced. The cameras are in a somewhat primitive stage of development and are still without facilities for remote (outside) or "dissolved" work. Breakdowns have been frequent. Electronic cameras do not appear likely seriously to compete with emitrons at any rate for a considerable time. They have advantages over other Baird apparatus in being instantaneous in action; in permitting good sound transmission; in mobility; and in the elimination of the mechanical scanning. At present their operation seems somewhat precarious.'

4 Telecine

'Originally, this apparatus gave the best picture obtainable by the Baird system and possibly (flicker apart) was better than the Marconi– EMI for reproduction from standard film. Line bending and twisting have since been noticeable due perhaps to mechanical scanning and the difficulty of maintenance in a first-class condition. Its contribution to programmes is limited by the present restricted use of 35 mm film.'

This very damaging report, from Baird Television's point of view, was discussed by the Television Advisory Committee[3] at its 34th meeting on 16th December 1936. They had to decide whether the time had now arrived for them to make a definite decision on the question of transmission standards. The contracts made with both companies contained a reference to the London Experimental Period which was defined as terminating on the day on which a decision was reached concerning the system to be employed at the London Station. In the light of Cock's adverse account of the working of the Baird system (with the possible exception of the telecine equipment) the committee had to consider whether further expenditure on the development of a system which appeared most unlikely to survive was prudent. Failing this, it could suggest that the system should be temporarily suspended until the apparatus could be handed over in a reliable and efficient state — and within a reasonable time limit. Cock foresaw that artists and celebrities might refuse to appear in Baird programmes, thus making the problem of producing programmes still more difficult. There was no doubt that the uncertainties and limitations of the equipment were having an adverse effect on the production staff, whereas with the

Marconi–EMI system the apparatus was sufficiently advanced and reliable to enable interesting and entertaining programmes to be devised and transmitted with a 'high degree of reality and with complete confidence on the part of the producers and the studio organisation'.

Cock's report was not the only one that the TAC had received on the operation of the two systems. A C Cossor Ltd, which had developed a velocity modulation television system in 1933, had produced, for the Television Advisory Committee's benefit, a number of constructive and helpful papers on various aspects of television. The company had prepared a most useful account of television at Olympia in September, and in November had written to the TAC on the Baird and Marconi– EMI waveforms[4]. Cossor was critical of the former but found that M–EMI's system had appeared to have settled down to a very satisfactory standard.

With Cock's and other reports in mind, the TAC quickly came to a conclusion: they would recommend to the Postmaster General the adoption of Marconi–EMI's transmission standards as the standards for the London Station, at any rate for the next two years. Bairds would be given the opportunity of making representations, before the public announcement, to the committee on 23rd December[5]. The discontinuance would date from 2nd January 1937[6]. Actually, because of some delay in the drafting of an agreement relating to the Ballard patent, the BBC's last transmission using the Baird Television Ltd system was sent out on 30th January 1937[7].

The BBC's decision probably caused no astonishment. Spotlight scanning had been employed in the USA, France, Germany, the UK, and elsewhere from the late 1920s and its limitations were well known. Again the intermediate film process had been much investigated and developed by Fernseh and others during the first half of the decade 1930–1940 and certain deficiencies of the process had been highlighted at the Berlin Radio Exhibitions. Furthermore, Farnsworth's electron camera was based on the image dissector tube which lacked charge storage and hence could not compete in sensitivity with camera tubes such as the Emitron and the iconoscope, which utilised the charge storage principle. These points were known generally at the beginning of 1937.

Effectively those countries which, in 1937 and subsequently, aspired to operating high definition television broadcast stations had to use studio equipment based on the Emitron and the iconoscope cameras. Only EMI and RCA, and their licensees, worldwide, manufactured such equipment: and so, during the formative years of television growth, prospective television administrations, perforce, had to found their nascent television systems on the work of RCA and EMI. Pre-war, the influence of these companies was particularly evident in France, the USSR, Japan and Germany.

The United States of America's Radio Manufacturers Association (RMA) was one of the relevant bodies which was influenced by the Marconi–EMI system, and, more particularly, by the Blumlein waveform. During the summer of 1937 engineers of the American Hazeltine Service Corporation established a temporary laboratory in England for the purpose of making a survey of television. The survey included observing the transmissions and making measurements of the received signals with special equipment designed for the purpose. Visits were made to Alexandra Palace and to receiver manufacturers, and talks were held with several television engineers. This work led to an important paper[8], by Lewis and Loughren, published in the magazine *Electronics* in October 1937.

The object of the authors was to determine whether certain aspects of the 'present practice of the proposed television standards in the United States [were] wise'. There were three US standards (on the polarity of the transmission, the transmission of the d.c. or background component, and the shape, amplitude and duration of the synchronising pulses), which were the exact reverse of the corresponding standards employed in the UK. Clearly it was highly desirable that the RMA's standards should be based on sound practical experience.

Lewis and Loughren were most complimentary about British television.

'First let it be said that the British pictures are remarkably good. They are steady; they are brilliant; they have an exceptional amount of detail. . . . That [the] British standards constitute a major improvement over present American practice is an inescapable conclusion because television is technically successful and an accomplished fact in England. . . . We cannot avoid the fact that the situation in the United States is much less favourable. Unless changes are made in the type of signal which is now being used for experimental transmitters, American receivers will be more expensive, more difficult to service and will give performance inferior to British receivers. . . . It will be in steadiness and control that the American picture will suffer. . . . In viewing the Alexandra Palace transmissions on a large variety of different receivers there were practically no cases of faulty synchronisation.'

As an indication of the stability of synchronisation of UK sets, Lewis and Loughren noted that reception of the Alexandra Palace transmission at 80 miles (which was beyond the optical horizon) gave an image that 'was not visible except as a hazy movement of light on a grille of noise', yet 'it was easily demonstrated that the grille [raster] was synchronised . . .'.

Furthermore, the authors found that even the most excessive automobile static which resulted in 'a snowstorm' on the screen failed to have any disturbing effect on the synchronisation.

In their paper Lewis and Loughren extolled the virtues of positive modulation and d.c. transmission and observed: 'It would not be putting it too strongly to say that the level of black is a definite foundation upon which the structure of British television is built.... A final confirmation of the practical nature of the British standard is that the other European countries are adopting its principal features...'.

Among the major industrial firms, apart from the Hazeltine Service Corporation, which sent representatives to Europe to collect data which could be helpful in the formulation of US national standards, were the Radio Corporation of America, the Columbia Broadcasting System, and Bell Laboratories[9]. Reports on these visits together with details of the latest endeavours being undertaken by organisations in the United States, were presented at the RMA–IRE Fall Convention held in Rochester, USA, in 1937.

Further work by the RMA Standards Committee in July 1938 led to several additional recommendations to its 1936 standards. D.c. transmission of the brightness of the televised scene was specified — black in the picture was to be represented by a definite carrier level, as in British practice; the radiated electromagnetic wave had to be horizontally polarised; and details of the line and frame synchronising pulses were delineated — the frame synchronising pulse had to be serrated and to include equalising pulses, again as in British practice.

Undoubtedly the engineering of the world's first, regular, public, all-electronic, high-definition television system was a superb achievement. And undoubtedly much of the achievement was due to Blumlein. It was his driving force, leadership, and skill in integrating many different talents into a coherent and effective team; his engineering genius and acumen; and his ability to make original contributions in many disparate fields of activity, which played such a crucial role in Marconi–EMI's success.

From November 1936 an extensive variety of programmes was produced, both in the studios on the premises of the Alexandra Palace and in the surrounding park. The studio programmes included extracts from West End productions, revues, variety, ballet and illustrated talks and demonstrations, as well as a weekly magazine programme of topical interest called *Picture Page*. From outside the studio came demonstrations of golf, riding, boxing and other sports. Because of the limited facilities for rehearsal, it was seldom possible, in the early days, to attempt original productions in the studio, but valuable experience was gained in methods of presentation over a very wide field. Transmissions were limited to two hours per day, excluding Sundays.

The programmes offered to British viewers in 1937 and subsequently can be classified broadly as studio-produced programmes, televised films and outside broadcasts (OBs). Of these, OBs were particularly noteworthy. With the mobile television units, the variety of programmes was much extended. Of course, the simultaneity between the actual event and the reproduced display in a viewer's home gave television a great advantage over the cinema.

The mobile outside broadcast television unit[10] comprised three large vans, each about the size of a large coach: one contained the control apparatus and scanning equipment, one housed the power plant, and the third was fitted with the VHF radio link transmitter (1 kW). Three Emitron cameras were provided.

Marconi–EMI's system was entirely self-contained and most comprehensive in its specification. The control room unit held two rows of equipment mounted along the sides of the vehicle so as to leave a clear aisle for the engineers operating the controls. Each row comprised six racks, each 7.5 ft high and 19.5 in wide: the total weight of the vehicle and apparatus was approximately 8.5 tons. The engineers were able to view the televised scenes on a monitor fitted into a compartment over the driver's head and, if necessary, make adjustments by means of controls mounted on the front panels of the equipment racks. In addition to the vision units, the vehicle also included sufficient faders and amplifiers to complement the four microphones which were used to pick up local sounds associated with the scenes being televised and the commentary.

The most outstanding outside broadcast television transmission during 1937 was the televising of the Coronation procession at Apsley Gate, Hyde Park Corner, London, on its return journey to Buckingham Palace from Westminster Abbey. Three Emitron cameras were utilised to televise the procession: two were mounted on a special platform at Apsley Gate and were fitted with telephoto lenses to obtain distant and midfield shots of the procession and the crowds to the north and south of the gate, and a third camera was installed on the pavement to the north of the gate to give close-range views of the royal coach and other important parts of the procession which passed through the gate. The cameras were connected by about 50 yards of special cable to the mobile television unit behind the park keeper's lodge, whence the sound and vision signals were propagated to the London Television Station.

The outside broadcasts of the Coronation procession represented an outstanding technical achievement by the Marconi–EMI Television Company. Great credit was due to Shoenberg and his research team for making the broadcast possible. Well over 50 000 people saw the scenes televised on Coronation Day and so became aware that 'television was a force to be reckoned with for the provision of entertainment and education in the home'[11].

From the inception of outside broadcasting on Coronation Day, the BBC's mobile television units enabled a section of the general public to see a wide variety of events in public, social and sporting life—the Wimbledon tennis tournaments, the Lord Mayor's Show, the Cenotaph ceremony on Armistice Day, the Prime Minister alighting from his plane at Heston after his visits to Berchtesgaden and Munich, the Cup Final, the Derby, the University Boat Race, Trooping the Colour, the test matches from Lord's and the Oval, boxing matches such as those between McAvoy and Harvey, and Boon and Danaha, and so on.

Two means existed for sending television signals from one place to another; it could be achieved by the use of a radio link or by the use of a cable. Prior to the opening of the London Station, EMI had been considering how the signals from the OB cameras could be transmitted to Alexandra Palace. Characteristically, Shoenberg and his research team had realised that OBs would take place, that radio link transmissions might be difficult, that a cable was probably the best solution, and that the Post Office had no experience of transmitting television signals. With typical foresight, he had initiated a research project to evolve a cable and associated terminal equipment which would send video signals from various locations in London, such as Westminster Abbey, Buckingham Palace, the Houses of Parliament, the Cenotaph, and the theatres, to Alexandra Palace. The estimated cost of such a cable, suitable for carrying video signals having frequencies up to 1.5 MHz was £300 per mile: in addition there was the cost of running the cable into a duct. The total expenditure was thought to be about £900 to £1000 per mile for a cable capable of handling signals up to 2 MHz[12].

The cable[13,14], designed by Blumlein and Cork of EMI, was of the twin or balanced pair type, and comprised two 0.08 in diameter conductors each located centrally within a tube of paper insulation having an external diameter of 0.91 in. The tubes were twisted together and had copper screening laid around them: the whole structure was encased in a lead sheath.

The cable could be used in lengths of up to 8 miles without the need to employ repeaters, and as the length of the cable route (Figure 7.1) from Broadcasting House to Alexandra Palace was 7.25 miles, the headquarters of the BBC became a convenient repeating station for OB transmissions into Alexandra Palace.

When the cable had been manufactured by Siemens, it was taken coiled-up to EMI's research and development department at Hayes and tested. Such was the excellence of the design that no significant difference

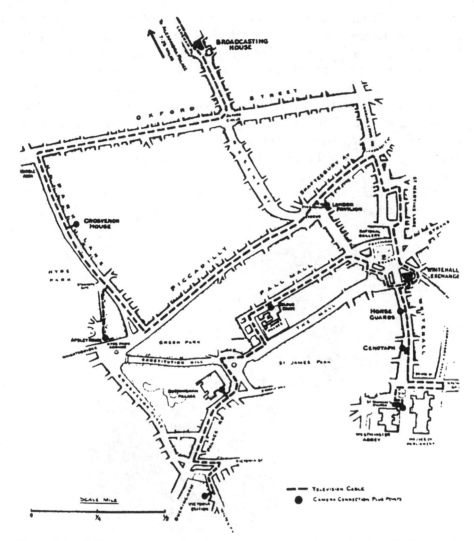

*Figure 7.1   Cable route in London. Camera connection points existed at St Margaret's
Church, the Cenotaph, Horse Guards Parade, St James's Palace, Buckingham
Palace, Victoria Station, Apsley House, Grosvenor House, the London Pavilion
and Broadcasting House*

Source: EMI Archives

between the signals applied to the sending end and those observed at the
receiving end of the cable could be detected.

Mention has been made of the six channels of the EMI vision system.
Two of these were used for telecine working and four for studio and outside
broadcasts. In 1937 Birkinshaw asked EMI if some of the channel outputs
could be combined to give overlay/inlay effects. The BBC had a few

creative producers and it was thought such a facility would introduce unusual and interesting images into televised programmes. Blumlein, with R E Spencer, quickly provided the means. Their patent[15] of 1937 was possibly, indeed probably, the first to be used for special television effects (see Figure 7.2), which are now such a common feature of television. On one occasion a producer devised a dance sequence which was televised by four cameras. A Bach fugue in four parts accompanied the dancer, each part being associated with a different camera. By mixing the outputs of the cameras, the images of three of the cameras could be superimposed on that of the fourth.

Other Blumlein patents which deal with television image enhancement were, *inter alia*, those on anti-ghosting, black-spotting, spot-wobble, and slot antennas.

Camera I produces a picture of the violinist.

Camera II a picture of the dancers

and the Overlay produces the combined effect to be seen on the screen

Figure 7.2   *An example of the use of the overlay technique*

Source: unknown, from the author's collection

In urban areas a received television signal may comprise two components, one of which has followed a direct path from the transmitter to the receiver, and another which has followed an indirect path due to reflection from a steel structure. Since the latter path is longer than the direct path the 'indirect' signal is delayed. Consequently a ghost image, or echo appears on the television tube screen. Patent no. 507 417 deals with the elimination of these ghost images[16].

On this patent Mr I J P James, a colleague of Blumlein from 1937, has written the following recollection[17].

'I joined EMI in January 1937 as a technical assistant in Isaac Shoenberg's patent department and was thus involved with ADB in preparing the technical write-ups, investigating the prior art, and acting as an advisor between the scientists and legal staff in particular, Freddie Cackett, the main Chartered Patent Agent, a man with a ready turn of phrase and logical mind in finding the essence of an invention and then "sewing it up" in the broadest possible claim. I well remember Blumlein's great interest in contributing to the formulation of the claims, and in building-up a formidable patent position in the television world.

'It is interesting to recall how ADB's knowledge and interest in filter networks were to develop from their use as low pass filters in the design of wideband video amplifiers and as delay networks in the timing of pulses, to their use in actually forming pulses. Subsequently he applied the use of delay in the reduction of echoes in television reception by a process of simulation of the echo characteristic, and later [to] the extension of the idea, in association with Kallman and Percival, to the transversal filter (patent no. 517 516). He [Blumlein] would have been highly intrigued with the modern application of these two ideas to self-adaptive digital echo suppressors. How ADB would have enjoyed working with the echo killer chip of Bell Labs. containing 34 397 transistors, which suppresses the 0.5 s delay echo encountered in satellite telephone links.'

In paragraph 48 of its report of January 1935, the Television Committee envisaged 'the ultimate establishment of a general television service in this country [the UK]' and the relaying of television productions by landlines, or by wireless from one or more main transmitting stations to substations in different parts of the country. It was obvious that the creation of a network of stations to serve all but the most sparsely populated areas would occupy a good many years and require the building of at least 12 to 15 stations.

The TAC gave much thought to this matter and in December 1938 forwarded to the Postmaster General a report on the expansion of the service[18]. Based on the technical experience which had been gained from more than two years' working of the London Station and the degree of acceptance of television by the general public, a plan of action had been formulated which would lead to the foundation of a semi-national service. The committee foresaw the setting up of four regional stations, each having a power output of approximately four times the power of the London television station, which would probably be located at or near Birmingham, Moorside Edge near Huddersfield, Westerglen near Falkirk, and Clevedon near Bristol. It was contemplated, not unduly optimistically, to anticipate that about 25 million people would be within the effective range of one or other of these stations and that the first of them, at Birmingham, would be ready after the end of 1941. Two other stations might be constructed simultaneously and become operational in 1943, and the fourth regional station might follow in late 1944 or early 1945.

Apart from the TAC, provincial traders also wanted an extension of the television service. In a concerted effort, a joint committee of dealers and manufacturers (known as the Television Extension Committee) was formed in March 1939 with the objectives of '[bringing] home to the authorities the need for a speed up in the extension of television into the provinces because of a new importance at home; because of its export capabilities; and because it [was] so desirable for Britain to maintain the lead which it at present [enjoyed] over all other countries'[19].

From a technical viewpoint, the most interesting, and surprising, aspect of the TAC's deliberations on relaying was its advice on the need to use a cable, even though the manufacturers of television equipment were quite unanimous in agreeing that radio links rather than cable links should be used. Indeed, so convinced were the manufacturers of the merits of radio links that it agreed that 'if the Government [proceeded] with the building of a Birmingham television station without delay, [the manufacturers would] be prepared to stand the loss of it should it not be a success'[20].

The TAC's advice seems curious when set against the opinion of the manufacturers. In June 1938 Shoenberg had submitted a paper[21] to the TAC on 'Television relays' and had expanded the reasons why a radio relay link was preferable to a cable link for the propagation of television signals.

1 'A long cable link [while] theoretically possible... does not seem to be practically workable with the components at present available.'
2 '...with [the] valves at present available, sufficiently stable amplifiers cannot be made to cover the 2.5 MHz band.' (These were required in the repeater units.)
3 'The production of a suitable single sideband signal [was] not known to be feasible yet with a 2.5 MHz band.' (This type of signal was needed for a coaxial cable link.)

4 '... [the] equalisation [of the cable had to] be performed with extreme accuracy using extremely stable components.'

Shoenberg concluded by saying that a cable link of 100 miles or more was commercially workable but 'at present it could only handle a band some 1.0 to 1.5 MHz wide'.

It seems highly likely that these points had been drafted by Blumlein, since he had in the 1920s gained much experience of narrow bandwidth telephone cable circuits. Blumlein and Collard were probably the only members of Shoenberg's staff with such expertise. Moreover, Blumlein had with Cork designed the eight-mile television cable link between Central London and Alexandra Palace.

On the possibility of using a radio link, EMI had examined the transmission of radio signals in the 60–100 MHz range and it appeared that the London to Birmingham link could be established by three radio hops. The link would comprise a 100 W, 175 MHz transmitter and antenna at Alexandra Palace, and intermediate relay stations sited at elevated positions in the Chiltern Hills and in the Cotswold Hills. Each of these stations would include a receiver and a 100 W transmitter, and the transmitter frequencies would be 185 MHz and 175 MHz, respectively. Shoenberg estimated the cost of the complete chain (excluding spares and buildings) as £20 000 and thought that it could be installed in about 18 months.

Shoenberg's opinion was based on extended tests on the characteristics of radio transmissions, including their susceptibility to interference signals. Both horizontally polarised and vertically polarised electromagnetic radiations having frequencies in the 6 m to 8 m band had been partially investigated prior to the inauguration of the London television station. For EMI, Blumlein[22] in February 1934 had informed the TAC that 'although they had not obtained conclusive experimental data ... they had found that the increase in noise level [due to automobile ignition] resulting from the use of a vertical aerial as against a horizontal aerial was not appreciable'. He had submitted that over flat country a better signal strength was obtained at ground level when vertically polarised waves were used. Also, he had opined that from a practical point of view it was probably much easier for the average viewer to erect a vertical aerial, 'the more so on account of directional effects associated with some types of horizontal aerials'.

Later, in June 1938, Shoenberg[23] told the TAC that further 'recent' tests had shown the advantages of utilising horizontally polarised radiations compared to vertically polarised radiations 'both from the points of view of signal strength and a comparative reduction of the effect of interference'.

Unfortunately the horizontal half-wave dipole antenna does not radiate uniformly in a horizontal plane and uniform coverage in this plane can only be achieved by employing an array of such dipoles. Ideally, for television use, a simple structure is required which will radiate horizontally polarised electromagnetic waves, and which will have a uniform horizontal polar diagram and a directive vertical polar diagram. This was a new engineering problem, of a type which appealed to Blumlein's enquiring mind, and he determined to find a solution. His work led to a new radiating structure — the slot antenna.

Blumlein was not an academic[24], he had no aspirations to being appointed to a chair at a university, and he had little interest in producing learned society papers. His prime interest was to solve engineering problems of the form encountered in a large industrial research laboratory. Thus, problems would arise in the development of a complex system; possible solutions would be suggested; investigations — both theoretical and experimental — would be undertaken; the best solutions meeting the various constraints — financial, technical and general — would be adopted; and then the system would be engineered.

Until Blumlein undertook his experimental investigation[25] in 1938 no one anywhere had considered that a longitudinal slot cut in an electrically conducting cylinder and suitably excited would have an engineering application. His work must have seemed revolutionary at that time but, since 1938, slot antennas have been extensively developed and employed in ground-borne and airborne communication systems, radar, navigational systems, and television. One writer has said: 'The most important technical advance in the 1930s [in the field of radiating structures] was . . . the invention of the resonant slot. This was a device which was both an aperture radiator and a resonant structure. Its novelty was major. Nothing so important had appeared since Hertz invented the dipole and the loop [antennas], and Lodge and Bose experimented with open-ended waveguide radiators. . . . The resonance slot was as significant an invention as the resonant dipole. . .'

Blumlein's experiments were conducted with a slotted tube excited either magnetically by a coupling coil, or electrically by direct connection of a balanced line to the edges of the slot. The tube diameter was approximately one-thirtieth of a wavelength, and the radiated wave was polarised perpendicular to the length of the slot.

In the patent an open-ended slot excited at one end is shown. However, Blumlein noted: 'Alternatively connections may be made to the opposite edges of the slot at the centre of the conductor, short circuit connections being provided at the opposite ends if desired'. This sentence describes a line-fed resonant slot as it is now known.

As usual with most of Blumlein's efforts, no paper was published, but his 1938 patent depicts distinctive feed and tuning arrangements. Investigations were undertaken of the structure's polar diagrams and gains at different frequencies, and of various slot configurations. An array of slots was proposed. The phenomena of travelling waves and of standing waves, of phase velocity control by shunt-inductive and shunt-capacitive loading, and of attenuation and polarisation were all understood by Blumlein. His work in this field typifies Abraham Lincoln's statement: 'Towering genius disdains a beaten path. It seeks regions hitherto unexplored.'

The slot antenna was much developed in the 1940s and 1950s. For aircraft the flush-mounted slot became the standard 'suppressed' antenna. And for the BBC's frequency modulated very high frequency sound broadcasting service the broad-band matched slot antenna was extensively used (Figure 7.3).

Blumlein's investigation of slot antennas highlights not only his genius but also, until recently, his anonymity. It has already been stated that until

Figure 7.3   Photograph of a modern multi-slot television antenna

Source: BBC

the 1950s the sound-recording engineers at the Bell Telephone Laboratories were ignorant of Blumlein's extensive 1931 patent on stereophonic recording and reproduction. And during the discussion in 1949 of an IEE paper on the transformer ratio-arm bridge one member, after extolling the advantages of Blumlein's 'brilliant' 1928 conception, observed: 'This brings me to a curious point: Blumlein's idea was conceived over 20 years ago, and it has these very attractive features. Why has it not made more of a mark on the general development of bridge measurements during this intervening period?' Similarly, Blumlein's work on slot antennas appears to have been unknown for many years. The author has not yet located a single book on the theory and practice of antenna design which makes reference to his 1938 researches.

These examples show Blumlein as an industrial R&D engineer and not as an academic engineer. An academic would almost certainly have written papers, for learned society publication, on his work.

References

1 BURNS, R. W.: 'British television, the formative years' (Peter Peregrinus Ltd, London, 1986), chapter 18
2 Television Advisory Committee: Minutes of the 18th meeting held on 18th July 1935, Post Office bundle 5536
3 Television Advisory Committee: Minutes of the 20th meeting held on 24th July 1935, Post Office bundle 5536
4 Television Advisory Committee: Minutes of the 19th meeting held on 23rd July 1935, Post Office bundle 5536
5 CARPENDALE, Admiral Sir C.: letter to Marconi–EMI Television Co. Ltd, 6th August 1935, Minute Post 33/5536
6 CARPENDALE, Admiral Sir C.: letter to Baird Television Ltd, 6th August 1935, Minute Post 33/5536
7 BURNS, R. W.: 'Television, an international history of the formative years' (Institute of Electrical Engineers, London, 1998)
8 LEWIS, H. M., and LOUGHREN, A. V.: 'Television in Great Britain', *Electronics*, October 1937, pp. 32–35, 60, 62
9 Ref. 7, chapter 22
10 BBC announcement: 'Television in Coronation week', 22nd April 1937
11 ANON.: 'The future of television. Coronation success', *Observer*, 23rd May 1937
12 Ref. 7, pp. 584–586
13 COLLARD, J.: 'London's television twin cable links', *POEEJ*, 1937, **30**, (3), pp. 215–221
14 BLUMLEIN, A. D., and CORK, E. C.: 'Improvements in cables for the transmission of high frequency electric currents', British patent no. 452 713, 6th December 1934, and British patent no. 452 772, 25th February 1935

15 BLUMLEIN, A. D., and SPENCER, R. B.: 'Improvements in or relating to television systems', British patent no. 501 966, 7th June 1937
16 BLUMLEIN, A. D.: 'Improvements in or relating to radio receivers', British patent no. 507 417, 8th December 1937
17 JAMES, I. J. P.: letter to the author, 9th March 1981, personal collection
18 Ref. 7, pp. 604–610
19 ANON.: 'Provincial traders urge television speed-up', *Wireless and Electrical Trader*, 8th April 1939, **58**, (764), p. 45
20 ANON.: 'Television manufacturers explain their offer', *Wireless and Electrical Trader*, 29th April 1939, **58**, (767); p. 127
21 SHOENBERG, I.: 'Relay of television by radio links', 15th June 1938, Post 33/5536, file 52, Post Office Records Office
22 Technical Sub-committee: Minutes of the 4th meeting, 26th February 1935, Minute Post 33/5533
23 Ref. 21
24 NIND, E. A.: taped interview with the author, personal collection
25 BLUMLEIN, A. D.: 'Improvements in or relating to high frequency electrical conductors or radiators', British patent no. 515 684, 7th March 1938

Chapter 8
Personality

Of all the great engineers and inventors—Brunel, Tesla, Steinmetz, Edison and others—there is a remarkable parallel between the characteristics of Blumlein and Brunel, probably the greatest engineer of the nineteenth century. Besides being outstandingly prolific, far-sighted and versatile, and being possessed of brilliant intellects, memories and powers of concentration and stamina (topics which are dealt with in Chapter 17), they both had similar personalities. Their standards of integrity, honesty and fair-mindedness, and their detestation of sloth, affectation and humbug were in consonance. And their everyday enjoyment of smoking, boyish fun, music, the theatre and other after-work interests were similarly in accord.

When Brunel's son, in the late 1860s, decided to write a biography[1] of his father, he found no shortage of associates of the great engineer willing to attest to his delightful nature. Typical of the comments given were the following:

'He was a joyous, open-hearted, considerate friend, willing to contribute to the pleasure and enjoyment of those about him, well knowing his own power, but never intruding to the annoyance of others, unless he was thwarted or opposed by pretentious ignorance.'

'His professional friends . . . well knew the genius, the intense energy, and indefatigable industry with which every principle and detail of his profession was mastered.'

'His influence among [his friends and associates] was unbounded, but never sought by him; it was the result of his love of fair play, of his uniform kindness and willingness to assist them, and of the simplicity and high-mindedness of his character.'

'He could enter into the most boyish pranks and fun, without in the least distracting his attention from the matter of business in which he was engaged. . . .'

'His light and joyous disposition was very attractive. At no time was he stern, but when travelling or off work he was like a boy set free. There was no fun for which he was not ready.'

Most, if not all, of these eulogies could be used to described Blumlein. He was a delightful[2], quiet[3], modest[4], very sensitive[5] person possessed of limitless energy[6], a remarkable memory[7] and an outstanding creative engineering ability. He had an inexhaustible patience for those who were receptive to learning and was 'very good at explaining anything'. He had a good sense of humour, much enjoyed practical fun and was 'very human indeed'. He was not a person to seek admiration, or to advertise his achievements. He was a generous and loyal colleague and friend.

Apart from his joining in a game of darts, Blumlein (Figure 8.1) would also participate in more boyish activities. EMI had an aerial installation at

Figure 8.1   A D Blumlein, date unknown

Source: Mr S J L Blumlein

Dawley Manor, near the Hayes Central Research Laboratories, where Cork and Blumlein carried out measurements on a tilted wire aerial. The grounds of the manor included a cherry orchard and here Blumlein would join in the rush to pick the cherries when they were ripe[8]. He was not a person to adopt an aura of pre-eminence, or to consider certain types of behaviour as being inappropriate to his station, if they led to enjoyment.

Sometimes his fun-making had an intellectual dimension. On one occasion, following an evening lecture given by Dr V K Zworykin at the IEE, Kaye, Blumlein and L H Bedford, (Chief Engineer of A C Cossor Ltd), decided to dine at Stone's Chop House, which was close to Leicester Square:

> 'On the route we called at a fun fair in the Haymarket. [It] had little racing cars on tracks.... And [they] were run two at a time because one raced against the other. Blumlein and Bedford were just on a par with their explanations of the mechanics, and the mathematics, and the way to get these cars round [the track] in the shortest possible time. It wasn't the maximum speed of the wheels [which was important] because if you overdid it the wheels spun on the track. You had to keep [the cars] on the track going at the maximum speed [possible].... [So there were these two brilliant engineers, thoroughly enjoying themselves, playing with gusto and hilarity a boy's game.] It was quite a merry evening. I should think [the owner] thought [he] had got three prize lunatics in.'[9]

Given that humour, in modern usage, means the comic or laughable, what is it about a situation that makes it humorous: How can Blumlein's sense of humour be delineated? Theories of humour are attempts to elucidate this question. They can be divided into three groups[10]: theories of superiority or degradation; theories of incongruity, frustrated expectation, and bisociation; and theories of relief of tension.

In theories of the first group, the laugher is in a position of superiority and the object of the laughter is in a position of degradation. The drunkard, the glutton, the henpecked husband, and the miser are all stock figures of comedy, as is the person who is hit by a custard pie. Laughter, also, arises from schoolboy howlers, sloppy speech, dialect and poor grammar.

Thomas Hobbes (1588–1679) was probably the originator of this theory. 'Laughter', he noted, 'is a kind of sudden glory.' For Hobbes, joy is the cause of laughter because of the sense of triumph that the laugher feels. 'Men laugh at mischances and indecencies, wherein there lieth not wit or jest.... Men laugh often... at their own actions performed never so little beyond their expectations.... Also men laugh at the infirmities of others, by comparison wherewith their own abilities are set off and illustrated.

Also men laugh at jests, the wit whereof consisteth in the elegant discovery and conveying to our minds of some absurdity of another. . . . For when it, jest, is broken upon ourselves, or friends of whose dishonour we participate, we never laugh thereat.' From such observation, Hobbes concluded that 'the passion of laughter is nothing else but sudden glory arising from some sudden conception of some eminence in ourselves, by comparison with the infirmity of others, or with our own formerly.'

In theories of the second group, the cause of the laughable is the sudden realisation of the incongruity between a concept and an actual situation or object. Schopenhauer, an adherent of this theory, used the following example to illustrate his views: some prison guards allowed an inmate to play cards with them; when they discovered he had cheated, they kicked him out — of jail. The humour here is based on two rules — 'offenders are punished by being locked up', and 'cheats are punished by being kicked out' — both of which are acceptable and self-consistent, being at variance in a given situation.

'Laughter', said Sir Philip Sidney, 'almost ever cometh of thinges moste disproportioned to ourselves, and nature.' However, for Blaise Pascal it was not incongruity but frustration of expectation which was important. 'Nothing makes people laugh so much as a surprising disparity between what they expect and what they see.'

Theories of the third group describe laughter as a manifestation that arises when the restraint of conforming to conventional decency is relieved. Bawdy, men-only, rugby songs and stories enable situations to be aired which, normally, society would repress. Again, laughter can erupt when people who have been subjected to considerable stress are suddenly relieved of that stress. Thus, the central element is neither a feeling of superiority, nor an awareness of incongruity, but the feeling of relief that comes from the removal of the restraint.

Monro has summarised these three theories as follows:

'Superiority theories account very well for our laughter at small misfortunes and for the appeal of satire, but are less happy in dealing with word play, incongruity, nonsense, and indecency. Incongruity theories, on the other hand, are strong where superiority theories are weakest, and weak where they are strong. Relief theories account admirably for laughter at indecency, malice and nonsense (regarded as relief from the "governess reason") but are forced to concede that there is an intrinsic appeal in incongruity and word play that is quite independent of relief from restraint.'

Blumlein, although of high intellect, never used his superiority to laugh at the determined endeavours of others. He never ridiculed a young engineer who was trying, but could ridicule people who really deserved it and who were capable of responding if they so wished. Blumlein did not swear, he didn't waste words, but he could be 'extremely witty at times without being cutting'[11]. On one occasion, following a discussion on passive and active networks with a Mr X at an IEE meeting, Mr X telephoned Blumlein, the following morning, at EMI to continue the argument. He was heard to say: 'Mr X, I may not know the difference between a passive network and an active network, but I do know the difference between a passive network and an active nit wit. Good morning.'

Blumlein was a 'grockle', as fans of the great Swiss clown, Adrian Wettach[12] (1880–1959), who was known as Grock on stage, were called. Blumlein's admiration for him was boundless. If, when Blumlein was working on a cable job for ISEC on the Continent of Europe, Grock was billed at one of the theatres in the region, Blumlein would make every effort to be there.

Grock's comic sketches, which caused so much mirth and merriment, were wholly dependent for their effect on their visual impact; written descriptions cannot easily convey the quick changes of facial expression, from bathos to pathos, and the exaggerated movements which characterised his clowning. He played the part of a simpleton among musical instruments — wondering where the strings had gone when he held a fiddle the wrong side up; or throwing his bow up, on three consecutive occasions, and bungling the catches, until after some practicing behind a stage screen, he succeeded in making a perfect catch at the fourth attempt.

These sketches, in narration, appear trite and lacking in humour or novelty, but such was Grock's genius that he could give pleasure to countless audiences in many countries using what seems to be most unpromising material.

J B Kaye has recorded an example of Blumlein's clowning:

'One day at EMI some studios were being built, or some building was going up, and workmen were there in the main yard, [when] suddenly a figure appeared at the end of the yard. He was carrying a colossal beam, so enormous that it [seemed] unlikely that any normal human being could carry it: it was about 20 feet in length and about 2 ft square. [Anyway] the figure had it on his shoulder and he was staggering along, lurching from side-to-side. The workmen dropped their bricks and everything that went with them and stood open-mouthed. He [Blumlein] went straight across the yard and disappeared through a gate still carrying the [enormous] beam. It was quite beyond all human comprehension. The workmen had obviously not heard of balsa wood!'

Blumlein and Kaye had a love of the ridiculous situation. The lyrics of Edward Lear would be read aloud to the accompaniment of much laughter. On occasions the two close friends would make up their own comical descriptions of events.

'At one time, I think after he was married, I happened to have a boil just to one side of my tummy; and he had a rash on his back. He rang me one evening and said: "Is that the Tumboil of Trim?" I said: "Yes", and he said: "This is the Backrash of Oufe." And for some time after that whenever we rang one another up . . . he always referred to me as the Tumboil and I referred to him as the Backrash. So I would ring up and if anyone . . . within earshot heard me say: "By the way, is that Backrash?, Tumboil here, how are you?", they would think [we were] a couple of [idiots].'

Blumlein had a penchant for inventing names. His units of 'flab', and of the 'decibeelze' have already been noted. Other names have been recalled by E A Newman[13], one of the members of EMI's Central Research Laboratories who was recruited by Blumlein:

'One of my first tasks was to make and test some large monitors [voltage measuring oscilloscopes]. The design of these — due to Blumlein — was, I think, quite brilliant. But the monitors were very large, and one usually took jobs to them. But there came a time when this was not practical and the first of them had to be carted up to the top floor. Blumlein spotted this and remarked that Mohamed had at last had to go to the mountain. From then on that monitor was Mohamed. Those that came afterwards were Abu Behr, Yerzid, Ali, etc, the Caliphs that followed Mohammad.'

Some of Blumlein's original circuit designs were given names such as Featherstonehaugh's follower, Cholmondley's tweeter, St John's X, and Marjoribank's Y; but he always insisted that these names should be correctly spelt. They were not pronounced, however, as written, being Fanshaw, Chumley, Sin-jin, and Marchbanks when referred to in speech.

M G Harker, formerly of EMI, has recollected that Blumlein 'had his own particular brand of humour based on hyperbole, which he often used to make a point. . . . He was a compulsive pipe smoker and the metal NO SMOKING notice in Clark's laboratory was removed from the wall and bent up to form an ashtray for him to empty the contents of his pipe into!'

F Charman, another member of EMI, who was generally known by his nickname 'Dud' (because he labelled his components cupboard with the word 'Dud' to stop the components being pinched), has written that Blumlein 'always worked a bit of humour into his expositions whenever he could'.[14]

' One day one of his colleagues said: "I hear Scophony have made a 3 ft cathode ray tube." "No", says Alan, "there can't be that much vacuum about." Then he went on to discuss the effect of [the] mass production [of the tubes] raising the atmospheric pressure and the fact that we would be able to land aircraft at low speeds. Then again, when somebody asked him a question on say audio screening, he would answer "ten thou of mumetal, a foot of lead, or a mile of sea water." But all this nonsense was effective. The recipient always learned something.'

In 1933 Blumlein married Doreen Lane, the daughter of William Lane, an auctioneer, and his wife (Figure 8.2); they had known each other since 1930. In that year Miss Lane was teaching at the school, owned and run by the elderly Miss Chataway, where Blumlein had been a pupil in his early childhood. One day, in 1930, Miss Chataway informed the future Mrs Blumlein[15]: 'Oh, Alan Blumlein came to see me last night.... His mother is going to South Africa, to see her people, and they are giving up their flat. They have a very nice Beckstein piano and don't want put it into store, [so] they thought that we might house it in the school.'

Figure 8.2   Mr and Mrs W Lane, the father and mother of Mrs Doreen Blumlein

Source: Mr S J L Blumlein

Miss Chataway was keen for Doreen Lane to see the piano because she played the piano at the school when taking some of the classes. Since the Blumleins were living in a flat in Lyncroft Gardens, which was near the school in Belsize Road, both headmistress and teacher subsequently visited them. The occasion was one Monday night in June 1930.

'We had quite a nice evening and Alan sat with me while Mrs Blumlein and Miss Chataway [talked] away. He took some pipe cleaners and made me three little dogs . . . father, mother, and co-respondent . . . [which had little patches which Alan made with his fountain pen]. That was Alan's sense of fun (he had a great sense of fun), and that was that. We said we would have the piano.'

A few days later, on Thursday, Miss Chataway and Miss Lane were having dinner when Miss Chataway remarked: 'Oh, there's Alan in his car again.'

'So in he came and chatted away and [then] he said: "Well I have got to post a letter, will you come with me?" [However) when we got outside he said: "I haven't got a letter to post, I only wanted to get you out".'

At this time Blumlein had obtained his pilot's licence and was flying Tiger Moths, of the London Aerodrome Club, from Stag Lane aerodrome. He invited Doreen Lane to fly with him 'some day'. The invitation did not have a great appeal, for she was a very travel-sick person, but as Blumlein had said some day, she thought she could 'put that off' and duly accepted. 'Right', said Blumlein, 'I'll [call for you] at three o'clock.' Taken aback, Doreen enquired what he meant and was told: 'I'll fetch you at three o'clock on Sunday.' The 'Sunday' had been misheard as 'some day'. 'Then he said to me: "How old are you, because I've got to put it down for flying?" So I said: "How old do you think?" He said: "22 or an experienced 21." I was 22 the week before so I was a bit annoyed about that. I went [into the school] and told the others I didn't want to go because I'd be sick flying.' Faced with what might be an unpleasant experience, Doreen prayed for rain. It did not arrive, and a flight in a yellow Tiger Moth was duly made. 'I suppose I was so scared of being sick in front of a strange man that I was all right . . .'.

Further embarrassment was felt by Doreen after the flight when Alan asked her where she would like to go, since she had arranged to meet someone else at 6 o'clock. Alan took her to a hotel in Radlett for afternoon tea. 'He was very annoyed with me [because] I couldn't go out [with him], so I made him drop [me] at the corner where, round the corner, I was meeting another young man. He said: "We'll go and have a drink." And I'd be blowed if he didn't take me to Radlett, same hotel. Of all the other hotels in the Home Counties he could have picked, he picked the same one.

I thought to myself, the waiter will think I am a tart. So, I took my hat off hoping he wouldn't recognise me, but when I went in he said: "Good evening, Miss." He almost said: "Are you here again?" I nearly said to him: "Yes, I'm bringing you custom", but I didn't and that was the start [of my going out with Alan].'

Some time later, Blumlein's mother decided to visit her brother and other relatives in South Africa. She gave up her flat, put the furniture into storage, and Blumlein moved into lodgings at 57 Earls Court Square (London SW5) where he stayed until 1932. A further move was made in that year to digs, looked after by a Mrs Upton, at 32 Woodville Road, Ealing (London W5). Presumably the new abode was more accessible to the recently established Electric and Musical Industries at Hayes. 'I used to go there [to No. 32] and [Alan] used to come and see me two or three times a week; he stayed there until we got married.' Doreen was living at Miss Chataway's school during this period.

J B Kaye was, of course, introduced to Doreen. The meeting occurred a day or two after Alan and Doreen had announced their engagement.

'When he told me he was engaged I was a bit anxious. I thought, he may be damned good on circuitry and that sort of thing, and especially on recording and television, but I wondered whether he was any good at choosing a wife. . . . Even the best men can make mistakes.

'But I was delighted when I met Doreen. She was a character and had a sense of humour, and was very charming. I thought, yes, this girl will be able to keep Blumlein under a sufficient degree of domestic control that will avoid any serious rift. [It seemed to me that] there would probably be a few brick ends flying about the kitchen, but that is all part of life's rich pageant.'

Naturally, Kaye was asked to be best man at the wedding, which took place on Saturday, 22nd April 1933, at St John's Church, Penzance, Cornwall (Figures 8.3 and 8.4).

The Blumlein's first son was born on 26th March 1935, but tragically died after a short time. Their second and third sons Simon John Lane and David Antony Paul were born on 11th May 1936 and 21st February 1938 respectively. J B Kaye became godfather to Simon. On the marriage, Kaye has stated:

'It really was a very successful married life; I think I can say that without any contradiction from anybody. They were very, very happy, and Doreen, with her sense of humour and fun, recognised that she was coping with someone who was not the average cabbage that you get in industry, but that he was effervescent. She gave him all the latitude he required for his studies, his reading and his fun.'

Figure 8.3   Mr and Mrs A D Blumlein on their wedding day

Source: Mr S J L Blumlein

Curiously, Kaye's effusiveness on the happiness of the marriage is not echoed in Mrs Blumlein's taped personal reminiscences of Alan Blumlein[16]. He was 'very difficult', and 'very temperamental', and there were times when she said to her husband: 'I'm going out and I'll come back when you are fit to speak to.'

> 'I think anyone would have found him difficult. I remember J B Kaye, his great friend, rang me up one day and [told me] he had married an artist. A bit later I said [to him]: "Oh, you are lucky, Millicent's got such a talent — she's a wonderful artist". And I said: "I haven't got any talent." [But he disputed that and mentioned] "You are the only person in the world who has made HB [Kaye's name for Blumlein] happy: that's your talent".'

Shoenberg, too, was aware of the beneficial effect which Mrs Blumlein had on her husband. During the early part of the war Mrs Blumlein and the two boys, Simon and David (Figure 8.5), lived in Cornwall to escape the heavy bombing attacks on London. From time to time visits were made by them to their home in London, and, as Shoenberg said to Mrs

Figure 8.4   A group wedding photograph showing J B Kaye (Best Man), A D Blumlein' great friend

Source: Mr S J L Blumlein

Figure 8.5   Mr and Mrs Blumlein with their two sons, Simon and David

Source: Mr S J L Blumlein

Blumlein, 'We always know when you're back in London. [Alan hasn't told us] but he's [more] human when you're here.' And when she returned to the capital 'there would be a box of [her] favourite violet creams waiting. He had great consideration and generosity for people. He was a very thoughtful man.' 'He was generosity itself.'

> 'Alan had a great joy in little things. I once, before we were engaged, bought him a picnic case — we were going to his sister's in Somerset for the weekend. I had packed a picnic and when we stopped he said: "Where did you get this?", and I said: "It is yours." His joy was unbounded, he even slept with it by his bed so that it should be the first thing he saw in the morning.'
>
> 'In the early days I once in fun complained that the glove shelf in the open Morris two-seater was so full of maps that there was no room for my gloves. We were bowling along a country road and he took the maps in his left hand and scattered them to the four winds. He had a great sense of fun — rather than a sense of humour.'

'Alan *loved* chocolate and in the war-starved years his uncle sent him some real chocolate from South Africa — a whole slab. He would only eat one square a day so as to make it last.' Van Warrington, who was a demonstrator with ADB at the City and Guilds (Engineering) College also remarked on his fondness for this sweet. And one of the few letters to have survived from the war period (from his mother, dated 18th August 1941), says only: 'Thanks for your letter. This is not a reply but just a line to send my love with this chocolate which I got by a lucky chance.'

In her recollections, Mrs Blumlein recalled occasions when her husband had got up at 5 o'clock in the morning and had gone off to work in a huff because some event, unbeknownst to her, had annoyed him in the middle of the night. Happily, Blumlein, as his colleagues knew, was usually apologetic whenever he felt he had upset someone and by mid-morning Mrs Blumlein would receive a bunch of white lilacs.

She has related that her husband 'got fits of depression'. 'He would go to bed and say he wasn't going to get up for Christmas, and he wouldn't eat anything. And if I caught him crawling downstairs [for some food] he'd get very annoyed if I found out.'

At other times he could be stubborn:

> '[On one occasion] he left his dressing case downstairs — he usually kept it in the loft — and I remember saying: "Alan, do put that case away in the loft." And he got annoyed: "I'm not going to." I said: "Do dear, it's in the way." And he took the case and put it in the middle of the landing, and said: "[The] dressing case stays there."
>
> 'We had a cook-general at the time, so I said to Florence: "Don't move it. We'll move it every morning and put it back every night before he comes in."

This went on for some days. At last he said: "Alright I'll give in and put it away." But it was all treated as a sort of joke.'

The pressures on Blumlein must have been quite excessive at times, particularly during the engineering of the 405-line television system, and the various radars he was associated with during the first three years of the war. He was *primus inter pares* in Shoenberg's research department and carried great responsibility for ensuring that the various systems being developed were properly implemented. These pressures seem to have led to a degree of absent-mindedness. Mrs Blumlein has given some examples of this trait. If her husband hadn't come home one evening or was a little late she 'would ring up his secretary and say: "Miss Bull, remind Mr Blumlein he's got a Ford V 8 and not a Vauxhall [his previous car] will you?" And he would later say: "I'm very glad you rang up because I was looking for the Vauxhall."' 'If he had to bring someone home to dinner I used to get Miss Bull to put a note in his hat. "Don't forget to bring Mr Dutton or whoever home tonight."'

> '[He was] very absent-minded. When he used to go to Petty France there was a lift and he said he [had] been known to throw his gloves in the gutter and hand his bus ticket to the commissionaire.'

On another occasion:

> 'I was in bed and he [hadn't] come in from his dressing room, and I thought 'What is he doing?' So I went in and there he was with a book propped up on the chest. I said: "Hurry up darling". And he [replied]: "Oh, I'll hurry up, where's my tie?"And I said: "You're not dressing, you're undressing." He didn't know which he was doing, his mind would be so [engrossed with other matters] . . .'

Again:

> 'I remember once we went out to dinner and we were sitting in our friend's drawing room afterwards, and he said to me after dinner: "Have we had the meal?" I said: "Yes, don't you remember?"'

Blumlein's preoccupation with his work led to him being reliant on his wife for many of the routine matters of married life—from arranging their holidays and buying the railway tickets, to looking after his clothes and ensuring that once a year new suits, shoes and other clothes were purchased. 'The only thing he worried about were shoes. He had one [shoemaker's] last at the Army and Navy [stores] and his shoes had to be bought there. But he would never clean his shoes. He said it was a habit he had acquired when he was a child in South Africa. Unless the maids did it, I would have to do it.'

Blumlein, like everyone else, had his faults. He did not suffer fools gladly and was not always tactful. He could be rude[17]. According to Kaye:

'Blumlein was never intentionally rude unless he was deliberately intentionally rude. [This is] perhaps drawing a fine [line]. But some people have caused him to come back at them. He was never rude of manner at all in my experience; [except] I suppose [on one occasion] one might have said [he was rude]. [This was after] an eminent gas-bag had spoken at the IEE on binaural reproduction. The man was not terribly conversant with it personally but from his position had managed to gather quite a lot of information together. Blumlein rose to his feet and said that Mr X's system was not only one-eared but one-eyed.'

Mrs Blumlein has confirmed that her husband's rudeness with some people was unintentional 'because he didn't seem to get his brain down to their level'. He couldn't tolerate small talk and this caused much difficulty. 'We lost many friends because, if I had friends of mine in, and they just talked small talk as people do, he'd say: "Don't ask them again."'

Despite these failings, he was a generous colleague[18]. Although Turnbull and Blumlein fell out (almost fisticuffs!) when they were second-year students at the City and Guilds Engineering College, Blumlein never bore Turnbull any malice and later asked him to take over the recording work (at EMI) when he, Blumlein, in 1933 began to concentrate on television matters. Turnbull, remembering the earlier association and altercation said: 'No, we had enough of that when we started [in 1921]'. Blumlein's response was to say that that was a long time ago and insisted Turnbull should join the recording group, which he did.

Blumlein was generous to others with his ideas and never expected any acknowledgement in return. McGee has commented[19] on this aspect of Blumlein's nature. 'I'd say this, that I can hardly imagine a case where you had a man dominating, or being so prolific in ideas, and [who] has caused so little jealousy or resentment. This was because he was so generous, so honest and [so] lacking in any kind of attempt to grab at other people's ideas.... Probably, Blumlein gave away more ideas which he was not credited with ... inventions which were largely due to him or to his inspiration.'

Condliffe, EMI's Research Manager, met this desirable attribute when he was preparing a patent on a film scanner. Blumlein made various suggestions for improving the scanner, and as a consequence, Condliffe proposed that the patent should be in both of their names, but Blumlein would not agree. 'It is your patent', he told Condliffe[20].

At times, Condliffe was placed in a difficult position. Shoenberg would send for Blumlein, developments would be discussed, and actions initiated, without the Research Manager knowing anything about them.

Hierarchically, Blumlein was third in line after Shoenberg and Condliffe but, as McGee has recorded, in fact Blumlein was well above Condliffe[21].

> 'This was, of course, a potential source of friction, of trouble between Condliffe and Blumlein, but nevertheless they remained quite good friends, right through. I think, Condliffe recognised Blumlein's genius and although he got a bit peeved about this by-passing of him all the time to Shoenberg he more or less had to take it. Had he really objected I think he would have been the one who went. Blumlein was quite happy about this because he didn't have to do the chores of administration.'

Fortunately, Blumlein never took advantage of the situation; he was only too anxious to put Condliffe in the right position and as a result a good rapport was maintained between the *de jure* research manager and the *de facto* research 'boss'.

Financially, Shoenberg acknowledged Condliffe's seniority to Blumlein. From July 1939 to December 1942 Condliffe's salary was £150.00 per month; that of Blumlein was £145.66 per month[22]. For comparison, the average weekly wage in 1942 for manual workers in all industries was £5.57, and a university professor of physics' annual remuneration was about £1000 to £1200. According to Shoenberg, Blumlein was the only member of his staff who never asked for a rise.

Part of Blumlein's, and of course EMI's, success stemmed from his ability as a project leader. F Charman, who worked at EMI for 40 years, has recalled: 'I gather from other reports that he was rather abrupt and difficult, but by the time I came to know him he was becoming more approachable. We got on well together. I spoke his language to some extent. He had an extraordinary store in his head and a very rapid address system. He was tolerant and would listen to a tyro and help him if he could. If one went to him with a problem, which nobody else could solve, he would come up at once with an answer and very quickly three or four more, and at least two of them would work.'

According to P B Vanderlyn[23]:

> 'Blumlein professionally had two cardinal virtues: he would keep in daily touch with all the work in the laboratories, and he would talk with the most junior of us as equals without ever making us conscious of our lack of status, which was good for our egos and even better for our technical education. In this task of training his staff he had unbounded patience. If he had one fault it was that he did not always appreciate where the problem lay, and would go to some lengths in explaining what was already clear and gloss over the more difficult bits that to him were obvious.

'It has been said of him that he could fly into a rage over unimportant details: if this was so I never saw evidence of it. He certainly had a sense of humour. One of his favourite devices was to bring a discussion to its close by quoting Thévenin's theorem, and he would often be halfway down the corridor, puffing his pipe and shouting Thévenin over his shoulder before we were able to gather our wits together. On one occasion, he missed an opportunity to carry out this manoeuvre, and the following day was presented with the Thévenin Medal fabricated from the lid of a cocoa tin. After a moment of hesitation it was gracefully accepted. Such are the touches that turn geniuses into great men. They also speak more eloquently for their relations with their staff than any mere statement of fact.'

In his dealings with EMI's R&D staff, Blumlein never lost his temper. 'To be looked at by [him] when he was annoyed was enough.'[24]

Blumlein was a modest person. He was always quick to correct anyone who called him Dr Blumlein, as happened during his wartime visits to the Telecommunications Research Establishment. And when, in 1941, M G Harker (who was working on air interception radar at Hurn airfield), drove specially from Hurn to pick up Blumlein from Christchurch railway station, his modesty was such that it had not occurred to him that he should be met, because he had not thought others regarded him as a VIP.

Notwithstanding his senior position in the company, Blumlein was not aloof [25].

'During one period of the war we were frequently working too late to go home and slept on beds in the laboratory. Blumlein used to join us about 6.00 p.m., take off his jacket, work on the bench beside us and sleep in one of the beds. On one such occasion, when we retired to the work's canteen for supper we found that the 6 ft high gate leading to it was locked, so that we had to scramble over it, led by Blumlein!'

Blumlein had many interests outside his work. Several of these — the Chelsea Arts Balls, dancing, swimming for the London Otter Club, driving his open-tourer, bull-nose Morris, Vauxhall, and Ford cars, flying Tiger Moth aircraft at Stag Lane and Hatfield aerodromes, pistol shooting under the arches at Charing Cross, and, occasionally, darts have already been mentioned. Other interests included bridge, the theatre, the cinema, music, and sometimes snooker.

On these interests Mrs Blumlein had several memories to relate. 'He took up horse riding when we first married. It was [at] the Seven Sisters Riding School at Northolt and it really was kept by seven sisters. . . . When we went down to Cornwall he used to go cub hunting, at the end of summer, with my father. Father had two hunters and they used to go cubbing.'

Blumlein never lost his love for aircraft, motor vehicles — on one occasion he managed to persuade a bus driver to let him drive his bus from Penzance to Land's End — trains and railways. 'He once saw me off on the night train from Paddington [and] told me he [hadn't] got home until 2 o'clock in the morning. I said: "What did you do from [the time] of the 10 o'clock train?" He said: "I went up into the signal box and I spent an hour or two helping the man up there." I was always told the name of every engine and generally introduced to the driver.'

Notwithstanding Blumlein's dislike of small talk, he and his wife were sociable and invited friends to their house. Card games, particularly bridge, were played. The Prestons and the Spencers (R E Spencer was also with EMI and had joined the HMV company in September 1929), but not the Kayes, were among those who enjoyed playing bridge with their hosts. Kaye had no interest in bridge.

Music was one of Blumlein's great loves. He particularly liked the works of Beethoven, and those of J S Bach because they were mathematical in conception[26]. 'He loved Caesar Franck's *Symphonic Variations*. Whenever he thought he had done anything especially well he would go around humming this.'

Blumlein never mastered the playing of a musical instrument, but for some time in the 1930s he received tuition in piano playing. The lessons were abandoned in the late 1930s. McGee has recalled[27] a conversation when he expressed to Blumlein some surprise that he'd given up his piano lessons, since piano playing was a good form of relaxation. Blumlein responded by saying: 'Well, my object in life is to be just as good an engineer as I can be; that is my number one object, everything else I regard as just relaxation. The thing that I find takes me right away from everything is just to play bridge.' So piano practising came to an end. Blumlein relished the intellectual challenge of bridge and the joking and lively banter that went with the game.

Theatre-going was another relaxation. Unlike his wife, he adored Bernard Shaw's plays, and possessed all the first editions of the plays although Mrs Blumlein has said: '. . . when I married him he could never read poetry or books and I used to read to him. We used to lie in bed at night and he used to say: "Say that Shakespeare speech over to me again", and I would say it two or three times and he would know it. That is the way he got to love poetry, through me reading it to him. We spent hours [with] me reading to him but he could never take it in if he read it himself because he never really learnt to read properly.' He had a great liking for musical comedies. Noel Coward's *Bitter Sweet*, the Cochrane Reviews, the

Immortal Hour, and Gilbert's and Sullivan's operettas were among the productions which gave him much pleasure. Blumlein 'was high-brow in his scientific mind but not exactly in [his] other [cultural interests]'[28].

Among the films which Blumlein enjoyed watching were those of Rene Clair's pre-Hollywood period. Notable among them were *A nous la liberté* and *Sous le toits de Paris*. They were quite a change to the usual run of films of those days and Blumlein found them delightful.

The German film *Metropolis* (which was a vision of the future), was one which he and Kaye found entertaining and thought-provoking[29]. 'There is only one thing wrong with this film', he told Kaye, 'and that is they [the population of the future] have too much reciprocating machinery. There won't be any reciprocating machinery of the type driving the wheels [of transport] at the time this film is supposed to apply.' Blumlein was of the view that mighty steam engines would be things of the past and that all engines would be rotary in operation.

Blumlein had, as W S Churchill said of T E Lawrence, 'a full measure of the versatility of genius. He held one of those master keys which unlock the doors of many kinds of treasure houses.' Aspects of Blumlein's versatility are considered in the following chapter and a comparison of his genius with the genius of other great engineers and inventors is given in Chapter 17.

References

1 BRUNEL, I.: 'The life of Isambard Kingdom Brunel, Civil Engineer (1870)' (David Charles Reprints, Newton Abbot, 1971)
2 NIND, E. A.: taped interview with the author, personal collection
3 TURK, W.: letter to the author, 11th May 1991, personal collection
4 HARKER, M. G.: letter to the author, 17th May 1991, personal collection
5 KAYE, J. B.: taped interview, National Sound Archives, London, UK
6 GREENHEAD, M.: letter to the author, 28th May 1991, personal collection
7 SPENCER, H. C.: notes of telephone conversation with the author, 24th November 1996, personal collection
8 Ref. 6
9 Ref. 7
10 MONRO, D. H.: 'Theories of humour', *Collier's Encyclopaedia*, 1991, Vol. 12, pp. 356–358
11 Ref. 7
12 WETTACH, A.: 'Grock, King of Clowns' (Methuen, London, 1957)
13 NEWMAN, E. A.: letter to B. J. Benzimra, 17th March 1962, personal collection

14 CHARMAN, F.: letter to the author, 20th August 1991, personal collection
15 BLUMLEIN, Mrs D.: taped interview, National Sound Archives, London, UK
16 Ref. 15
17 COLLARD, J.: quoted in BENZIMRA, B. J.: 'A. D. Blumlein—an electronics genius', *Electronics and Power*, June 1967, pp. 218–224
18 TURNBULL, I. L.: taped interview with the author, personal collection
19 MCGEE, J. D.: taped interview with the author, personal collection
20 Ref. 18
21 Ref. 19
22 Personnel files, EMI Music archives
23 VANDERLYN, P. B.: 'In search of Blumlein: the inventor incognito', *Journal of the Audio Society*, September 1978, **26**, (9), pp. 660, 662, 664, 668, 670.
24 Ref. 13
25 Ref. 3
26 Refs. 7 and 15
27 Ref. 19
28 Ref. 15
29 Ref. 5

Chapter 9

Blumlein's technical achievements in electronics

(Non-technically minded readers may omit this chapter.)

The development of the world's first, public, all-electronic, high-definition television system was a most remarkable and splendid achievement. On 12th January 1934 N Ashbridge, the chief engineer of the BBC, had been given a demonstration of EMI's latest 150-line television system, which used mechanical scanning of film images. Yet just over a year later, in the spring of 1935, 'a full scale studio equipment answering the requirements for a [405-line] television service [had been] completed. . . . [It] was [according to C O Browne], apart from a number of small improvements, essentially the same in fundamental design as the equipment installed at the London Television Station and in the mobile television scanning van which was subsequently constructed for outside broadcasts'[1]. The all-electronic system was first mentioned in C O Browne's diary on 6th February 1935 and stemmed from the meeting, held one Sunday in November 1934, when Blumlein and a few of his colleagues drafted the specification of the 405-line waveform and planned the necessary circuits to implement their design.

Blumlein's circuit and other contributions to the evolution of the final engineered system extended from the Emitron camera to the final stage of the main modulator. (Some of these contributions have been mentioned in Chapters 6 and 7.) Dr E L C White, who worked closely with Blumlein from April 1933 until his death on 7th June 1942, has described the influence of Blumlein on the television R&D team[2].

'Many of the [EMI R&D] team were physicists, with some knowledge of valve circuits but no formal training — there wasn't much available then, as the subject of electronics had not been established — but although ADB's first-class honours degree was in heavy electrical engineering, he had subsequently had a year demonstrating on a telephone engineering course followed by five years experience in the telephone industry. This was where most of the knowledge of circuit designs existed at that time, so ADB became our mentor on such matters as filter design, including artificial delay lines, negative feedback (for which the theory of stabilisation had only been published by Nyquist in January 1932) and his favourite circuit theorem, Thevenin's.

'Reverting to inter-valve couplings, Blumlein showed us how these can be derived from low pass filter theory, introducing us to such refinements as m-derived terminations. W S Percival's distributed amplifier is based on filter theory; it was first used in the modulator for the radio transmitter of the second outside broadcast unit we made. Nearly all ADB's circuits utilize negative feedback as a means of linearising amplifiers dealing with large signals, and also as a means of reducing the effects of variation of valve characteristics, thus reducing the need for pre-set adjustments. All our circuits had to follow this lead and be as far as possible designable.

'Circuits can be designed in many ways to perform a required function. It is always worth thinking again after finding a first solution, and going on exploring alternatives until eventually one is found which not only does the required job, but does it economically in terms of components used, and also reliably in the sense of not being unduly sensitive to errors in the values of components.

'These best solutions have an aesthetically satisfying element of symmetry; in this sense circuit design is an art as well as a science, and Alan Blumlein was a supreme artist in circuit design.'

Blumlein's influence on good circuit design was not limited to EMI's R&D team but extended to the staff at the Air Ministry Research Establishment, Swanage. One of the senior members of its staff was Dr (later Professor) F C Williams. In a paper entitled 'Introduction to circuit techniques for radiolocation', presented at the 1946 Radiolocation Convention, he wrote: 'First and foremost the author wishes to express his indebtedness to the late Mr A D Blumlein, whose contributions to circuit technique were very great, and from contact with whom the author derived enormous benefit during the early days of the war.'

In considering the breadth of Blumlein's interests and the fertility of his genius in electronics, it is important to appreciate that prior to about 1933 almost the only known applications of the thermionic valve were, in the main, to generating, amplifying and rectifying narrow bandwidth sinusoidal signals. Some relaxation oscillators circuits had been described and were being used to generate time-base signals for cathode ray oscilloscopes, but ideas on deliberately shaping waveforms were 'extremely rudimentary'. When it is borne in mind that EMI's camera control equipment included approximately 500 valves, the extent of EMI's task and

Blumlein's involvement in it can be appreciated. The engineering of such a complex, challenging and innovative system was an outstanding accomplishment.

White's description of Blumlein as 'a supreme artist in circuit design' is most apt. Creative artists, whether painters, sculptors, composers, or whatever, develop idiosyncratic styles of execution; their works can be recognised by various distinctive features and traits. Many of Blumlein's systems have a personal stylistic character and employ recurrent devices. He had a penchant for using closely coupled coils, negative feedback, constant-impedance networks, cathode followers and delay lines. Above all, as White has said, Blumlein's circuits had the exemplary characteristic of being designable. Empirical, or trial and error, methods were anathema to Blumlein. His insistence on good engineering design certainly had its reward. After the London Television Station was inaugurated in November 1936, Marconi – EMI's studio and outside broadcast equipment gave excellent service: it proved reliable in practice and was liked by producers, artists and technicians.

On the distinctive style of Blumlein's work, McGee has recalled[3]:

> '[Some] years ago a patent question came up about a radar patent [in the name of J Collard] on moving target indicators. I'd been involved in this, not as one of the inventors, but [as a member of the staff] working on the actual project later on. Preston [the Patents Manager] and I were discussing the technical merits of this invention and we were reading [the] patent specification [when] we both stopped in our tracks. "This really sounds like Blumlein", [we said], and we came to the conclusion that almost certainly the idea was very likely to have been really Blumlein's, but that, because Collard was involved, Blumlein had perhaps left it entirely up to Collard's proposal . . .'

The circuits outlined below illustrate Blumlein's 'remarkable combination of physical insight into what goes on in the circuits together with a most fertile imagination'.

The energy conserving scanning circuit[4]

Blumlein's first television patent was for an efficient line frequency scanning circuit for supplying the magnetic deflection coils, of a cathode ray tube, with a linear sawtooth current waveform, (Figure 9.1). With previous circuits, based on the use of a linear amplifier fed with a sawtooth voltage, the transient oscillation, caused by the self-capacitance and self-inductance of the coils, during the flyback period, had to be damped by a shunt resistor, thus wasting power. Blumlein's solution was to remove

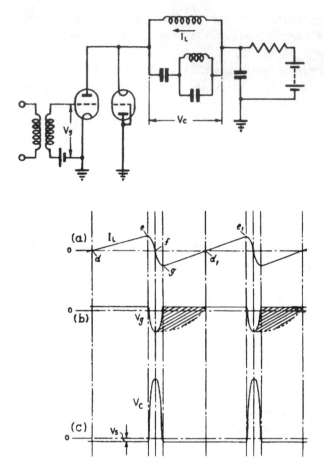

Figure 9.1   The energy conserving scanning circuit

Source: *BKSTJ*, July 1968

the damping and allow the circuit to execute just one half-period oscillation. When the current in the coils reverses, further oscillation is prevented by the diode. By this means the average current in the driving valve is much reduced. The circuit is known as the resonant return scanning circuit and is now used universally in semiconductor form.

The cathode follower[5]

Blumlein was the first person to appreciate the properties of this most useful circuit and employed it extensively in the equipment of the London Television Station. In its most basic circuit configuration, the cathode follower comprises just two components, a valve and a resistor. Nevertheless, the circuit's important properties have led to its universal adoption in electronics, including solid-state electronics where the circuit is known as the emitter follower.

From Figure 9.2 the output impedance of the triode between cathode and earth, with the grid and anode effectively earthed, is $m/g(1 + m)$] where m is the amplification factor and g is the mutual conductance of the valve. The open-circuit voltage developed between the cathode and earth when a potential difference, E, is applied between the grid and earth is $mE/(1 + m)$. Thus, if $m > 1$ the open-circuit output impedance is approximately equal to the reciprocal of g, and the open-circuit output voltage is almost equal to the input voltage applied to the grid. It follows that, if the cathode load resistance, R_k, is large compared to $1/g$, the cathode potential will follow the grid potential both in amplitude and in phase. Effectively the negative feedback introduced by the potential difference across the load increases the linearity of the valve and enables it to handle peak-to-peak voltage variations which are limited only by the h.t. (high tension) supply voltage. Furthermore, if an impedance Z_{gk} is connected between grid and cathode, the input impedance between grid and earth is approximately mZ_{gk} when $m > 1$, $R_k > m/g$, and $Z_{gk} > 1/g$. It is these properties of high input resistance, low output impedance, and practically unity voltage gain which make the cathode follower such a valuable circuit.

Some of Blumlein's applications of the cathode follower were outlined by C O Browne in his 1938 IEE paper on the vision input equipment of the Alexandra Palace station. His descriptions[6] of these are given below.

Case 1. High input impedance: Emitron-camera input circuit

'The first valve in the head amplifier is a high-slope triode connected as a cathode follower with large cathode resistance, and it constitutes a low-impedance source for feeding the succeeding amplifier stage. By virtue of the fact that the cathode potential follows the grid potential, the grid-to-cathode capacitance is effectively removed. The capacitance to earth of the signal plate of the emitron (which is connected to the grid of the first valve) is also reduced by surrounding it with a shield connected to the cathode. The emitron is worked with a high load resistance, and compensation is subsequently introduced for the loss of the higher frequencies. A higher input resistance is made possible by the use of a cathode follower, and the overall result from this arrangement is a better signal/noise ratio than would be realized by a lower input resistance and the conventional amplifier circuit.'

THE CATHODE FOLLOWER AS A MEANS
OF CORRECTING CHARACTERISTIC FOR INPUT CAPACITY.

$$i_a R_o = E_s + (v_z - v_c) + \mu(v_g - v_c)$$

$$i_a R_o = E_s + v_z + \mu v_g - (\mu+1)v_c \quad -------(i)$$

$$v_c = Z(i_a + i_i) \quad -----------(ii)$$

$$i_i = \frac{(e - v_c)}{G + g} \quad ---------(iii)$$

$$v_g = v_c + i_i\, g \quad --------(iv)$$

CASE I. $V_z = 0$

Multiplying (ii) by $R_o :-$

$$v_c R_o = i_a R_o Z + i_i R_o Z$$

Substituting from (i)

$$= E_s Z + v_z Z + \mu v_g Z - (\mu+1)v_c Z + i_i R_o Z$$

Substituting for v_g from (iv)

$$= E_s Z + \mu v_c Z + \mu i_i\, g Z - (\mu+1)v_c Z + i_i R_o Z$$

$$= E_s Z - v_c Z + i_i(\mu g Z + R_o Z)$$

Substituting from (iii) for i_i

$$v_c R_o = E_s Z - v_c Z + \frac{e - v_c}{g + g}(\mu g Z + R_o Z)$$

$$v_c\left\{R_o + Z + \frac{\mu g Z + R_o Z}{G + g}\right\} = E_s Z + e.\frac{\mu g Z + R_o Z}{G + g}$$

Divide both sides by $Z :-$

$$v_c\left\{\frac{R_o}{Z} + 1 + \frac{\mu g + R_o}{G + g}\right\} = E_s + e.\frac{\mu g + R_o}{G + g}$$

Divide both sides by μ, writing $\frac{R_o}{\mu} = \frac{1}{g} = R_v$

$$v_c\left\{\frac{R_v}{Z} + \frac{1}{\mu} + \frac{g + R_v}{G + g}\right\} = \frac{E_s}{\mu} + e.\frac{g + R_v}{G + g}$$

Ratio of Shot Noise to Signal

$$= \frac{\frac{E_s}{\mu}}{e.\frac{g + R_v}{G + g}} = \frac{E_s}{\mu.e.\frac{g + R_v}{G + g}}$$

Figure 9.2   *Blumlein's analysis of the cathode follower*

Source: Mr I J P James

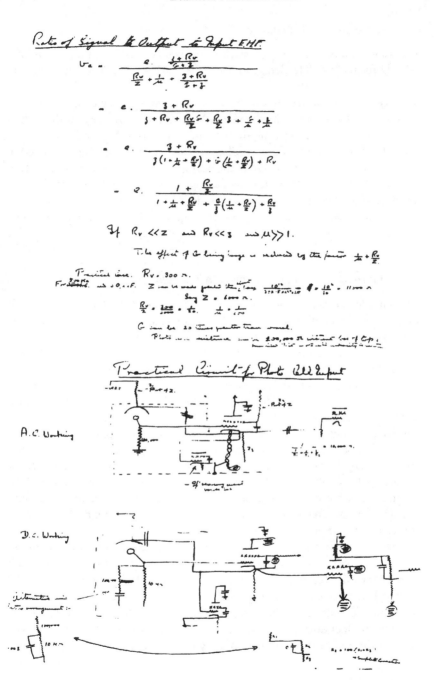

Figure 9.2   *(continued)*

Case 2. Low output impedance

'Cables for interconnecting the various units used in the equipment usually take the form of small feeders, and in order that these shall be of good practicable construction so that they may be handled without fear of breakage the inner conductor consists of a wire of comparatively heavy gauge. In consequence, the characteristic impedance is low, and it is obvious that difficulties will be encountered in feeding lines of this type from the anode circuit of a valve if a direct coupling is used. The practice has therefore been adopted in [the EMI television] equipment of including a cathode follower in the output circuit of each unit, so that the disadvantage of a high steady potential superimposed upon the output signals is eliminated. This arrangement has the advantage of simplicity, and the circuit is, in general less susceptible to anode-supply impedance and voltage fluctuations.'

Case 3. Cathode follower feeding a capacitive load

'An example of this condition arises in the case of the CAT 6 cathode follower stage of the modulation amplifiers; this stage feeds the television signals into the input capacitance of the radio transmitter and connecting line.

'In general, variations of potential at the cathode would lag behind the corresponding variations on the grid by an amount determined by the cathode circuit time constant. In the conducting state, the time constant of the cathode circuit of the valve will be $CR(1 + Rg)$, where C is the capacitance of the load in shunt with the cathode load resistance R and cathode impedance $1/g$. If, however, a rapid negative pulse of large amplitude is applied to the grid, so that the rate of change of grid potential is appreciably greater than that of cathode potential, the grid may depreciate in potential with respect to the cathode far enough to cut off the anode current. Momentarily, therefore, a longer time constant, determined by the product of C and R, is operative until the valve again conducts, after which the rapid time constant will be realized for the completion of the pulse.

'In the design of a stage in which this possibility exists, since the rate of discharge of the capacitance is proportional to the voltage across it, it is desirable to make the initial cathode potential as high as possible, by operating the valve with high anode current. In the particular case of the CAT 6 cathode follower, the picture signals are positive, so that it is improbable that the valve would cease to conduct during the reproduction of wave-fronts which are likely to occur in the amplitude range occupied by the picture signals.'

Case 4. Cathode follower as a voltage stabiliser of low regulation impedance

'In order to prevent instability or cross talk in the equipment, due to interaction between the various units through the HT supply circuits, it is necessary to provide a number of effectively independent supply sources of low internal regulation. For this purpose, stabilising circuits are provided consisting of cathode follower valves, in the cathode circuits of which the loads they are required to feed are connected. The grids of these valves are held positive with respect to earth by a bias battery.

'The open circuit voltage available at the cathode of the stabilising valve will be equal to that of the grid bias battery V_b plus the voltage V_g to reduce the anode current to zero.

'Current may be taken by the load from this source, which has, therefore, a regulation impedance approximately equal to $1/g$; a limit (imposed by grid current) to the amount of current available being reached when the open-circuit cathode voltage has been depreciated by V_g to the voltage of the bias battery only.

'A variation of supply voltage applied to the anode of the stabilising valve will be reproduced at the cathode diminished in the ratio $1/(1 + m)$, so that the voltage as applied to the load is effectively stabilised, and is independent of load fluctuations of the HT supply due to other parts of the equipment.'

Negative feedback circuits

The origin of negative feedback is universally attributed to H S Black, who worked at Bell Telephone Laboratories. His paper[7] on 'Stabilised feedback amplifiers' was published in the January 1934 issue of the *Bell System Technical Journal* (*BSTJ*).

Blumlein's earliest patent, with H A M Clarke, on a negative feedback power amplifier (with both current and voltage feedback) was taken out in September 1933, and their 26-page internal report[8] (no. F 7) on 'The use of negative retro-action to increase the distortionless output of a power amplifier' is dated 23rd August 1932. Consequently their work considerably predates that of Black.

By this date the only important paper on regeneration theory which had been published was that written by H Nyquist[9], also of Bell Telephone Laboratories. Nyquist's highly mathematical paper (of January 1932) was not concerned explicitly with the application of negative feedback theory to the design of amplifiers; rather it defined the general conditions under which a linear amplifier could be stable when feedback was applied.

It is not known whether Blumlein saw the paper, or whether EMI subscribed to specialist journals such as the *BSTJ*. What is indisputable is that EMI had an excellent patent department: as Shoenberg mentioned to Lord Selsdon, in June 1934: 'I might also tell you that we have a pretty large patent department that watches all patents which have any relevance to our business and that no piece of apparatus can be built by the company in any way whatsoever without the approval of that patent department.'

In 1932 Blumlein was endeavouring to produce a portable recording apparatus based on his studio recording equipment. The latter was too heavy to transport easily to recording locations, so the company required a much lighter version which could be comfortably handled. It appears that during the design and the engineering of the portable model EMI noted a relevant patent, of a competitor, which had to be circumvented. The outcome was an internal paper, dated 19th July 1932, by Blumlein on 'Notes on proposed portable recording system output stage with special reference to infringement of British patent no. 323,823 — Philips'.

The Philips company's patent[10] was submitted on 18th October 1928 and related to circuits for 'obtaining amplification without any appreciable distortion'. Voltage feedback was described and the usual formula, $(V_2/V_1) = mx/(x + m)$ where m is the stage gain and $1/x$ is the fraction of the output voltage (V_2) fed back, was deduced. No experimental results were given in the simple two-page patent, which contained just two claims.

In his paper Blumlein gave the full theory of 'the use of current feedback for impedance matching and harmonic reduction'. He noted that 'with modem valves it is advantageous to work into a load of several times the valve anode impedance in order to limit the harmonic introduction. By the use of a current feedback circuit it is possible to put an impedance in the anode circuit which may be several times as great as the valve impedance and yet maintain an output impedance equal to the actual load in order to work under matched conditions'. Blumlein's theoretical considerations were supported by experimental work on a two-stage amplifier in which the 25 W output stage valve was either a DO25, or a LS6a, or a PP5/400.

In their F 7 report Blumlein and Clark showed that with voltage feedback, the apparent output impedance is reduced with increase of the feedback voltage. Again experimental confirmation of the theory was undertaken. They deduced that voltage feedback has the disadvantage that as the anode load decreases, the feedback voltage is reduced correspondingly. With current feedback the feedback voltage decreases as the anode load increases. So, Blumlein and Clark reasoned that by using a combination of current and voltage feedback the feedback voltage could be made independent of variations of the load impedance. They verified their theoretical analysis by experimental tests on a five-stage amplifier, (Figure 9.3) and demonstrated that with feedback an output of 6 W, at 1% total harmonic production, was realisable, which was 'much greater than

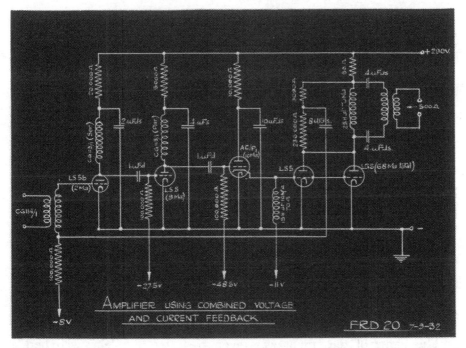

Figure 9.3   Blumlein's negative feedback amplifier circuit, of 1932, which incorporated both voltage and current feedback

Source: EMI Archives

could be obtained without both feedback and a gridding output valve'. Interestingly the amplifier utilised a cathode follower stage (no. 3) even though this configuration was not patented by Blumlein until two years later, in September 1934. The theoretical analysis of this most important circuit was, of course, given in Blumlein's and Clark's report.

Blumlein's enthusiasm for negative feedback led him to design several novel and valuable circuits including the long-tailed pair, the ultra linear output amplifier, and the Miller integrator.

The long-tailed pair[11]

Blumlein's long-tailed pair circuit (Figure 9.4) was devised originally for the eight-mile video cable network which linked various potential Emitron camera sites in central London to the television transmitter at Alexandra Palace. It was first used during the transmission of the televised images of the Coronation procession following the crowning of King George VI. As noted previously, the cable was not of the now familiar coaxial type but

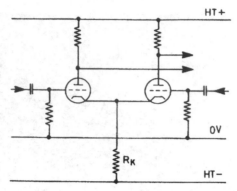

Figure 9.4   The long-tailed pair circuit in its basic form

Source: *Engineering Science and Education Journal*, **2** (3) 1993

comprised a balanced shielded pair of conductors. The problem was to amplify the signals applied in push–pull while rejecting common mode (push–push) interference. In telephone practice a transformer serves this purpose, but in 1936 transformers able to handle the wide video bandwidth were not available.

The name of the circuit was given to it by Blumlein, who had a penchant for coining unusual names, and was so descriptive that it has persisted.

An important application in the post-war years was in d.c. amplifiers, because of the circuit's almost total insensitivity to changes in HT supply voltage and other causes of drift, such as those produced, for example, by

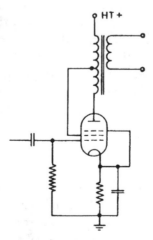

Figure 9.5   The ultra linear amplifier

Source: *Engineering Science and Educational Journal*, **2** (3) 1993

temperature changes. The circuit configuration was the basis, *mutatis mutandis*, of transistorised operational amplifiers. Other applications include mixers in communications equipment, high-speed non-saturating switches (such as Emitter Coupled Logic, ECL), and push-pull output stages for driving the deflection plates of cathode ray tubes.

In some applications the 'tail' is replaced by a third valve, with a cathode feedback resistor, to make the impedance seen at its anode very high. By combining two long-tailed pairs in this way, various logic circuits can be implemented. These were typical of the National Physical Laboratory's Automatic Computing Engine (ACE). According to White: 'Perhaps one can trace this to the fact that two or three of the team designing ACE had worked with Blumlein'.

The ultra linear amplifier[12]

In the circuit given in Figure 9.3 voltage feedback was obtained from the output stage by means of a voltage divider, consisting of the 3090 Ω and 236 000 Ω resistors connected across the primary of the output transformer. The junction of the resistors was connected back to the grid transformer of the first valve via an 8 μF capacitor. By using a tapping on the primary of the transformer a similar effect can be achieved.

An interesting modification of this circuit arises when the output triode is replaced by a pentode and the tapping point is connected to the screen-grid electrode of the valve (Figure 9.5). The circuit is deceptively simple, but has important properties. If the tap is at the anode end of the primary winding, the valve is configured as a triode, but if the tap is at the supply line end, the valve is connected as a pentode. Now when a pentode valve is connected as a triode it becomes less efficient and has a lower power output. It has the advantage that its odd order harmonic production is lower than for the pentode load connection, and the distortion is less affected by variations in the load impedance. Such variations occur when the load is a wide-frequency range loudspeaker. It is clear that by suitably positioning the tapping connection, the output impedance of the valve can have any value between the impedance of a pentode or a tetrode and the impedance of a triode.

This property of the circuit, known later as the ultra linear amplifier, formed the basis of Blumlein's 1937 patent and enabled the output impedance of the valve to be more nearly matched to its optimum load impedance. Blumlein found that a suitable tapping point, in practice, gave a voltage swing on the screen grid electrode of between 25% and 50% of the voltage swing on the anode.

The ultra linear output stage in its push–pull form was for many years widely used in post-Second World War high fidelity amplifiers. They had excellent characteristics. Apart from the ease by which the load could be matched to the output stage, the ultra linear configuration prevented damage to the valve or output transformer if the load was inadvertently disconnected. Furthermore, the circuit was very effective in reducing harmonic distortion. The Mullard three-stage 20 W amplifier (about 1959), for example, had a power versus frequency characteristic that was flat, within ±0.5 dB, from 30 Hz to 20 kHz, with a total harmonic distortion output, with overall feedback, of less than 0.05%. Blumlein did not consider the reduction of distortion in his patent, but it is most likely he was aware that the distortion characteristic would be improved, since distortionless amplification was the main topic of his 1932 laboratory work with Clark.

The Miller integrator

During the early days of radio communication, receivers commonly used triodes as radio frequency amplifiers. It was sometimes found that these would give an oscillatory output voltage even when they were not receiving a signal: the necessary condition for this to happen was an inductive anode load. Investigation by E Armstrong showed that the oscillations were due to positive feedback, from the anode circuit to the grid circuit, via the inter-electrode capacitance, C_{ag}, between the anode and grid of the valve.

In 1919 J M Miller[13], of the US Bureau of Standards, gave a mathematical analysis of the triode as an audio frequency amplifier. He demonstrated that when the anode load was inductive—for example, at frequencies below the resonant frequency of an anode tuned circuit load—the positive feedback gave rise to a negative real component of the grid–cathode circuit's admittance. If this negative resistance were in parallel with a high Q grid tuned circuit, sustained oscillations could occur. Later, various means were developed for negating this effect. They included the use of neutralisation circuits, tetrodes and pentodes.

For an amplifier stage having a purely resistive load in the anode circuit, it can be shown that the effective input capacitance—between the grid and cathode of the stage—is approximately equal to $[C_{gk} + (1 + |A|) C_{ag}]$, where C_{gk} is the grid–cathode capacitance, C_{ag} is the anode–grid capacitance and $|A|$ is the magnitude of the voltage again. A typical triode might have values of these parameters equal to 8 pF, 3 pF and 30 pF respectively, thereby giving an input capacitance of 101 pF. This effect is known as the Miller effect. It imposes a constraint on the ability of *RC* amplifiers to function at frequencies above a few hundred kilohertz.

With his remarkable insight into the operation of valve circuits, Blumlein saw that the feedback feature provided by C_{ag} could be put to good use in the design of integrators, i.e. circuits which develop a voltage which represents the time integral of a given voltage. If the voltage to be integrated is applied across a series connected resistor *R* and capacitor *C*, the voltage v_c developed across *C* is proportional to the time integral of the applied voltage v_i provided v_c is never allowed to exceed a value which is small compared with v_i. Blumlein sought to remove this limitation.

If now *C* is connected across C_{ag}, on the application of a constant voltage v_i, current flows into the integrating circuit to charge the capacitor *C*. In the absence of the valve, v_c rises towards v_i but as soon as v_c exceeds a small fraction of v_i the charging current *i* ceases to be proportional to the applied voltage and the integration ceases to be accurate. Due to the presence of the valve, however, any tendency of *i* to decrease due to v_c is opposed by negative feedback via *C*, since any decrease of i will reduce v_R thereby making the control grid of the valve more positive. Consequently the anode current will increase, the anode voltage will fall and effectively v_c will increase. The charging circuit behaves as though it has a very long time constant compared to *CR*.

Blumlein, in his patent[14], mentioned that the voltage developed between the anode and cathode of the valve would be substantially proportional to the time integral of the applied voltage provided the angular

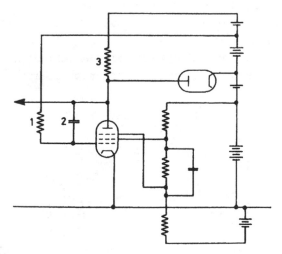

Figure 9.6   Blumlein's Miller integrator time-base circuit

Source: *Engineering Science and Educational Journal*, **2** (3) 1968

frequencies of the Fourier components of the applied voltage satisfied the inequality $G \gg 1/\omega CR \gg 1/gR$ where G is the gain of the valve in the absence of C, and g is the valve's mutual conductance. 'As in practice the product GRg may usually be as large as 10^6 accurate integration can be effected over a wide range of frequency with good amplification so that it is possible to generate integrated voltages of amplitude substantially greater than the amplitude of the applied voltage without appreciable error in integration.'

Figure 9.6, taken from the patent, shows a sawtooth waveform generator. During the integration period, the anode potential of the pentode falls linearly until it reaches the knee of the anode voltage/anode current characteristic. Then the screen grid current begins to increase at the expense of the anode current, so causing the screen grid voltage to fall. This drop is coupled to the suppressor grid which further increases the screen grid current. The action is cumulative: very quickly the anode current is cut off and the anode voltage rises rapidly at a rate determined by the anode load resistor and C. The rise is coupled to the control grid and at some point the valve begins to conduct again and the cycle of linear voltage rise followed by a rapid fall is repeated. In Figure 9.12 the diode is used to clamp the anode voltage at a suitable value.

The Miller integrator, or more correctly the Blumlein integrator, and its variants the sanatron and the phantastron found many applications during the Second World War as time bases and ranging circuits in radar equipment.

An important application of the circuit was in post-war electronic analogue computers. These employed low-drift operational amplifiers based on Blumlein's long-tailed pair amplifiers, and operational integrators of the Blumlein type.

Although Blumlein's patent application was submitted on 5th June 1942, two days before his tragic death in an aeroplane crash, the notion of

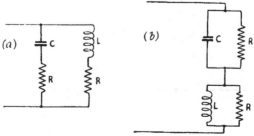

Figure 9.7   Constant impedance circuits

Source: *J.IEE*, 1938, **83**

using negative feedback, to linearise a time-base scan of the sawtooth form, had been used by Blumlein a few years earlier. His television frame scan circuit (patent no. 479 113 of 29th April 1936) utilises the basic idea of the Blumlein integrator in a somewhat elaborate form.

Constant impedance networks

It was well known, in the 1930s, that the circuits shown in Figure 9.7 have a constant impedance which is purely resistive, and equal to R, at all frequencies, provided $L = CR^2$. Blumlein adapted the circuits, particularly Figure 9.7b, as a means of removing from critical points in a circuit the stray capacitances to earth of, for example, floating battery supplies.

Using the networks he made 'very valuable contributions'[15] to the design of the TV modulator. This had to provide an output voltage (containing frequency components up to 2.5 MHz) of 2000 V peak-to-peak to the input of the grid-modulated Marconi 17 kW, 45 MHz transmitter output stage, the grid current of which, with the total shunting capacitance of the input circuit, constituted a non-linear load. Blumlein's design solution was to use a water-cooled triode as a cathode follower. However, the valve had a directly heated tungsten filament which required a supply of 18 V at 30 A and this had to be insulated from earth with as small an effective stray capacitance as possible. In practice a d.c. generator, mounted on insulating stilts, was driven via an insulating coupling by a suitable motor. The actual stray capacitance of the bulky generator was too large to enable the design specification of the modulator to be met, so it was reduced by using one of Blumlein's stand-off constant resistance circuits. White has said: 'This whole arrangement has always struck me as a really heroic bit of engineering'.

The design of the final stage of the modulator is far from obvious (see Case 3 later) and well illustrates Blumlein's supreme ability to envisage the physical limitations of a circuit and, with his most creative imagination, to obtain an engineering solution.

The application of the constant impedance network and others were treated by C O Browne in his 1938 paper[16]. His descriptions follow.

Case 1. Stray capacitance of a 'hold-off' supply

'Referring to Figure 9.8a the capacitance of the hold-off supply, here indicated as a battery connected between the anode circuit and the following grid, may introduce serious loss of the higher frequencies, due to the diminishing impe-

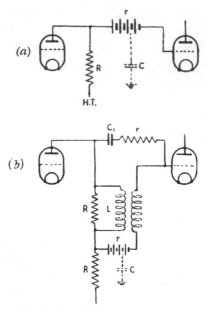

(a)

(b)

Figure 9.8   Use of constant impedance circuits to eliminate the effect of the stray capacitance of a battery

Source: *J.IEE*, 1938, **83**

dance of the stray capacitance in shunt with the anode load R with increase of frequency. In Figure 9.8b the anode resistance R shunted by the stray capacitance of the battery (C) is connected in series with an inductance L shunted by a second resistance R. The value of the inductance is determined by $L = CR^2$, so that the anode load now presents a constant impedance R at all frequencies.

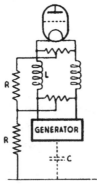

Figure 9.9   Use of constant impedance circuits to remove effectively the capacitance of a filament generator

Source: *J.IEE*, **83**, 1938

'The negative battery voltage is applied to the grid through a second winding tightly coupled to the first, the mutual inductance between the two windings being made equal to L. Assuming negligible grid input impedance, any alternating e.m.f. on the anode will be transmitted to the grid, but the impedance facing the grid will be increased by the leakage inductance l of the two coils. By comparison with the circuit of Figure 9.7a if r is the resistance of the battery the impedance of l_1 and r in series can be made constant by connecting a condenser and resistance in series between anode and grid of the two valves, where the condenser capacitance is given by $C_1 = l_1/r^2$.

'The circuit of Figure 9.8b will now be equivalent to that of Figure 9.8a except that the effect of the stray capacitance of the battery has been completely removed.'

Case 2. Filament-heating generator for cathode follower

'A circuit shown in Figure 9.9 is used by which the capacitance of the filament generator may effectively be removed from the cathode of a cathode follower valve. In this case the inductance consists of a bifilar winding through which the filament heating current is passed. The coil is shunted by a resistance equal to the cathode load resistance R, and has inductance determined by $L = CR^2$, where C is the capacitance of the generator to earth.

Case 3. Hold-off supply and filament-heating generator

'In the case in which it is required to include a hold-off voltage supply S_1, in addition to a filament generator S_2, the capacitance to earth of the hold-off supply (C_1) is included in a circuit of constant impedance equal to R, as shown in Figure 9.10 where the inductance $L_1 = C_1 R^2$, and R is given as the cathode load resistance. The effective impedance R of this circuit is shunted to earth by the capacitance C_2 of the filament generator S_2, and the combination is made part of a further constant impedance circuit by the generator inductance L_2 shunted by R. The value of L_2 is given by $L_2 = C_2 R^2$.

'In order to pass the negative voltage of the hold-off source to the following grid, a third winding is provided which is tightly coupled to L_2. As in Case 1, the leakage inductances l_1, and l_2, which are effectively in series with the resistance r of the hold-off supply S_1, are corrected by the resistance r and the condenser C_3 in series, where $C_3 = (l_1 + l_2)/r^2$.

'In the particular case of the filament generator and hold-off for the main modulation amplifier stage, the capacitance between the tightly coupled windings was such as to resonate within the working frequency range with the leakage inductance. This capacitance, together with the leakage inductance, made an effective transmission line and was treated as such.

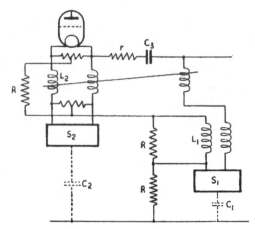

Figure 9.10   Blumlein's hold-off supply and filament heating generator circuit

Source: *J.IEE*, **83**, 1938

Case 4. HT smoothing circuits

'The circuit shown in Figure 9.11 is typical of that used for HT smoothing in the equipment. Each section of the circuit is made to appear as a constant impedance equal to R, where R is the regulation resistance of the rectifier, by choosing values for the chokes and condensers such that $L = CR^2$. The values of R and L may, of course, be chosen to account for regulation due to the reactance of the supply mains, transformers, etc. The final stage of smoothing has been shown inverted with respect to the preceding stages; an alternative arrangement which still preserves the constant impedance properties of the network.

'The cables connecting the HT supply circuits to the corresponding amplifiers present considerable capacitance between the positive conductor and earth. This capacitance is shunted at the supply unit end by the constant impedance of the smoothing circuit, and the cable is made to present a constant impedance to the amplifier by connecting, in series with it, the appropriate inductance and resistance in parallel.'

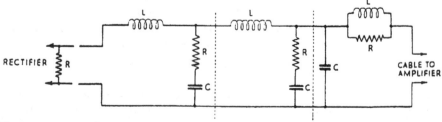

Figure 9.11   HT smoothing circuit

Source: *J.IEE*, **83**, 1938

Oscillator stabilisation[17]

Blumlein's fondness for circuit design which minimised the dependence of a circuit's specification on uncertain external factors, such as production spreads, supply voltage changes and valve characteristic variations, is exemplified by his patent, no. 563 464 of 1940, which describes a simple method of stabilising the output of an oscillator by means of negative feedback, (Figure 9.12).

Previously the achievement of equilibrium in the operation of a valve oscillator had been dependent on the flow of grid current, 'with its attendant disadvantages'. But the effect of the grid current '[could not] be easily calculated' and Blumlein sought a method which '[lent] itself to accurate design'. In the circuit of Figure 9.12 capacitor C (15) shunts the cathode resistor R (14) (thereby providing a low-impedance path for the generated oscillatory anode current) until the amplitude of the generated oscillations exceeds a predetermined value. Then, the diode (5) becomes non-conducting, C ceases to shunt R, causing negative feedback to be applied, and the amplitude of the oscillations is stabilised at such a value that grid current does not flow. Blumlein found that under these conditions the waveform was substantially rectangular and had a peak-to-peak anode current amplitude 'substantially equal to the potential of the [diode's] anode divided by $[R_{14}]$'.

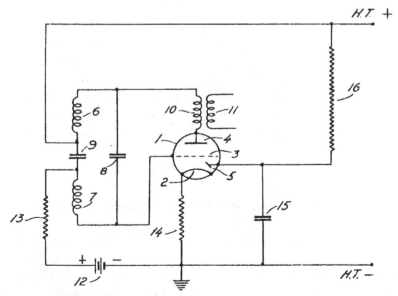

Figure 9.12   Method of amplitude control for a valve oscillator

Source: British patent no. 563 464

Potential divider

Freedom from convention is necessarily a trait of geniuses. They are endowed with imagination, that most important gift which the great Albert Einstein, of general relativity fame, considered was more important than knowledge itself. Many of Blumlein's circuits illustrate this trait. During the discussion of an IEE paper titled 'Instruments incorporating thermionic valves' he observed[18]:

> 'All the devices referred to in the paper have one point of similarity — the signals are applied at the grid and come out at the anode. It might be worthwhile considering where it might be advantageous to depart from this convention. One possible application of valve voltmeters is to the measurement of very high voltages. For this purpose one constructs a valve where the grid is earthed and used as a shield between anode and cathode. A very high voltage is put on a very small anode remotely situated from the grid, and in the cathode lead is put the inevitable feedback resistance, which is given a high value. In these circumstances the valve cathode rises to such a potential that there is practically no anode current. The valve acts as a potentiometer, the cathode voltage being about $1/(1 + \mu)$ of the anode voltage [μ is the amplification factor of the valve]. At first sight the arrangement would seem to have no advantage over a resistance-type voltmeter, but the usual very high voltage voltmeter resistances are cumbersome, and this arrangement will follow variations of the anode voltage quickly, so that measurements of very high voltage can be made in the comparative safety of a shielded cathode.'

Blumlein's comments were based on a patent[19] (with C S Bull) dated 18th November 1937. In this a circuit arrangement for measuring or monitoring, for example, the anode potential of the output stage of a radio transmitter, was described. The circuit arrangement seems to have been inspired by Blumlein's cathode follower patent of September 1934 and also his patent (no. 477 392 of May 1936), with E L C White, for a diode–triode valve voltmeter which essentially was another application of the cathode follower configuration.

Voltage measurement using a cathode ray tube (c.r.t.)

Blumlein's work demanded the ready availability of suitable laboratory apparatus, for, as Leon Brillouin once said: 'Experiments are the only means of knowledge at our disposal. The rest is poetry — imagination.' Ideas and thoughts on the way forward, and more particularly the designs which originated from them, had to be correlated with the products and practices which were the tangible implementation of those ideas.

In the 1930s much of the specialist equipment needed for this purpose was commercially unobtainable. Until EMI commenced its work on wide-band television, only narrow bandwidth circuits were a feature of radio communication systems; electronics as it is known today hardly existed as a separate discipline. Consequently Blumlein and his colleagues had to design, develop and construct many of their testing facilities. Signal generators, impedance bridges, amplifiers, power supplies and cathode ray oscilloscopes all had to be built in-house to enable EMI to achieve its excellence in engineering and to advance the practice of television by dramatic steps.

Some of EMI's test apparatuses were characterised by basic but novel features. In 1938 Blumlein, J Hardwick and C O Browne patented[20] a method of measuring the amplitude of a waveform, traced on the screen of a cathode ray oscilloscope, by moving the trace past a datum line, using a calibrated variable bias voltage. The technique is elementary—almost trivial—but has been used universally for very many years. Until 1938 no one had sought to determine waveform amplitudes in this way, even though low-voltage laboratory cathode ray oscilloscopes had been available from about 1924.

'Black spotting' and 'anti-ghost' circuits

The delay line was one of Blumlein's circuit 'favourites' in his engineering compositions. His utilisation of this circuit entity dates from about September 1934. According to White, Blumlein 'was fascinated by the great variety of ways in which they could be used'. Delay lines were employed by him in 'black spotting' and 'anti-ghost' television circuits, in transversal filters, in pulse generation and in radar systems. Whether he was the first person to incorporate the device in general electronics, as distinct from telephone engineering practice, is not known.

Blumlein's first patent in this field was submitted after EMI had given its evidence to the Television Committee, and when a detailed consideration was being given to the specification of the transmitted television waveform. This depended on many factors previously mentioned: on whether to use positive or negative modulation; sequential or interlaced scanning; d.c. or a.c. modulation; on the line and frame synchronising standards; and on the nature of the line and frame synchronising and their combination with the picture signals. With positive modulation the effect of interference is to produce white spots on the television raster, whereas with negative modulation the effect of such interference is to produce black spots.

In patent no. 446 663[21] Blumlein sought to reduce or eliminate the white spots, whether caused by interference or whether due to whiter-than-white synchronising signals, by inverting the peaks of the whiter-than-white unwanted signals and amplifying them to produce black rather than white marks on the television screen.

After the inauguration of the London Television Station it was found that, in some areas, the received and desired television images were accompanied by ghost images. Blumlein, in patent no. 507 417[22] of December 1937, proposed two methods by which the 'ghost' signal could be either attenuated or substantially annulled provided it was of smaller amplitude than the desired signal. If the latter signal is delayed with respect to the 'ghost' signal and then the two signals are suitably combined, the undesired signal can be partially or completely neutralised.

Delay-line circuits

In both television and radar engineering it is necessary to generate pulses for a variety of purposes; for example, line and frame synchronization in television, modulation of a continuous wave (c.w.) source in a pulse radar transmitter, and timing of the various circuits in complex transmitter— receiver systems. These pulses may be produced by propagating a step voltage/current waveform along a delay network and utilising the applied and delayed waveforms to establish the commencement and termination of the pulse.

The only circuit element which will delay an infinitely steep wave-front, such as a step function, without distortion, is a loss-free transmission line. For an air-cored line the velocity of propagation of an electrical signal along the line is the velocity of light in free space, 3×10^8 m/s. Consequently a line 150 m in length will delay a wavefront by 0.5 μs. Such lines are much too bulky to be used in practice. Fortunately, it is possible to simulate a low-loss line by a cascade connection of LC network sections having lumped inductances and capacitances (Figure 9.13). These artificial lines behave as low-pass filters. In their simplest form their characteristic impedances and time-delays are functions of frequency, and considerable distortion of a wavefront will occur. By changing from a simple (constant k) section to a more complex (m-derived) section much improvement can be obtained.

If an artificial line is terminated in a short-circuit or an open-circuit, reflection of a step function voltage, applied at the input, will take place, as in an actual line. The reflected wave will be out-of-phase, or in-phase, respectively with the incident voltage wave. On reaching the input of

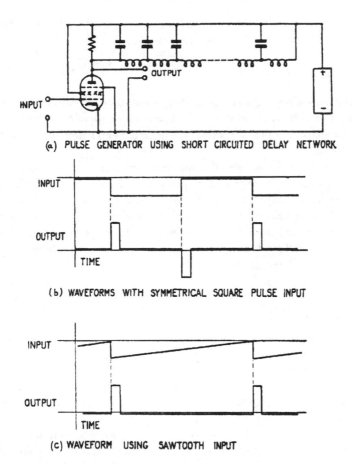

(a) PULSE GENERATOR USING SHORT CIRCUITED DELAY NETWORK

(b) WAVEFORMS WITH SYMMETRICAL SQUARE PULSE INPUT

(c) WAVEFORM USING SAWTOOTH INPUT

Figure 9.13   (a) Pulse generator using short-circuited delay network. (b) Waveforms with symmetrical square pulse input. (c) Pulse production in short-circuited delay network

Source: *J.IEE*, **93**, Part IIIA, (1), 1946

the line, the reflected wave will either be absorbed or further reflected depending upon whether or not the input is connected to a resistor having a resistance equal to the characteristic impedance of the line. An applied current step function will also give rise to a reflected wave but the phase relationship of this wave to the applied wave will be the opposite of that for a step voltage wave. At the input to the line, the resultant time variations of voltage or current are separated by a time interval of twice the one-way delay time of the network.

Using these principles, various pulse forming circuits can be designed. Some of these, based on Blumlein's patents, were given in a paper[23] which Dr E L C White read at the 1946 IEE 'Radiolocation Convention'. He mentioned that the paper would have been presented by Blumlein but for his untimely death. White's descriptions follow.

Generation of single pulses

'Figure 9.13a shows a practical circuit[24] for generating a positive potential pulse from a negative potential step. A shunt-fed generator of high impedance is used in the shape of a pentode valve, the negative step being applied to the grid of the valve.

'The input will usually be periodic, and the level must return in some pre-determined manner to its initial value prior to each step. Figures 9.13b and 9.13c show the input and output wave forms for two usual cases.

'Figure 9.14 shows a circuit making use of the current into an open-circuited network. The drive is from a cathode-follower padded out to match the network at the input end, and the current pulse is most conveniently utilized in the anode circuit of the same valve.'

Generation of groups of pulses

'In radar applications it is often necessary to generate a group of pulses bearing constant time relations with each other. For instance, three pulses of equal widths may be required, (i) and (iii) contiguous and (ii) overlapping (i) and (iii) equally, with the object of using (i) and (iii) for differential strobe pulses in an automatic range-tracking circuit, and (ii) for a marker pulse to indicate on a display unit the position of the group.

'Such a family may be produced from one delay network by a circuit similar to Figure 9.13a, except that the matching resistance is replaced by an extra delay network twice as long as the original, and itself matched at its far end. Since it is matched, its impedance at the junction with the valve anode is the

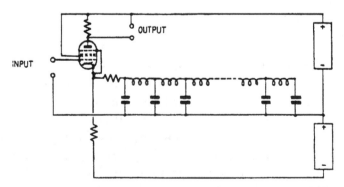

Example of use of open-circuited delay network.

Figure 9.14   Example of the use of an open-circuited delay network in pulse formation

Source: *J.IEE*, **93**, Part IIIA, (1), 1946

same as though the matching resistance were still placed directly at this point. Hence the pulse produced at the anode is identical with that in Figure 9.13b or 9.13c. This waveform now travels along the extra delay, and is absorbed at the far end, so a similar waveform but delayed by the desired amounts can be obtained from suitable taps along the network. The taps should not be loaded appreciably, i.e. in general they should be connected to the grids of further valves.

'Sometimes two or more pulses of different widths may be required. These can be obtained from taps on the short-circuited section[25], but the start of pulses from such taps will always be delayed with respect to the start of a pulse taken from the anode tap. To secure coincident starting, the latter pulse may be delayed by combining this arrangement with that previously described.

Use of 'inverted' and 'twisted' delay networks

'It is possible to produce pulses from a non-reflecting delay network by a circuit[26] such as Figure 9.15a. The network is matched at the far end, so that no reflection occurs. Hence matching at the input end is unnecessary, and if used

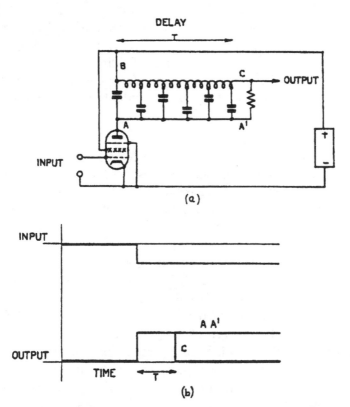

Figure 9.15   Example of the use of an inverted network in pulse generation

Source: *J.IEE*, **93**, Part IIIA, (1), 1946

will halve the potential for a given current. The network is "inverted", i.e. the side at the input end which is earthed is the side connected to the first coil, instead of the common connection to all the condensers [capacitors] as is usually the case. Since it is matched, it appears to the input as a simple resistance, so the potential waveform at A is similar to the current input waveform. Initially, the whole of the network to the right of AB, including the output point C, is carried with A, but the step waveform across AB travels along till it reaches A'C, when C returns to its original potential. The waveforms at A'A and C are shown in Figure 9.15b.

'A disadvantage is that stray capacitances to earth come across points on the "live" side of the network, and do not form part of the shunt capacitances as in the normal arrangement.

'A complex pulse which consists of a series of pulses with gaps between them, the pulses and the intervals being each of any desired width, can be produced from a "twisted" network, in which successive portions of the network, each of delay equal to the length of a particular pulse or interval, have reversed connections. This circuit makes a useful marker pulse generator for "bracketing" a radar echo pulse.'

In his paper White mentioned that other applications of delay networks used in conjunction with valves were the generation of more complex waveforms, the separation of input pulses of different widths, the stabilising of relaxation oscillation frequencies, and numerous other purposes. 'The delay network as a circuit element has undoubtedly come to stay.'

The transversal filter

Using linear network theory, the output of a wave filter, having given amplitude versus frequency and phase versus frequency responses, may be determined for any input. Blumlein, with H Kallmann and Dr W S Percival, showed that the problem of wave shaping could be treated from a radically different point of view from that used in wave filter theory.

Figure 9.16 shows a delay line with a series of taps along its length, a resistive network which enables weighted signals to be obtained from the tapping points, and a sub-system '21' which inverts the signal applied to it. The summation of the weighted signals (some of which may have negative coefficients), following the application of a signal to the input of the line, forms the output signal. By proper choice of the weighting of the successive taps the transversal filter can have the same amplitude versus frequency and phase versus frequency characteristics as a given filter.

According to White[27]: 'a particular virtue of the transversal filter is that it is easy to make the equivalent of a linear phase filter, simply by ensuring that the weights are symmetrical about the centre tap. In this case it is only necessary to provide half the delay line, and the first half of

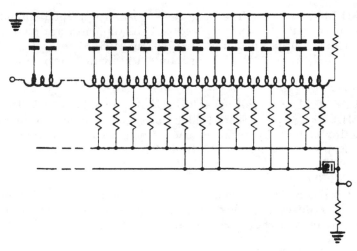

Figure 9.16   The basic transversal filter circuit

Source: *BKSTJ*, July 1968

the set of taps and weights: if the line is left open-circuit the reflection will do the rest. Rudimentary forms of this are often used for aperture correction circuits'.

References

1 BROWNE, C. O.: 'Vision input equipment', *J.IEE*, 1938, **83**, pp. 767–782
2 WHITE, E. L. C.: 'Blumlein and television', *Engineering Science and Education Journal*, June 1993, **2**, (3), pp. 125–132
3 MCGEE, J. D.: taped interview with the author, personal collection
4 BLUMLEIN, A. D.: 'Scanning circuit (suitable for hard valves, though a thyratron version is also described)', British patent no. 400 976, April 1932
5 BLUMLEIN, A. D.: 'Cathode follower for increasing input impedance of valves', British patent no. 448 421, September 1934
6 BROWNE, C. O.: 'Vision input equipment', *J.IEE*, 1938, **83**, p. 780
7 BLACK, H. S.: 'Stabilised feedback amplifiers', *Bell System Technical Journal*, 1934, **14**
8 BLUMLEIN, A. D., and CLARK, H. A. M.: 'The use of negative retroaction to increase the distortionless output of a power amplifier', EMI Report no. F 7, 26pp., EMI Central Research Laboratories archives
9 NYQUIST, H.: 'Regeneration theory', *Bell System Technical Journal*, 1932, **11**, pp. 126–147
10 Philips Co.: British patent no. 323 823, October 1928

11 BLUMLEIN, A. D.: 'The long-tailed pair (push–pull circuit with undecoupled cathode: negative feedback discriminates against common-mode signals)', British patent no. 482 740, July 1936

12 BLUMLEIN, A. D.: '"Ultralinear" amplifier circuit (screen tapped in on output transformer, to improve linearity)', British patent no. 496 883, June 1937

13 MILLER, J. M.: Scientific paper 351, US Bureau of Standards, 1919

14 BLUMLEIN, A. D.: 'The Miller integrator (circuit to linearise a sawtooth waveform by means of a valve with a capacitor from anode to grid)', British patent no. 580 527, June 1942

15 Ref 2, p. 128

16 Ref 6, pp. 780–782

17 BLUMLEIN, A. D.: 'Method of amplitude control for a valve oscillator (negative feedback at cathode is made operative by means of a diode, when amplitude reaches a certain value)', British patent no. 563 464, June 1940

18 BLUMLEIN, A. D.: discussion on IEE paper 'Instruments incorporating thermionic valves', *J.IEE*, September 1939, **86**, pp. 381–410

19 BLUMLEIN, A. D., and BULL, C. S.: 'Valve used as potential divider for high voltages (grid earthed and input to anode: a reduced output appears at the cathode)', British patent no. 507 665, November 1937

20 BLUMLEIN, A. D., HARDWICK, J., and BROWNE, C. O.: 'Measuring waveforms on a cathode-ray oscilloscope by moving the trace past a datum line, using a variable bias', British patent no. 515 044, March 1938

21 BLUMLEIN, A. D.: 'Inverting tips of whiter-than-white signal peaks, whether caused by interference or due to whiter-than-white synchronism signals, and amplifying them to produce black marks on screen rather than white', British patent no. 446 663, September 1934

22 BLUMLEIN, A. D.: 'Anti-ghosting device for television receiver (main signal delayed and fed back as required, so as to neutralise ghosts)', British patent no. 507 417, December 1937

23 WHITE, E. L. C.: 'Generation of very short pulses', Proceedings of the Radiolocation Convention, *J.IEE*, March–May 1946, **93**, Part IIIA, (1), pp. 312–314

24 BLUMLEIN, A. D.: 'Production of pulses using reflection from the open-circuit or short-circuit end of a line', British patent no. 528 310, February 1939

25 BLUMLEIN, A. D., WHITE, E. L. C., and WILLIAMS, F. C.: 'Radar system for searching (in range) and locking onto any echo found', British patent no. 582 503, October 1943

26 BLUMLEIN, A. D.: 'Improvement to previous patent (517 516) for designing filters of desired characteristics using delay lines', British patent no. 574 133, June 1942

27 WHITE, E. L. C.: 'Contributions to circuitry', *BKST Journal*, July 1968, reprint

Chapter 10

Air defence

Of all the battles of the Second World War, two of the most important were the Battle of Britain and the Battle of the Atlantic. Both battles directly threatened the freedom of existence of the United Kingdom; both battles were won by narrow margins; and both battles were won because of the support given to the fighting forces by technology. An indispensable element in both battles was the Allies' ability to detect and locate hostile air and sea forces by means of radar.

On the importance of radar in air battles, the German General Adolf Galland has written[1]: 'The British had, from the first, an extra-ordinary advantage, never to be balanced out at any time in the whole war: their radar and fighter control network.' And on the Battle of the Atlantic, Hitler himself, in his 1944 New Year's Day address, said[2]: 'The temporary setback in the U-boat offensive can be ascribed to a single invention of the enemy'. That invention was 10 cm radar, more precisely ASV Mark III radar.

The need for means to detect and locate offensive aircraft (and ships) did not originate in the Second World War. In 1913–14, just before the outbreak of the First World War, the German air force comprised five battalions of aircraft totalling 232 machines and a military airship organisation, also of five battalions[3]. This was set against the total effective strength of 63 aeroplanes in the Royal Flying Corps, and 41 aircraft and 52 seaplanes in the Royal Naval Air Service[4].

When the war started, the RFC was sent to the Western Front as part of the British Expeditionary Force. Because the Army had no aeroplanes left to defend Britain, from September 1914 the Admiralty, at the request of Lord Kitchener, accepted responsibility for the air defence of the country[5].

In 1914 the means to detect and locate enemy aircraft were limited to unaided visual sighting by day, searchlight-aided sighting at night, and, initially, unaided listening by day and night. The inadequacy of these means, especially for night attacks by Zeppelins, is exemplified by a few statistics relating to 1915: there were 37 airship journeys in 20 separate raids; the defending aircraft made 81 sorties; they sighted the enemy three times, but failed to engage them[6]. Despite the immense size of the German airships (about 200 m long by about 25 m diameter) and their slow speeds (50–70 miles per hour), they proved to be difficult to find on a dark night.

The Admiralty soon appreciated the importance of the contributions that scientists and engineers could make to the war effort, and in July 1915 it established the Board of Invention and Research[7]. Admiral of the Fleet Lord Fisher was appointed Chairman of the Central Committee of the Board, which included six sections. Section 1 was concerned with 'airships, aeroplanes, seaplanes and aeronauticics generally'; its committee members were Professor the Honourable R J Strutt, R Threlfall, R E Glazebrook and F W Lanchester.

Lanchester gave some thought to the problem of night air attacks and on 23rd September 1915 he submitted to the Aeroplane Committee 'A note on the general illumination of a belt or region for the detection of enemy aircraft'[8]. His idea involved illuminating an area of the country 100 miles (160 km) in length by 20 miles wide to form a 'veritable carpet of light against which hostile aircraft would be visible, in silhouette, from aeroplanes flying at a higher altitude. Lanchester's scheme was discussed by the BIR in October 1915, and quantitative model tests were initiated at the National Physical Laboratory. By 27th November, Lanchester reported that these were 'highly satisfactory'.[9,10]

Following this early success, a consultant, Charle Mertz, was asked to produce an estimate for illuminating an area of country 200 miles by 10 miles. He assumed a power density of 3.86 kW/km^2 and one lighting tower (25–100 ft high, depending on the terrain) for each two square miles. The total power requirement would thus be 20 MW, and this could be supplied from 1000 petrol-driven generators (110 V, 182 A) to the 1000 towers. Mertz estimated the cost at £1 002 000. About 1200 men would be needed to operate the scheme, and it was thought that it could be constructed in approximately eight months[11].

Full-scale trials of Lanchester's plan took place at Upavon on 16th July 1917. Four light units had been erected at the corners of an imaginary square, the sides of which were one mile in length, and two types of lighting had been installed at each point. Two aircraft were employed during the trial, one flying at 3000 ft (914 m) and the other at 3500 ft. The object of the test was to determine whether the observer or pilot of the higher flying aircraft could see the other aircraft.

The trial was not a success[12]. Lanchester was most unhappy about the conduct of the experiment, which he felt had been carried out 'under conditions which were foredoomed to failure'. The night air had been clear and free from ground mist (an essential prerequisite to scatter the light) and hence the experiment—was no experiment at all', Lanchester felt. The scheme was abandoned in favour of a plan for general illumination[13].

The reason for this change of strategy was the knowledge that enemy aircraft were flying at a higher altitude than was assumed when Lanchester's idea was first proposed. Zeppelins in 1917 were operating at up to 6000 m (20 000 ft) in raids over England and had to be attacked mainly from below. Hence Lanchester's initial (1915) premise was invalidated.

Although the original scheme came to nought, nevertheless the notion of a plan of general illumination found favour with Lieutenant General Sir David Henderson, Director General of Military Aeronautics, for, on 28th October 1917, he wrote[14] that 'At present there is no method of defence against hostile aeroplanes by night which gives promise of final success or even much hope of improvement. I agree with Mr Lanchester's view that an illumination scheme is fundamentally and inevitably a first and vital necessity for any successful scheme of defence'. It is interesting to note that during the Second World War the Germans successfully used defence methods known as Wilde Sau and Zahme Sau, in which searchlights and flares were used to illuminate a cloud base. German fighters were then able to see the silhouettes of the Allied aircraft against the bright cloud background. A belt of land from northwest Germany to the Low Countries was provided with illuminants for this purpose[15].

The significance of Henderson's statement is illustrated by the figures given in Table 10.1. In 131 flights, 58 of them in efficient fighting aircraft, the defenders only caught sight of an enemy machine on eight occasions; the eight sightings led to three combats, only one of which was rewarding.

The difficulty in accurately locating hostile air targets is further evident from the sheer number of shells that had to be fired to endanger aeroplanes. A survey showed that, for the period September 1917 to May 1918, 14 540 shells had to be discharged at night to bring down each enemy aircraft. For daylight attacks the figure was a mere 2335[16].

For night defence, it was obvious that pilots needed urgent assistance from the ground. However, towards the end of 1917 all that could be done was 'to continue the reorganisation of the air patrol lines, to limit the

Table 10.1   Air raids

Raid	Defending aircraft			Hostile aircraft		
	Efficient fighters	Others	Total	Came well overhead	Seen by our pilots	Combats
Raid of 31st Oct.	19	31	50	20	4	0
Raid of 6th Dec.	12	22	34	13	0	0
Raid of 18th Dec.	27	20	47	16	4	3 (1 successful)

height zone to be watched by erecting balloon aprons, and to improve the services of the searchlights'. A similar situation prevailed in the early stages of the Second World War.

Apart from Lanchester's plan, the BIR also considered and implemented development work on sound detection and sound location of aircraft. The earliest Admiralty work was undertaken in 1915–16 by a Lieutenant Richmond and the Western Electric Company, using trumpet-shaped collectors[17]. In 1916 Professor Mather, of the City and Guilds College, presented a report to the Munitions Inventions Board on the properties of a 16 ft (4.9 m) diameter sound mirror cut in chalk at Bimbury Manor, near Maidstone, Kent. He recommended the use of concrete as a suitable material for the construction of sound mirrors, and subsequently one of his concrete parabolic sound reflectors detected an aeroplane at a distance of ten miles.

The first of the fixed concrete mirrors, which were operational during the later stages of the war, was the 15 ft mirror at Joss Gap, North Foreland, operated by the Signals Experimental Establishment, Woolwich. It used a hot-wire microphone and a Helmholtz resonator to explore the focal surface of the mirror and to detect any received sound waves. Calibrated azimuth and elevation dials enabled the directions of the sound sources to be determined[18].

A great deal of work was carried out on 20 ft, 30 ft and 200 ft concrete mirrors, in the 1920s and early 1930s, particularly by the Air Defence Experimental Establishment at Biggin Hill. The 20 ft and 30 ft track plotting mirrors had a range of 10 miles under average weather conditions, extending to 15 miles for still air and favourable weather. Pairs of mirrors used together could give the course and ground speed of an aircraft. The 200 ft sentry mirrors were able to detect aircraft at a range of 20 to 25 miles, depending on local weather conditions[19].

In 1926 it was suggested that, if listening could be conducted from a ship, Britain's early warning line could be moved away from its shores.

Hence it was thought desirable to investigate the feasibility of listening for aircraft at sea. HMS *Lunar Bow* was fitted with three sound collectors: a conical horn, an exponential horn and a parabolic mirror. The ranges obtained with these devices were compared with the performance of the unaided ear and of a microphone–amplifier system[20].

Despite the considerable efforts that were made, the experiments showed that the use of collectors did not improve greatly on the unaided ear. As an average of 67 readings, the mean range with the parabolic mirror was just over 5 miles (8.9 km); the corresponding figure for the unaided ear was 4.3 miles (6.9 km).

Experiments were also undertaken in the 1920s with trumpets carried aloft on kites. The idea was that the increased distance to the horizon would extend the detection range (cf. the modem AWACS system); however, the trumpets were more exposed to winds and disturbing sounds, which completely outweighed any advantage from getting above the acoustic horizon[21].

By 1930 the characteristics of sound mirrors were well documented and a plan for an early warning system based on 30 ft and 200 ft mirrors was evolved. Briefly, 20 200 ft mirrors were to be built between Yarmouth and St Alban's Head at a cost of £3000 per mirror, and 12 30 ft mirrors, at re-entrant parts of the coast such as the Thames Estuary and districts near the Isle of Wight, to supplement the larger mirrors, at a cost of £600 each. The total cost of £67 200 would allow 350 miles (560 km) of coast to be provided with listening means for a cost of £118/km[22].

As well as the many investigations on trumpet, horns and parabolic mirrors, much work on listening wells and discs was carried out from around 1918[23,24]. A pilot disc system built at Romney Marsh comprised two rows of 16 discs each, the two rows being 3 miles apart. In each row the 105 ft (32 m) diameter 2 in (5 cm) thick discs were based at distances of half a mile apart. A November 1932 report[25] on the operation of the system mentioned that the location of an aircraft could be given to an accuracy within 0.1 mile, the direction to within two degrees of true course, and the ground speed to within 4 mph: the height of an aeroplane could be measured (in 1927) to an accuracy of about 900 feet. The cost of such a system was estimated to be about £500 per mile.

During the development period of sound detection, aircraft speeds gradually increased. By the mid-1930s monoplane fighters and bombers were being designed and built having speeds of up to 300 mph — an appreciable fraction of the speed of sound. Consequently the 20 to 25 mile maximum range of the 200 ft mirrors could provide a warning time of only three minutes.

Of the other methods of detecting aircraft, only those based on the measurement of infrared radiation showed any promise of success. The earliest known example of the detection of aircraft by infrared apparatus dates from 1918 when Hoffman, of the US Signal Corps, undertook some experimental work in this field[26].

Trench warfare on the Western Front in the First World War necessitated the frequent use of night patrols. Early in 1918 Hoffman began work on the problem of detecting, in the dark, men and other objects that were at a higher temperature than their backgrounds. He found that men could easily be detected at 180 m, with apparatus comprising a Hilger thermopile, mounted at the focus of a 36 cm silvered parabolic mirror, and a D'Arsonval galvanometer. Other tests showed that aircraft could be detected, on clear nights, at ranges 'well over a mile'. Hoffman observed that, as long as clouds did not drift across the field of view, the galvanometer reading was admirably steady, but the 'slightest whiff of cloud gave a warm indication as large as a plane would give'.

The Admiralty, during an extensive programme of work on infrared science and technology undertaken in the 1920s, confirmed Hoffman's observations. Following one series of night trials using a 36 in mirror, a Moll thermopile and a Paschen galvanometer mounted at distances of up to 1.25 miles from a tethered aircraft, the Admiralty noted[27]: 'Definite deflections were obtained but disturbances (probably due to cold and gusty night winds blowing over the thermopile) were frequent and of the same order of magnitude as the aeroplane effect.... Under the conditions of the experiment it would not have been possible to detect the aeroplane without some previous knowledge of its approximate position'.

Further investigations on an infrared detector were carried out, by Dr R V Jones, from 1935 to 1938[28]. He evolved an airborne infrared receiver and achieved a ground-to-air detection range of approximately two miles. However, because of the increased noise level from microphonic disturbances, the in-flight range was under half a mile on single-engined aircraft. A Vickers Vincent (with a 635 hp engine) could be observed broadside on at 500 m. The disadvantages of Dr Jones's equipment were an inability to work through cloud and a lack of provision of range information; activity in this field ended in March 1938.

On 30th January 1933 Adolf Hitler was appointed Chancellor of the Third Reich. His trusted friend, Herman Goering, was rewarded for his loyalty during the lean years of the 1920s and was given four posts in the new administration. One of them was Special High Commissioner for Aviation. In April 1933 the Commissariat was upgraded to the status of Air Ministry with Goering as Minister and Erhard Milch, who on 6th January 1926 had been installed as Chairman of Deutsche Lufthansa, as Secretary. Together, Goering and Milch embarked on a course of action which would lead to the reformation of the German air force.

No German combat aircraft were constructed in 1933 but in 1934 840 were built, followed by 1923 in 1935. Berlin informed the world on 1st March 1935 that the German Luftwaffe had been recreated. With its 20 000 officers and men and 1888 aircraft, the new service became a potential threat to peace[29].

Early in 1934 the Director of Scientific Research, Air Ministry, H E Wimperis, with his personal assistant, A P Rowe, investigated the situation on the air defence of the United Kingdom, with particular regard to the means which could be adopted to counter attacks made against the country by an enemy possessing a large modern air force. Wimperis and Rowe located 53 files on the subject of air defence, but none of these were felt to contain ideas which were especially suitable for the establishment of a modern air defence system. As a consequence, Wimperis considered that the Air Ministry should be warned of the dangers inherent in such a situation. He urged the setting up of a Committee for the Scientific Study of Air Defence. The Secretary of State for Air agreed and the committee was constituted under the Chairmanship of H T Tizard, and comprised Professor P M S Blackett, Professor A V Hill and H E Wimperis; A P Rowe was the Secretary. Its terms of reference were: 'To consider how far recent advances in scientific and technical knowledge can be used to strengthen the present methods of defence against hostile aircraft'[30].

The potential threat, posed in 1934 by the renewed militarisation of Germany, led to speculation on the prospect of constructing equipment which could produce damaging radiation, popularly called 'death rays'. Following a discussion, early in 1935, between Tizard and Watson-Watt, Superintendent of the Radio Research Station, Slough, Watson-Watt wrote a memorandum in which he dealt with quasi-fixed and rapidly moving targets and the practicability of their being destroyed by death rays[31].

For the first type of target, Watson-Watt calculated the amount of power which would be needed to produce physiological disablement in a man remaining in a radio beam for as long as ten minutes, at a distance of 600 m. He could be treated as composed simply of 75 kg of water with a projected area of one square metre (2 m high by 0.5 m wide). Making reasonable assumptions, Watson-Watt calculated that it would be necessary to deliver 1.5×10^4 g cal/min to raise the man's temperature by $2\,^{\circ}C$. Using a simple half-wave radiating antenna, approximately 5000 MW would have to be radiated to give this flux per square metre at 600 m distance.

On the question as to whether rapidly moving targets could be immobilised, Watson-Watt assumed that the bombing aircraft of the immediate future would be all-metal monoplanes with cowled engines and screened ignition systems. Thus pilots, magnetos and the rest of the ignition systems would be enclosed in very effective Faraday cages and would be immune from external electromagnetic wave radiation. Watson-Watt ended his memorandum by stating that 'attention was being turned to the still difficult, but less unpromising, problem of radio detection as opposed to radio destruction'.

He submitted, to the Air Ministry, his next memorandum on 12th February 1935 and a definitive version was put before the Committee for the Scientific Study of Air Defence on 27th February 1935. With his assistant A F Wilkins, Watson-Watt calculated the electric field strength which would exist at a position 'T' if radiation from T were reflected back by an aircraft to T. He assumed that the typical night bomber would be a metal-winged craft, well bonded throughout, with a span of the order of 25 m, and that the wing structure could, to a first approximation, be considered as a linear oscillator, with a fundamental wavelength of 50 m, and a low ohmic resistance. For an aircraft 6 km from the transmitter and flying at a height of 6 km, Watson-Watt and Wilkins deduced that a received field at T of 0.1 mV/m could be expected, after making generous allowances for losses[32].

The correctness of their views was borne out, on 26th February 1934, when a Heyford aircraft at an altitude of 1800 m was detected flying up and down the centre line of a radio beam radiated by the BBC's Daventry transmitter. This had a power output of 10 kW at a wavelength of 50 m, the beam width of the antenna was ± 30° at an inclination of 10° above the horizontal. A c.w. system of transmission and reception was used[33].

In March 1935, the Tizard Committee recommended that large-scale experiments should be performed, and to initiate these the Treasury made £10 000 available. Furthermore, a new Air Defence Research Sub-committee of the Committee of Imperial Defence was set up to include senior representatives of the three Services and other departments, so that all the various bodies concerned with radiolocation could constructively and speedily advance Radio Direction Finding (RDF), now known as RADAR (RAdio Detection And Ranging). A special laboratory was prepared at Orfordness and, by 16th June 1935, Watson-Watt and his co-workers were able to demonstrate to the Tizard committee a pulse radar system which could detect aircraft at distances up to 64 km. By mid-September the height of an aircraft, at a distance of 19 km and flying at 2100 m was measured successfully and, in January 1936, the bearings of aircraft at 40 km could be determined with reasonable accuracy[34].

Larger premises, at Bawdsey Manor, near Felixstowe, were purchased for the research group and from March 1936, the new establishment was known as the Bawdsey Research Station. Watson-Watt was the Superintendent. With grounds of about 73 ha in extent there was ample space to construct the large transmitter and antenna towers needed for further experimental work[35]. Towers 73 m high were erected and, on 13th March 1936, an aircraft flying at 406 m at a distance of 120 km was located. This range was very much greater than could be achieved using acoustic methods of detection.

On 7th March 1936 Hitler ordered his troops into the Rhineland. About 35 000 men crossed into the demilitarised zone and occupied the main towns. In the same month a secret UK report[36] was prepared on 'Appreciation of situation vis-à-vis Germany. Condition of our forces to meet possibility of war with Germany'. The report stated that the total numbers of guns and searchlights which could be deployed, to protect all the defended ports (including the great naval bases of Plymouth, Portsmouth and Chatham) from Milford Haven south about to Harwich were just 16 and 25 respectively; that the naval base at Scapa Flow was without 'any defence'; that all available AA ammunition had been sent to the Mediterranean and 'very little remained at Home'; and that the numbers of guns ready for use for the safeguard of London and the Thames was only 26, for those fitted with the then latest fire-control gear, and 42 without but capable of firing. Furthermore, of the total of seven squadrons of fighters which could be made available after mobilisation, two squadrons would be unable to operate at night and three squadrons were equipped with obsolescent aircraft possessing an inadequate performance. Hence the RAF had just two squadrons of fighters to safeguard the whole of the United Kingdom, at night.

Fortunately Winston Churchill, ably assisted by some friends and informants, had been alerting Parliament and the general public to the dangers inherent in the rise of the Nazi Party and the re-armament of Germany[37]. At this time the UK Secretary of State for Air was Sir Phillip Cunliffe-Lister (later Lord Swinton). He had succeeded Lord Londonderry in 1935 and was determined to improve the RAF's strength.

RAF Fighter Command was formed on 6th July 1936 with Air Marshal Sir Hugh Dowding as the first Air Officer Commander-in-Chief One of his earliest tasks was to consider the forces needed to provide an 'ideal defence scheme' for the protection of the UK. He showed that, broadly, 45 fighter squadrons would be essential for this purpose. The number was not large, but in 1936 it was not envisaged that in the event of a war France would fall and bring Britain within range of enemy fighters[38].

Dowding worked tirelessly to integrate the growing numbers of men (and later of women), airfields, aircraft, radar, balloons, headquarters and communications into a coherent and efficient organisation. This was first tested in the summer of 1937 when an 'enemy' force of Bristol Blenheim bombers had to be intercepted by a force of RAF fighters. It was found that none of the biplane fighters was able to intercept the new, fast, twin-engined Blenheims without being in a position of advantage above them. Even if the defending fighters chose precisely the right moment to dive, the Blenheims drew away steadily until the fighters were obliged to give up the chase and return home. In his report Dowding observed[39]: 'The exercise again emphasised the self-evident fact that fighters which cannot overtake enemy bombers are quite useless'. The main feature which emerged from an analysis of the exercise was 'the great superiority of speed of Blenheims and Battles giving them the ability to refuse combat at will, by use of the throttle only'[40]. In 1938 Fighter Command had 759 fighters (including reserves), of which 93 were Hurricanes. No Spitfires were in service and the Hurricanes could not fight above 4570 m, even in summer, because they were without heating for their guns[41]. Thus one year before war broke out the majority, 666 planes, of Fighter Command's force comprised outdated biplane fighters, namely Gloster Gladiators and the like. The German Luftwaffe had 1200 modern bombers in 1938.

Clearly, it was essential, if hostilities commenced, for the RAF's few modern fighters to be vectored to positions which would enable attacks on enemy aircraft to be initiated. A highly effective detection scheme was thus of prime importance in the Air Defence of Great Britain (ADGB).

The Bawdsey RDF system became the prototype of a chain of stations (Figure 10.1) built to cover the entire seaboard of the country. On 16th September 1935 the Air Defence Research Committee of Imperial Defence recommended the provision of a 20-station radar chain extending from the Tyne to the Solent[42]. The Treasury subsequently agreed and approved the expenditure of £10 million. In December 1936 the Bawdsey establishment[43] comprised just 13 scientific officers of various grades supported by 15 assistants (grades II and III) and three laboratory assistants. Most of the officers were physicists without experience of large-scale engineering. The construction of the new RDF stations and the manufacture of the necessary apparatus were quite beyond the resources of the research station; extra support from outside contractors and manufacturers had to be employed to effect the implementation of the 20-station plan. Actually, steps had already been taken, by 1935, to involve British industry in secret government work.

During the 1920s the design and development of new Service equipment had been entirely the responsibility of three Service establishments, namely HM Signal School, the Signal Experimental Establishment, and the Royal Aircraft Establishment (RAE). The Halahan Committee, which in 1924 had considered the organisation of RAE, thought that much of the work of the Wireless Section of the estab-

Figure 10.1   Chain Home (CH) transmitter antennas

Source: DERA, Malvern

lishment could with advantage be given to private firms. However, the Committee recognised that this was not feasible, since the problems of weight, size and multiplicity of wavelengths were quite distinct from those of the commercial manufacturers. The latter were concerned more with quantity than quality, whereas for the RAE quality had to take precedence over quantity. In the pre-radar period 1930–1935 RAE designed and developed a complete series of ground and airborne equipment. No project was handed over to industry for subsequent manufacture until fully developed drawings had been prepared by RAE[44].

This policy was not in accord with the view of the Air staff who wished to encourage some linkage between industry and the needs of the Service. Marconi and STC were approached to undertake the design of some new equipment, but they were unwilling to participate. Since they were the two firms with the most extensive design facilities, the gap between commercial design capacity and Service requirements was apparent. The Air Ministry accordingly adopted a strategy which would narrow the gap before it became a dangerous obstacle in the rearmament programme.

The policy had two objectives: first, to enlist the support of firms capable of originating the design, and subsequent manufacture, of major items of new equipment from a detailed theoretical specification; and secondly, to have available a large number of industrial organisations which would be prepared to undertake design work of a minor nature. Success was quickly achieved. In 1932, 5% of the development effort formerly undertaken by RAE was carried out by industry; by 1935, the year rudimentary radar was demonstrated, the proportion had risen to between 50% and 60%, and by 1939 to 90%. Nearly 60 firms, out of the total of 80 to 90 possible equipment manufacturers, were able to carry out such work. Several companies—Marconi, STC, GEC, Plessey, E K Cole, and Siemens— were now able and willing to design radio equipment from a technical specification. By 1939 the number of approved firms had risen to 17.

This policy did not at first embrace radar generally because of the highly secret nature of the development. Watson-Watt referred to this matter in a paper on radar[45].

'In 1938 a number of radar instruments were manufactured in Britain in pre-paration for war. This caused [a] problem. The government could not tell the manufacturers what they were making in case they should give away the secret of radar. This meant that no manufacture could make a complete radar instal-lation. So different parts of the equipment were made by different manufac-turers, and these were then put together in secret by scientists.

In January 1937 Watson-Watt recommended that RDF should be disclosed to Metropolitan Vickers so that the firm could carry out some research work for the Bawdsey Research Station. A few days later, on 22nd January, it was agreed to describe only the transmission aspects to Metropolitan Vickers, and the receiving side only to A C Cossor Ltd[46]. Magnificent pioneer work was undertaken by Metropolitan Vickers on the transmitters (Type T 3026), and by A C Cossor Ltd on the receivers (Type RF 7). In a relatively short period of time, following the basic work of the few scientists at Orfordness and Bawdsey, the CH early warning system was planned and the equipment engineered, manufactured, installed and commissioned, just in time to be of central tactical importance in the defence of the nation against the onslaught of the German air force during the Battle of Britain.

Figure 10.2 illustrates the principles of the CH system[47]. The volume of space to be kept under surveillance is illuminated (floodlit) with radio frequency (r.f.) pulsed energy by a vertical array of half-wave dipole radiators strung between two steel lattice towers. Radiation reflected from aircraft within this volume is received back at the ground station by a set

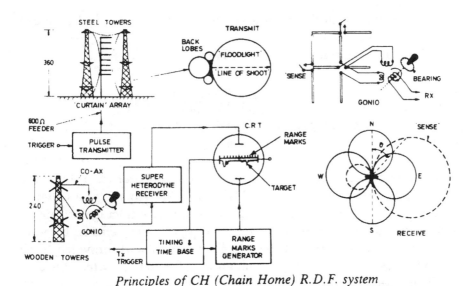

Principles of CH (Chain Home) R.D.F. system

Figure 10.2   Principles of Chain Home Radio Direction Finding (RDF) system

Source: *GEC Journal of Research*, **3**, (2), 1985

of crossed dipoles, mounted on a wooden mast, connected to a low-noise, high-gain receiver. The detected output is displayed as a Y-deflection along the linear time base of a cathode ray oscilloscope. If the transmitter-target-receiver distance is $2R$, the time taken taken for a pulse of energy to travel this distance is $2R/c$ where c is the speed of electromagnetic waves, (3.10^8m/s in free space). Using a calibrated time base, the elapsed time between the transmitted pulse and the received echo, and hence the range R, can be easily measured.

Bearing information is provided directly by comparing, using a goniometer, the outputs from the E−W and N−S dipoles. The horizontal polar diagram of a horizontally mounted dipole has a figure-of-eight shape. For the E−W dipole the output from it is proportional to $\cos\theta$, for radiation received at an angle θ, where θ is measured from the direction of north. Similarly the output from the N−S dipole is proportional to $\sin\theta$. These two outputs are fed to the two orthogonal stator coils of a goniometer and establish a resultant field the bearing of which, relative to north, is θ if the stator coils are aligned N−S and E−W respectively. This resultant field is detected by the goniometer's rotor coil.

Height finding is indirectly determined by comparing, again using a goniometer, the outputs from two similarly aligned horizontal dipoles situated at different heights above ground. The ratio of these signals is related to the angle of elevation of the reflecting target. A 'gap-filler' antenna was used with the main array to improve the vertical coverage pattern of the radiated field.

Typical operating conditions of the CH transmitter were:

carrier frequency	20 to 30 MHz
peak power	350 kW (later 750 kW)
pulse repetition frequency (p.r.f.)	25 Hz and 12.5 Hz
pulse length	20 μs

The original CH plan was based on the utilisation, for each station, of four allocated frequencies in the 20 MHz to 50 MHz band, as a counter-measure to possible enemy jamming, and as a precaution against opera-tional difficulties caused by interference or propagation effects. For that purpose four transmitter and four receiver towers were provided, each pair of towers being dedicated to one allotted frequency. Later, this plan was abandoned in favour of a simpler arrangement which used a principal, and a standby, frequency in the 20 MHz to 30 MHz band. Various ingenious anti-jamming/anti-interference devices were incorporated in the CH system. These included intentional p.r.f. 'jitter'; double phosphor screens to improve the discrimination of synchronised signals against unsynchronised signals; narrow-band notch filters to reject CW interfer-ence; and 'black-out' circuits to prevent unsynchronised interference being displayed on a c.r.t. screen (Figure 10.3).

The second and third RDF stations were completed, in July and August 1937, at Dover and Great Bromley, Essex. By August 1938 five stations were available for participation in the Air Exercises[48]. The work involved in building, calibrating and operating the new stations required great efforts by both civilian and Service personnel, and imposed a consid-erable burden on the Bawdsey research staff. Research activity, in large part, came to an end and the staff spent much of their time with the difficult task of finding suitable sites and with the 'laborious and tricky business of calibrating the stations'[49].

The siting specification (about 1936) for CH stations detailed the condi-tions which had to be met before work could commence on a new RDF installation[50]. A site had to be well back from the coast, with a smooth slope between it and the sea, to provide good height-finding and good range-finding. Ground irregularities had to be avoided, if possible, since these could lead to permanent echoes and distort the height-finding prop-erties of the apparatus. In addition, the chosen sites had to be accessible to lorries carrying heavy engineering equipment, and have ground suitable for supporting 109.7 m steel masts; they had to be close to sources of elec-trical energy, secure against sea bombardment, and be inconspicuous from

Figure 10.3   CH receiver room, Bawdsey

Source: DERA, Malvern

the air. Above all, it was essential that they should not 'gravely interfere with the grouse shooting . . . '.

In May 1938 Rowe became Superintendent of the Bawdsey Research Station when Watson-Watt was appointed the first Director of Communications Development at the Air Ministry. By December 1938 Rowe[51] was complaining to Watson-Watt that 'The effort being put into the erection of the masts does not seem to be commensurate with the sacrifices in research made by Bawdsey'. These sacrifices did not affect the CH programme, the completion of which in time for use in the Battle of Britain was a most splendid achievement (Figure 10.4), but, it would seem, they restricted the progress of some radar developments notably AI, GL, and ASV radars. It will be seen, in Chapter 12, that in 1940, when the battle against the night bombers was being fought, the available AI and GL radars were largely ineffective and had severe operational limitations. Furthermore, no GCI radars, for vectoring night fighters to positions from which attacks could be initiated, existed.

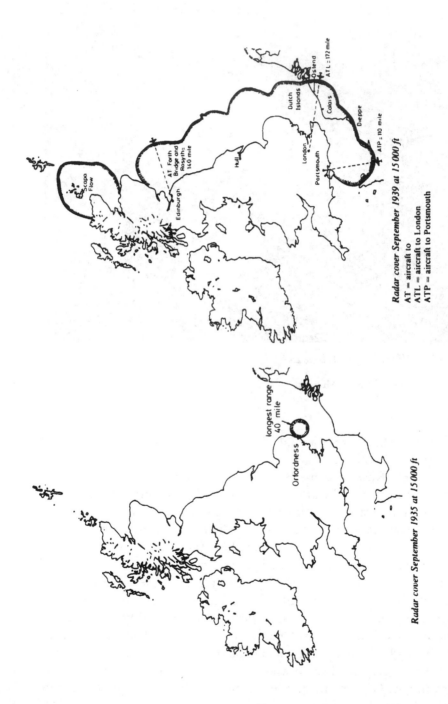

Figure 10.4   Maps showing the radar coverage of the United Kingdom in 1935, 1939, 1940 and 1941

Source: 'Science at war', by J G Crowther and R Whiddington (HMSO, 1947)

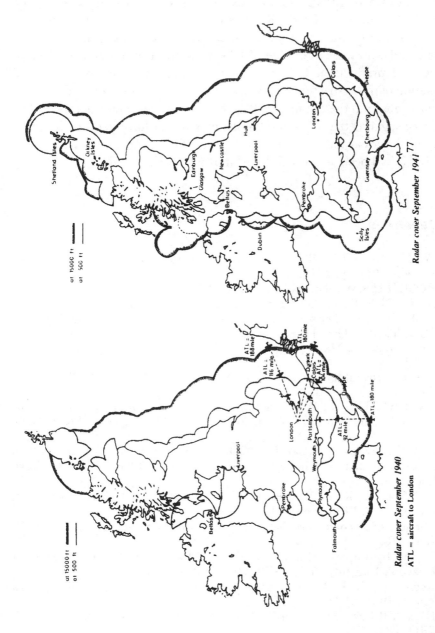

Figure 10.4 (continued)

Until early in 1939 the secret of British radar was restricted to a few people. Even within government circles most scientists outside Bawdsey were unaware of radar. Notable exceptions were the members of the Tizard Committee, Lord Rutherford, a defence adviser to the Government, and the few key staff at Metropolitan Vickers Ltd and A C Cossor Ltd who were in charge of CH transmitter and receiver developments and production. It is a matter for conjecture whether a partial relaxation of this restriction could have led to an amelioration of the devastation wrought in the autumn of 1940 by the German night bombers. The evidence suggests that considerable benefits would have accrued if one or more of the great industrial research laboratories, such as those of GEC and EMI, had been actively engaged in R&D work on radar.

References

1 WATSON-WATT, R. A.: 'Three steps to victory' (Odhams Press, London, 1957), p. 249
2 ROSKILL, S. W.: 'The war at sea 1939–45', (HMSO, London, 1954), p. 15
3 ASHMORE, E. B.: 'Air defence' (Longmans, Green, London, 1929)
4 NORMAN, A.: 'The great air war' (Macmillan, New York, 1968), p. 52
5 SCOTT, P.: 'The defence of London against Zeppelins 1915–1916', *Strand Magazine*, May 1919, pp. 523–528
6 Ref. 3
7 Board of Invention and Research 1915–17, ADM 212/158, PRO, Kew, UK
8 LANCHESTER, F. W.: 'A note on the general illumination of a belt or region for the detection of enemy aircraft', AVIA 8/473, PRO, Kew, UK
9 ANON.: 'Proposed remarks on BIR paper of 23rd September 1915', a paper of the Board of Invention and Research, AVIA 8/473, PRO, Kew, UK
10 LANCHESTER, F. W.: letter to T. H. Hoste, 27th November 1915, AVIA 8/473, PRO, Kew, UK
11 Minutes of a conference held on 6th April 1916, AVIA 8/473, PRO, Kew, UK
12 Report by Captain Gilling on 'Trial of Lanchester lighting scheme at Upavon', 16th July 1917, AVIA 8/477, PRO, Kew, UK
13 Reply by Mr F. W. Lanchester on report by Captain Gilling, 23rd August 1917, AVIA 8/477, PRO, Kew, UK
14 HENDERSON, D.: memorandum, 28th October 1917, AVIA 8/475, PRO, Kew, UK
15 STREETLY, M.: 'Confound and destroy: 100 Group and the Bomber Support Group', (Jane's Publishing, London, 1978)
16 Report, AIR 9/51, PRO, Kew, UK
17 ADM 212/158, PRO, Kew, UK

18 Report no. 87, 'Preliminary report on the Joss Gap Station of Acoustical Section SEE', W. S. Tucker, 22nd September 1920, AVIA 23/84, PRO, Kew, UK

19 Report on 'The employment of troops for long distance listening with sound mirrors', June–July 1932, AVIA 12/132, PRO, Kew, UK

20 'Report on Admiralty experiments with HM Drifter Lunar Bow', September 1926, AVIA 7/2961, PRO, Kew, UK

21 SEE Report no. 278, 'Observations of aeroplanes from kite balloons', October 1924, AVIA 23/269, PRO, Kew, UK

22 'Sound discs and mirrors development', 1931–, AIR 16/316, PRO, Kew, UK

23 Report no. 18, 'Experiments with the disc system at Hendon', 9th June to 23rd June 1919, AVIA 23/18, PRO, Kew, UK

24 SEE Report no. 19, 'Location of aircraft by sound using electrical methods, discs and listening wells', AVIA 23/19, PRO, Kew, UK

25 'Disc system at Romney Marsh', 9th November 1932, AVIA 7/2958, PRO, Kew, UK

26 HOFFMAN, S. O.: 'The detection of invisible objects by heat radiation', *Phys. Rev.*, 1919, **14**, (2), pp. 163–166

27 BURNS, R. W.: 'Aspects of UK air defence from 1914 to 1935: some unpublished Admiralty contributions', *IEE Proc.*, November 1989, **136**, Part A, (6), pp. 267–278

28 JONES, R. V.: 'Infrared detection in British air defence, 1935–38', *Infrared Physics*, 1961, **1**, pp. 153–162

29 WOOD, D., and DEMPSTER, D.: 'The narrow margin' (Hutchinson, London, 1961), pp. 35–51

30 Reports and Minutes of Meetings, the Committee for the Scientific Survey of Air Defence, 1935–38, AIR 20/80, PRO, Kew, UK

31 WATSON-WATT, R. A.: Document 3, 28th January 1935, T 160/80, PRO, Kew, UK

32 WATSON-WATT, R. A.: Document 6, 12th February 1935, T 160/80, PRO, Kew, UK

33 Ref 1, pp. 109–112

34 CROWTHER, J. G., and WHIDDINGTON, R.: 'Science at war', (HMSO, London, 1947), pp. 5–6

35 BROWN, R. H.: 'Boffin' (Adam Hilger, Bristol, 1991), pp. 4–5

36 Report on 'Appreciation of situation vis-à-vis Germany. Condition of our forces to meet possibility of war with Germany', 11th March 1936, AIR 9/73, PRO, Kew, UK

37 GILBERT, M.: 'Winston Churchill. The wilderness years' (Macmillan, London, 1981)

38 Ref. 29, p. 77

39 DOWDING, H. C. T.: 'Combined training exercise, 1937', 25th September 1937, AIR 16/67, PRO, Kew, UK

40 GOBLE, S. J.: 'Brief interim statement on sector and combined training exercise 9/8/37–12/8/37', 20th August 1937, AIR 16/60, PRO, Kew, UK

41 Ref. 29, p. 7

42 Ref. 1, p. 182

43 BURNS, R. W.: 'Radar development to 1945' (Peter Peregrinus Ltd, London, 1988), p. 25
44 'Development and production. Service radio equipment in pre-war period', CAB 102/640, PRO, Kew, UK
45 WATSON-WATT, R. A.: paper on 'Radar', Acc. 9343, No. 7, 1940–42, National Library of Scotland
46 Ref. 44
47 NEALE, B. T.: 'CH — the first operational radar', chapter 8 in Ref. 43
48 Ref. 1, p. 183
49 ROWE, A. P.: 'One story of radar' (Cambridge University Press, 1948), p. 24
50 Ref. 47
51 Ref. 1, p. 188

Chapter 11

EMI's '60 MHz job'

EMI's official involvement in the government's radar programme dates from 1940 when the Hayes based company was asked to solve an important minimum range problem associated with air interception radar. Subsequently, EMI's Central Research Laboratories and the Gramophone Company's manufacturing capacity were associated with the design, development and production of radar systems which played crucial roles in the air offensive over Germany, the defeat of the German night bombers and the defeat of the U-boats.

Before 1940 EMI had devised, as a private venture, a radar system which would enable the range, bearing and elevation of an aircraft to be determined. The method and the design of the apparatus of the '60 MHz job', as the system was known, were almost entirely due to Blumlein. Remarkably, when the 'job' commenced in September 1939, EMI was ignorant of the activities of the Bawdsey Research station on radar.

Blumlein's initiative followed on from another EMI private venture project, on sound location, which had attracted the attention of the War Office's Royal Engineers Signals Board (RESB). The Board and especially the Air Defence Experimental Establishment (ADEE) were much concerned with the detection of aircraft by means of sound location. From the end of the First World War various static and mobile sound detectors had been investigated by the ADEE and sometime in 1938 EMI and the Gramophone Company were given a contract for the manufacture of the Mark VIII sound locator[1]. It would appear that this contract stimulated Blumlein to inquire whether the principles of stereophony, which he had enunciated in 1931, could be applied to the problem of aircraft location.

In October 1938 EMI invited the Director of Scientific Research (War Office) and members of the RESB to a demonstration at the Hayes works of 'a new type of sound locator, in a trial model form, that [included] electrical listening and visual means of indicating the binaural effect'. During the demonstration[2] the Board had the opportunity of making tests on a fixed source of sound and also on aircraft. 'The results were sufficiently satisfactory, in spite of rather unfavourable background conditions, to warrant further investigation of [the method] with a view to possible improvements to service instruments'. Subsequently, contracts were placed for the provision of binaural equipment for two Mark IX sound locators and a single parabolic locator. Later trials elicited the opinion that 'the visual evidence of accurate listening, during operations with aircraft, under favourable weather conditions, [had] been convincingly supplied'.

The merit of this achievement by Blumlein, assisted by H A M Clark, may be gauged by the comment of Dr W S Tucker, Director of Research, ADEE, who noted: 'Visual methods have been tried with the Army electrical locator, but the results have not been of much value. If the EMI visual method — certainly the best of [the] visual methods — proves useful on the Mark IX then there [was] no reason why it should not be adapted to any electrical locator . . .'[3].

Blumlein's apparatus was known as VIE (Visual Indicating Equipment), and used two cathode ray oscilloscopes for the display of bearing and elevation information. In addition, electrical listening channels were provided so that complete protection was afforded to the equipment users. A salient feature of the VIE was the incorporation of principles advanced by Blumlein during his 1931 work on binaural recording. The application of these principles to the sound location and to the radar location of aircraft will be described in the section on the '60 MHz job'. The VIE filled a technological gap until appropriate gun-laying radars were manufactured in suitable numbers. By then many thousands of sound locators had been fitted with VIE[4] (Figure 11.1).

Sound locators suffer from a fundamental limitation. The speed of sound is about 330 m/s, and is appreciably affected by prevailing atmospheric conditions. Consequently by the time sound from a moving source has reached the locator, the bearing and elevation information provided are most inaccurate. For example: the sound from an aircraft distant 10 km from a sound locator will take about 30 s to travel the distance, but during this time the plane moving at, say, 300 m.p.h. will have travelled about 4 km: and the aircraft may have completely changed its direction. A further defect of the sound location method concerns the siting of the location equipment. The locators must be positioned with great care since traffic and other unwanted environmental noises may make normal operations completely impossible: even winds can cause difficulties. An additional disadvantage is the inability of a single sound locator to provide range information. The estimation of range requires two or more stations

Figure 11.1 EMI's (1939) anti-aircraft sound locator designed to incorporate Blumlein's visual indicator equipment

Source: EMI Archives

spaced some distance apart, with the attendant complication of a central control or plotting room, and the problem of ensuring that all the locators follow the same target. These difficulties effectively disappear when the location method is based on the use of electromagnetic waves, the speed of which is about 3.10^8 m/s.

All of this was, of course, well known to Blumlein and others and so, after Blumlein had completed his work on the visual indicating equipment, he investigated whether the same principles of binaural location could be adopted for the location of aircraft by means of electromagnetic waves. The investigation extended from September 1939 to March 1940 and led to several patent applications.

In Chapter 4, on stereophony, it was mentioned that Blumlein had advanced the proposition that the sense of the spatial location of a sound source could be provided by the difference in time of arrival of the sound waves at the listener's ears. At low audio frequencies this time interval can be interpreted in terms of a phase difference. His stereophonic system enabled the phase difference between the outputs of two similar microphones, spaced apart, by a distance small compared to the wavelength of the incident sound wave, to be converted into two in-phase signals of different amplitudes which could drive two loudspeakers separated by a distance greater than the spacing of the microphones.

Ideally, Blumlein required two co-located, directional (velocity type) microphones, having 'figure-of-eight' (cosine) polar diagrams, orientated so that their major lobes were orthogonal to each other. Such instruments were not available in 1931 and Blumlein had to utilise two pairs of omni-directional (pressure) microphones. He showed how the correct loud-speaker inputs could be obtained from such microphones by the use of suitable modifying circuits, which he named 'shuffling' networks.

Since a vertical half-wave dipole antenna is also omnidirectional in a horizontal plane, Blumlein appreciated that his theory of sound location by binaural reception could be applied to the location of an aircraft, provided the aircraft was an emitter, or reflector, of electromagnetic radiation. For the latter purpose a ground-based transmitter could be employed as in standard radar practice. Essentially, the '60 MHz job' was based on Blumlein's principles of binaural reception, *mutatis mutandis*.

Technically, the difference of two slightly displaced signal waves, of the same waveform and of equal amplitude (as received by, for example, a pair of spaced pressure microphones or a pair of spaced $\lambda/2$ dipoles), is a signal wave having a waveform corresponding to the differential of either of the waves and an amplitude dependent upon the time difference between the waves. If, now, this difference signal wave is integrated, the resulting waveform is substantially the same as the waveform of the sum of the two received signal waves, although the amplitudes are different and depend upon the time delay of the two original signals. It is this time delay which had to be determined.

In applying his ideas to aircraft sound locators, Blumlein obtained the necessary sum and difference signals directly, i.e. without frequency trans-lation or envelope detection, from the outputs of the two microphones. But in the extension of his principles to aircraft locators, based on the reflection of modulated high-frequency electromagnetic waves, Blumlein showed that the same method could be adopted if the envelopes of the received modulated signals were used for summing and differencing. The extension formed the basis of patent no. 581 920[5].

Three applications are delineated in the patent:

1 a system, Figure 11.2a, for determining the direction of a radio trans-mitter radiating a low-frequency modulated carrier wave.
2 a method, Figure 11.2b, of locating the absolute position of an aeroplane by means of reflected electromagnetic radiations (i.e. a form of radar); and

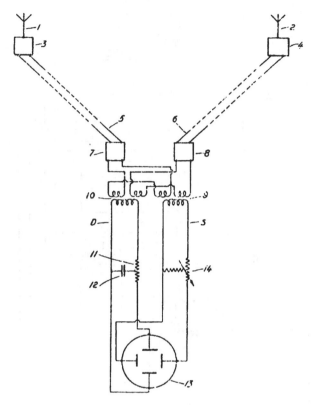

Figure 11.2 *(a) Schematic diagram of a system for determining the direction and/or elevation of a radiator or a reflector of modulated carrier wave signals*

Source: British patent no. 581 920

3 a method of finding the positions of objects at sea, such as submarines or icebergs, by using reflected modulated supersonic waves (i.e. a form of sonar).

Equipment to test Blumlein's ideas and methods was constructed by EMI during the autumn of 1939. The apparatus was set up on Lake Farm, an open area adjoining the EMI research laboratories. Three receiving antennas sited at the corners of an equilateral triangle of 70 m each side were utilised and the transmitting antenna was disposed centrally. The carrier wave of 66 MHz was modulated by 0.5 μs nominally rectangular pulses at a constant repetition frequency of 5000 Hz. A feature of the system was the use of continuously running oscillators, so that the carrier signal in successive pulses was coherent; that is, the carrier wave at the start of each pulse was the same as if no modulation had been applied.

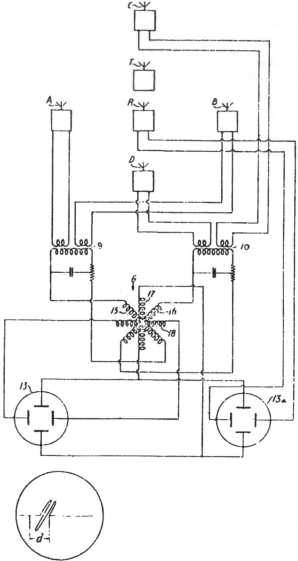

Figure 11.2   (b) Schematic diagram of a system for determining the azimuth, elevation, and distance of, for example, an aircraft

Source: British patent no. 581 920

In operation the three antennas received reflected pulses from an aircraft at different times, the differences in time indicating the direction of the aircraft. Signal delay networks enabled the received pulses to be coincident in time. By adjusting the signal delay in a pair of antenna circuits, a coincidence of echoes could be obtained. The delay introduced was a measure of the difference in the path lengths of the two rays

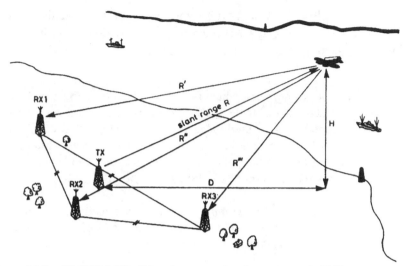

Figure 11.3   EMI 60 MHz GL radar proposal to the goverment in 1939

Source: *Engineering Science and Education Journal,* **2**, (3), 1993

reaching the two antennas from the reflecting aircraft. Hence, since the antenna spacing is known, the azimuth angle could be calculated. Ambiguity was annulled by repeating the measurement procedure on another pair of antennas. If two antennas were in line with the aircraft a similar process yielded its angle of elevation. Such alignments could be artificially produced by means of goniometers. (See Figure 11.3.)

A characteristic of the Blumlein system was the utilisation of resonance and integration (see Technical note 2). By heterodyning down the r.f carrier signal in several stages, and shock exciting the resonant circuit at each stage, the intermediate frequency carrier signals are prolonged until in the last stage the signal is almost continuous. As a consequence, the signal-to-noise ratio is enhanced since the wanted signals are phase additive whereas the noise is random.

The phase relationships of the signals produced from the several receivers are used as a measure of the phase relationships of the pulsed carrier signals received at the antennas. But since there are no pulses after the low decrement signal processing, ranging is not possible, although direction finding is still attainable.

A report on the partial demonstration model of the new form of radio locator was written in May 1940 by the Director of Communications Development (Air Ministry). He noted the 'novel principles' devised by Blumlein and stated: 'The principles involved are described in several patent specifications but the broadness of the latter is such that it is extremely difficult to obtain a clear idea of what the system, as at present established, really comprises'. The passage of more than 50 years has not eased this difficulty. Watson-Watt, who had had discussions with Blumlein seems to have seen EMI's equipment and described its operational working as follows[6].

> ' ... the aerials would be switched and time delays adjusted to produce simultaneity of echo trace by a process which though involved would be rapid and two controls would be rotated to define, after conversion, azimuth and elevation. A first cathode ray tube is used to locate the echo delay (range) approximately, the process of setting a pointer to any particular echo automatically setting the high accuracy stroboscopic sampler to locate the sampled pulse on the centre line of a second cathode ray tube having a much greater writing speed.
>
> 'The demonstration model is incomplete and does not include measurement of elevation. It has been set up to demonstrate the repetitive process of reception and the enhanced discrimination for range'.

Early in April 1940 Dr W B Lewis and Mr A F Wilkins visited EMI to assess 'the usefulness of the Blumlein system for GL [gun-laying radar] purposes'. 'Several ingenious devices which might find RDF [radar] applications' were noted: six were mentioned in the report[7] sent to the Air Ministry.

1 'A system of repetitive observations for pulses for increasing the signal-to-noise ratio. This system is sensitive to the Doppler effect from a moving target, and might therefore be developed to eliminate fixed echoes.

2 'A system of height finding depending on the measurement of time delays between reception at well spaced aerials. Given suitable sites, this might be useful for CHL [Chain Home Low radar] height finding.

3 'An economical transmitter, delivering very short pulses. This could find application in 1.5 m AI [radar].

4 'A capacity radio-goniometer. This might find application in other CHL height finding systems.

5 'A device for providing a pulse very accurately timed with respect to another pulse. This system is very similar to one we are developing for the electrical measurement of range.

6 'Increased accuracy of DF [direction finding] at long ranges afforded by the repetitive observation system.'

The report concluded with the suggestion that EMI should be 'encouraged to proceed with development of apparatus more particularly suited to some or all of these applications'.

The immediate effect of EMI's demonstration to Lewis and Wilkins was an invitation to the company to send two representatives to the Air

Ministry Research Establishment (AMRE), which was now (April 1940) based in Dundee, for technical discussions. At this time operational experience had shown that the minimum range of AMRE's Air Interception (AI) Mark III radar was too great for successful night fighter operations. Urgent steps had to be taken to decrease this range from about 1000 ft to about 400 ft. The engineering solution required a transmitter capable of delivering very short, accurately timed, pulses. Since such a transmitter had formed part of EMI's demonstration equipment, there was a possibility the design principles could be applied to the AI system.

Blumlein and White visited[8] Dundee in April and subsequently EMI was given a contract for a high-voltage anode modulation transmitter for AI working. This is discussed in Chapter 12.

There was also a possible application of some of Blumlein's innovations to the Chain Home (CH) and Chain Home Low (CHL) radars. Both of these surveillance radars radiated electromagnetic waves over the sea — a fairly uniform surface with no unwanted reflecting objects. Good discrimination could be obtained between the wanted echoes from hostile aircraft and the sea returns. But when enemy aircraft flew parallel with the coast, and at a comparatively short range out to sea, it was found that they were more difficult for the RDF chain to track than those aircraft which were well out to sea. Many CHL stations suffered from permanent coastline echoes when searching up or down the coast and their effective operational areas were perforce limited[9].

Similarly, inland CH radar working, in general, was impeded by fixed echoes. An increasing number of main CH stations were (in June 1940) being severely hampered by echoes from balloon barrages. Moreover, there was a need to extend the early warning radar chain to the west coast of Britain, but it was anticipated that the ruggedness of the terrain would produce troublesome unwanted echoes. Thus there was a requirement for a CH type of radar which would only be responsive to moving reflecting objects.

On 9th June 1940 AMRE (now at Worth Matravers, Swanage) wrote to the Ministry of Aircraft Production and suggested that EMI should be asked to develop a system, based on the Doppler change of frequency of electromagnetic waves reflected from a moving target, for RDF in a mountainous country, to meet the following specification[10]:

1 to detect aircraft having a velocity component radial to the station exceeding 20 m.p.h.;
2 to be insensitive to fixed echoes;
3 to provide as rapid a search as possible over 360°, and
4 to have a range of at least 40 miles on a single aircraft of the bomber class.

It was further suggested that this might be achieved on a wavelength of 50 cm; that the Blumlein system of repetitive observation should be used to achieve the range; that the transmitter should employ valves (e.g. type E.1046D) which were being developed for pulse applications on 50 cm in order to provide the necessary power; that coherence of the c.w. signals from pulse to pulse should be used; and that a method of presentation (plan position indicator, p.p.i.) similar to that of the radio lighthouse might be adopted.

The system of repetitive observation also offered the possibility of improving the accuracy of direction finding on a system such as the CHL (Chain Home Low) split switch system if this was coupled with electronic switching. AMRE recommended that EMI should be asked to submit proposals provided the modifications could be carried out as an addition to existing CHL apparatus. Unfortunately this apparatus could not be loaned to EMI for re-engineering: it was urgently required for items of CHL research of 'more immediate importance than increased access of direction finding'.

On these suggestions MAP (the Ministry of Aircraft Production) stated that the need for RDF in mountainous districts was 'now in abeyance'; but the CHL question could be raised again when the establishment had completed its CHL research.

Lewis pressed his case regarding the difficulties being experienced by some CH and CHL stations and opined: 'All these considerations should be taken into account before a system which promises relief from these disabilities [was] put in abeyance'. Subsequently, in September 1940 EMI was given a contract for the 'Reception of pulses from a moving object only', that is for a type of moving target indicator (MTI) radar.

As originally outlined, EMI proposed to develop apparatus which would detect objects moving with a radial velocity relative to the radar of greater than 60 m.p.h. It seemed that with this constraint the 'speed of indication could be made sufficiently rapid with a single integrating circuit to enable useful information to be obtained'. On this basis the execution of the design 'appeared reasonable'. However, such a design would not meet the given AMRE operational requirements.

After the contract had been placed, Blumlein and White visited Worth Matravers for discussions with Lewis and others of AMRE[11]. From these, two main points emerged. First, the angular coverage of the CHL antenna was very narrow so that a rapid search would be required; secondly, the minimum speed of 60 m.p.h. was too high for useful working as it would give too long a period during the transit of an aircraft in which its position could not be determined.

Consequently for the EMI method to be successful, either the wavelength would have to be shortened 'very considerably', or a scheme would have to be devised which employed a 'large number' of integrating circuits. The first possibility was not thought to be satisfactory because EMI's proposed modification would no longer be an addition to an existing equipment; an *ab initio* start on a new type of apparatus would be necessary and this would duplicate work already in progress at Worth Matravers.

The next morning at AMRE a possible system was advanced involving a homodyne type of reception and a multiplicity (about 30) of integration circuits. At that time the switching requirement had been solved theoretically rather than practically. So, when Blumlein and White returned to Hayes, some consideration had to be given to embodying the principles in a practical form. It was decided the implementation was possible if the switching was effected by a circular scanning electron beam scanning over 30 targets. Two such switches would be required and the whole system would be rather complex and need a certain amount of electronic development. It had the merit that adjustment of the integrating circuits would not be critical.

The task of making some preliminary experiments on the scheme was given to Collard, 'who later pointed out that there was a possible means of attack giving practically instantaneous reading which again involved "theoretical switching". On further examination it appeared that an electronic switch could be made to perform the function suggested by Mr Collard, only one such switch being required and the apparatus becoming much simpler.' The system would necessitate the design of a special cathode ray tube in which the beam made contact over 270 separate elements, which had to have a large capacitance to a fixed electrode. It was proposed to obtain this capacitance by using a very thin aluminium oxide dielectric[11].

Both Blumlein and Lewis were agreed that this conception would provide a real solution to the problem, though the development of the cathode ray tube would be a major operation even with the EMI electronic section. Unfortunately in September 1940 this section was fully occupied with another substantial development (possibly the infrared detector) for the Air Ministry. It was therefore suggested that the work on the Doppler radar should be deferred until the section was free from other very important activities.

Actually during the period when discussions between EMI and AMRE were taking place, there were many more urgent tasks to be accomplished than the provision of Doppler radar for inland RDF working. The situation which faced Great Britain in June 1940 was dire. France had fallen and in the retreat to Dunkirk and the subsequent evacuation of the

British Expeditionary Force (26th May to 4th June 1940) an immense amount of munitions had had to be abandoned. These included[12] 7000 tons of ammunition; 90 000 rifles; 2300 field guns; 82 000 vehicles; 8000 Bren guns, and 400 anti-tank rifles. The shortage of weapons in the UK was so desperate that in June 1940 in the St Margaret's Bay region near Dover, a prime invasion area for the German SEELOEWE assault, there were just three anti-tank guns to defend 6.4 to 8.0 km of coast and each gun had only six rounds of ammunition.

Much had to be done to rebuild the nation's defence capability. But just over a month after the evacuation from France the Battle of Britain commenced. It lasted from 10th July to 31st October 1940, a period in which the RAF lost 892 aircraft[13]. All these losses had to be made good and the UK's factories and production centres were hard pressed. Additional resources were necessary to advance not only the radar effort, but also all the other inventions which were thought to be meritorious.

In August 1940 Sir Henry Tizard and other members of a mission, known as the British Technical Mission[14], left the UK for the United States 'to enlist the help and powerful resources of the American radio manufacturing industry'. It was appreciated that the mission would not 'get their help adequately without disclosing our technical information', and therefore the disclosures which would be made had to be without any conditions attaching to the offer.

During the ensuing talks with US scientists all the UK's most precious and closely guarded secrets were handed over to the Americans. These secrets included the specifications, design plans, test reports and, where appropriate, actual hardware, relating to ASDIC, RDX explosives, proximity fuses (of all types), rockets, jet propulsion, gyro-gun sights, the cavity magnetron and micropup valves, and radar (of all forms), among others. The disclosures either initiated or spurred efforts in the development of centimetric radar, radio proximity fuses, and electrical predictors, which were to have a most profound effect against the 1944 Vergeltungswaffe (VI) offensive. Indeed the combination of the US SCR 584 and UK GL Mark III radars, the US M-9 electrical predictor, and the US T98 radio proximity fuse, with the British Mark IIC power-controlled AA gun proved highly effective in defeating the flying bombs[15]. Some gun batteries achieved an attrition rate of 40 shells per V1 destroyed, and on average just 156 shells had to be fired to destroy a V1. These figures may be compared to the figures previously given for the success rates in the First World War, namely 2335 and 14 540 shells fired to bring down one enemy aircraft during day and night attacks respectively.

It will be shown in Chapter 12 that the UK's defences, both gunnery and night fighter, were totally inadequate to counter the 1940 German night bombing Blitzes on British cities. As a result, a great deal of destruction and loss of life were suffered. The prime factor which led to this situation was the inability of AA gun crews and fighter pilots to locate night-flying enemy aircraft. There was an imperative need in 1940 to develop further radar equipment to aid the ground and air forces. Air interception (AI), gun-laying (GL), and ground controlled interception (GCI) radars all had to be progressed to the stage where they would be operationally useful. This stage had not been reached in the summer and autumn of 1940.

And so after EMI had given a partial demonstration of its radiolocator, and during the talks which were being held to consider what steps could be taken to advance salient aspects of the radar, the grave war situation demanded a reappraisal of the country's defence needs. The United Kingdom was alone and facing a formidable opponent. Essentially, at that time, research and development on equipment could only be undertaken if it led to a strengthening of the country's defensive capability. Inland CH radar coverage which could be enhanced by a development of Blumlein's repetitive observation method, using the Doppler effect, was clearly not of the first importance. The island realm could only be assailed by air and sea forces which crossed, or landed on, the coast. Hence the prime requirement was for effective AI, GL, CD (Coastal Defence), and CHL (Chain Home Low) radars able to locate hostile air and sea targets.

On 28th June 1940 Rowe, the Superintendent of the Telecommunications Research Establishment (TRE), wrote[16] to the Ministry of Aircraft Production and stated: 'Experience of AIH at the Fighter Interception Unit (FIU), Tangmere and our own analysis of the interception problem has shown that delay in providing the pilot with AIH information is now perhaps the biggest single factor affecting the success of the interception'. He mentioned that work had been in progress, at the Establishment, for some time on a visual indicating device which was expected to reduce this delay. However: 'Owing to the importance of the problem, the possibility of its use in single-seater fighters, and because of EMI's exceptional experience of this type of problem, it is felt that they should be making a parallel effort toward the same end. It is therefore suggested that a development contract be placed with EMI for the production of an automatic indicating device for attachment to existing AI equipment.' The suggestion led to Blumlein's crucial involvement in the AI Mark VI radar development programme: this is the dealt with in Chapter 13.

The government's failure to enlist the support of EMI's brilliant research and development staff at a sufficiently early stage in the advancement of radar was a monumental policy blunder. The company's R&D

team had engineered the world's first, all-electronic, high-definition television system and its engineers and physicists had acquired a wealth of knowledge and experience in the design of wide-band amplifiers (of all types), cameras, modulators, transmitters, cables, antennas, power supplies, oscillators, pulse circuits and electronic display tubes. There were very few, if any, other companies or establishments in the UK which in 1935–36 could have implemented such a system. Nearly all of the techniques and circuits developed for 405-line television were directly applicable to radar practice, especially the metric radars of the 1935–40 era. Significantly, EMI was the only British company which engineered—albeit partially—as a private venture, a pre-war radar system. And yet the first approach to EMI for assistance with a radar problem was in April 1940.

Certainly, there was a need for some additional engineering expertise among the (pre-war) staff at the Orfordness and Bawdsey radar laboratories. Professor Hanbury Brown, who was recruited in 1936, has given a fascinating account in his book *Boffin* of the early days of these laboratories, and of the chronic shortage of equipment and staff expertise and experience[17].

> 'The man whom Watson-Watt had put in charge of developing the new transmitter, although very bright, was a physicist with no experience of radio at all. He learned his radio engineering on the job, and by studying my copy of "Short Wave Wireless Communication" by Ladner and Stoner. I think we all began to fear for the future of that transmitter when, at tea one day, he turned to me and said; "Hanbury, how can I break down the sharpness of resonance?" You don't have to know much about radio engineering to realise that he was starting from scratch!'

On the facilities available, Brown has written:

> 'We also had a very little of the test gear which, even in those days, one might reasonably expect to find in a modest radio laboratory, let alone one engaged on important work. All we had in the receiver hut was a Cossor double-beam oscilloscope, an Avometer, and a wavemeter which would have been more at home in a science museum. There was no signal generator! . . .
>
> 'At first I couldn't understand why anyone in their right mind could allow this to happen when the work was so obviously urgent. . . .
>
> 'To get a simple thing like a resistor, worth perhaps one penny, one had first to establish its proper Air Ministry description and then fill up Form 674 in triplicate; not surprisingly, there were vast piles of coloured forms in every laboratory.'

No library or proper workshop facilities existed. One of Brown's first jobs in constructing a new antenna was to bore a large hole in a wooden mast, but no brace or bit existed. 'The only way of getting them in less than 24 hours was to walk across the airfield and negotiate a loan from the caretaker of the bombing range.' Two separate forms in triplicate had to be completed. 'I bored that hole, but it took me most of the day and permanently undermined my faith in the administrative ability of the people in charge of the early development of radar.'

Most of the early work on radar was carried out by young physicists rather than by radio engineers: there was not a single experienced radio engineer in the Orfordness group. As a consequence, the quality of design and construction of the electronic equipment at Orfordness and Bawdsey was inferior to that of the better industrial laboratories of that time, 'especially in labs. which were engaged in the development of television'.

'This was brought home to me [Brown] vividly in 1940 when we collaborated with EMI on the design of airborne radar; it was a great shock to find how far advanced their electronic circuit techniques and test gear were compared with ours. It made me wonder how much the early development of radar could have been improved and accelerated if more radio engineers had been recruited in 1936 when it became necessary to expand the original small group.'

On EMI's expertise in the field of electronic circuit technique, Professor F C Williams, who joined the Air Ministry's Bawdsey Research Station in March 1938 and remained with its successor, TRE, throughout the war period, has stated[18]: 'I would say that it [was] a very regrettable thing that the very considerable advances in circuit techniques which were made in connection with television were not in fact well-known in the radar sphere.' He regarded Blumlein, who was a 'frequent visitor' to TRE, as 'a real pioneer of electronics, as distinct from communications. ... There is no doubt about that. ... He was the only man I ever met from whom I ... really [derived] any benefit'.

Watson-Watt was aware of the need to strengthen the radio engineering side of the group. In a letter, dated 12th January 1937 to the Air Ministry, about staffing for the Bawdsey Research Station, he noted[19]: 'The greatest single weakness of the Bawdsey team is the absence of a senior scientific officer with a wide radio experience, and I have been able to find no-one to fill the place for which I had Dr Bartlett in view.' Watson-Watt's statement seems surprising given the extensive and rapid expansion of radio communications and broadcasting from the early 1920s, and the involvement of several major British companies in manufacturing radio, and later television transmitters and receivers.

The inadequate facilities and lack of experience of the Bawdsey team appear to have had an inhibiting effect on early radar development.

Following an unsatisfactory demonstration given in September 1936, Tizard sent a stern letter[20] to Watson-Watt.

'To say I was disappointed on Thursday is to put it very mildly. As you put it yourself you have to face the fact that very little progress in achievement has been made for a year. I am surprised that you encouraged, indeed proposed, the September exercises. Unless very different results are obtained soon I shall have to dissuade the Air Ministry from putting up other stations. The Secretary of State must have got a very bad impression.'

Rowe, too, referred to the Bawdsey staff in his recollections. Apart from three or four exceptions, the scientists were of average ability; but 'the percentage of genius was greatly increased by recruitment from the universities and from industry during the war'[21].

'The difficulty', according to Sir Stafford Cripps, the post-war President of the Board of Trade, was that 'in the pre-war days the paramount need for secrecy denied us some of the help we might otherwise have got from the industrial laboratories, but this temporary disadvantage was soon got rid of under stress of war'. In his address[22] Cripps mentioned the outstanding contributions, in the field of scientific research in the laboratory, of the Metropolitan Vickers, A C Cossor, Pye, and GEC companies. 'Outstanding, too, for their contribution in scientific research as well as development to the production stage were Dynatron Radio Ltd, Ferranti Ltd, and EMI Ltd. In developing to the production stage the designs and models that came from the labs., E K Cole Ltd, Murphy Rado Ltd, Bush Radio Ltd, Allan West, and BTH made particularly notable contributions'. And in the production field Cripps gave 'a particularly honourable' mention of STC Ltd and the Philips organisation.

The association of some of these firms, particularly EMI, in the pre-war advancement of radar would most certainly have had a profound effect on the country's ability to defend itself during the Blitz attacks of 1940. This offensive and Blumlein's contributions to the means of countering it form the basis of the following chapter.

Technical note 1

In a sound locator, which comprises two identical sound receivers A and B, spaced a distance d apart, the time delay (t) between the arrival of the sound at A and at B is $(d \sin \theta)/v$ where v is the velocity of sound and θ is the bearing of the sound source relative to the perpendicular line of symmetry of the two receivers.

Since the human brain appears to be able to detect a time delay of about 100 μs, or slightly less for a skilled listener, the smallest value of θ at which a sound source can be detected is about 9.5° from $\sin\theta = v10^{-4}/d$ where d is the spacing between the ears). Thus for an unaided observer any source bearing less than about 10° will appear to be either directly ahead or directly behind the observer.

If the distance between two sound receivers is, say, 8 ft (2.4 m) the value of θ is just 0.8° and greater precision in measuring the bearing of a sound source is obtained. However, this greater precision is accompanied by a narrowing of the 'angle of view' of the sound locator. This follows from the maximum time delay which an unaided observer can detect, namely, about 600 μs (from $t = d/v$, $\theta = 90°$, where $d = 20$ cm, the distance between the ears). For a sound locator with $d = 2.4$ m, a delay of about 600 μs corresponds to a value of θ of about 5°. Consequently no accurate directional indication can be determined for $\theta > 5°$.

The use of an observer's binaural faculty for sound location was limited in practice by fatigue and, for some observers, 'binaural squint'.

For these reasons Blumlein and his colleagues developed visual indicating equipment (VIE) in which the electrical signals from the microphones associated with the two pairs of orthogonally mounted paraboloidal sound collectors were presented on two cathode ray oscilloscopes. The method enabled the effective base length to be controlled without actually changing the distance between the sound receivers. 'This allowed the use of a wider "field of view" with reduced accuracy when first searching for an aircraft, while the base length [could] later be increased at will to raise the accuracy for final adjustment' (EMI report on 'The application of the EMI systems on binaural sound recording to aircraft sound location').

Summarising, the advantages of the VIE system (Figure 11.4) were expected to be:

1 'a more accurate location;
2 'less training for a given standard of reliability;
3 'less fatigue;
4 'independence of binaural squint;
5 'a method of checking an observer's work while under instruction.'

The addition of aural observation of the new type was presumed to give:

1 an 'initial wider angle of observation by reduction of the effective base length';

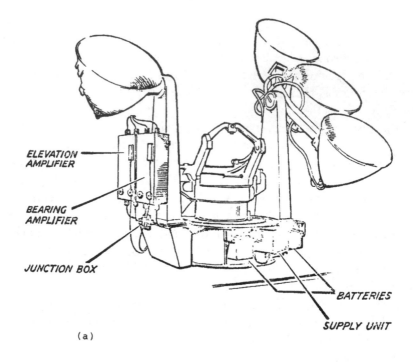

ELEVATION
AMPLIFIER

BEARING
AMPLIFIER

JUNCTION BOX

BATTERIES

SUPPLY UNIT

(a)

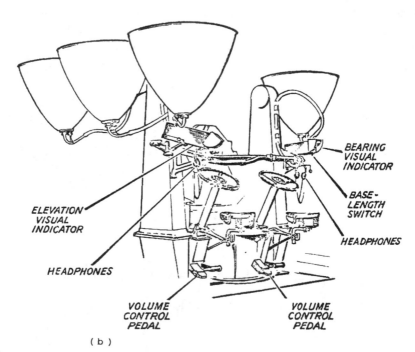

BEARING
VISUAL
INDICATOR

BASE-
LENGTH
SWITCH

ELEVATION
VISUAL
INDICATOR

HEADPHONES

HEADPHONES

VOLUME
CONTROL
PEDAL

VOLUME
CONTROL
PEDAL

(b)

Figure 11.4   EMI/Blumlein sound locators with VIE

Source: EMI Archives

2 an 'increased accuracy of final observation by increase of effective base length'.

The use of aural observation retained the useful features of:

1 'recognition of aircraft type by sound';
2 'separation of aircraft sound from interfering noises at extreme range'.

Technical note 2

As with many of Blumlein's patents, no. 581920 contains several important secondary ideas. These have been delineated by Dr E L C White in 'A review of the EMI 60 Mhz job', 9th October 1991. The following notes follow those of Dr White[23].

1 With two pairs of receiving antennas, arranged so that the axes of the two pairs are orthogonal, the difference signals, from the two pairs, when applied to the two pairs of deflection plates of a cathode ray oscilloscope result in a straight line deflection at the bearing angle (suitably referenced), independent of the elevation angle.
2 With two pairs of antennas, a goniometer (with two quadrature windings on the stator and two similar windings on the rotor), connected between the receivers and the display, enables a null method of measuring the bearing to be employed. The rotor is turned to bring the deflection on the c.r.o. screen to a reference direction (say, the Y direction); then the rotor angle gives the bearing.
3 With the arrangement of 2 above the two rotor outputs have zero and maximum values. If the latter (integrated) output is applied to one pair of plates of a second c.r.o. and the sum signal to the other pair, a straight line is traced whose deflection is a measure of the elevation angle of the target. However, because of the cosine law the method is insensitive at small angles of the elevation.
4 With the use of a third tube and a sawtooth range scan, a null method of measuring range can be used by adding a variable voltage to the sawtooth to bring the received echo to a reference position, for example, the centre of the display. Similarly, if the same biased scan waveform is applied to the elevation and azimuth tubes, separate observers utilising these will be in no doubt which echo (if more than one) is to be measured. Furthermore, the scan can then be arranged to stop for the duration of the echo, giving these latter observers a 'clean line' (apart from noise) to measure. This procedure would enable the manual

tracking of a target in three coordinates, by searchlight or gun crews, to be facilitated.

5 With the arrangement of 4, a range-dependent gain control for the receivers can be developed from the sawtooth scan waveform.

6 With the low duty cycle (pulse) operation of the transmitter, a much greater power could be radiated than would be possible with c.w. or modulated c.w. operation.

In operation the rectified echoes from each receiver were displayed in turn, and two adjustable signal delays were set to show coincidence of the pulses. The delay values allowed bearing and elevation to be calculated.

Since the maximum delay of the received echoes could not exceed $0.23\,\mu s$ ($70/3 \times 10^8$) the use of a timebase of $200\,\mu s$ ($1/5000\,s$) would not have allowed the pulses to be aligned with a sufficient accuracy. Accordingly a portion of the scan containing the echo was selected and expanded.

A better method was described in patents $585\,907^{24}$ and $585\,908^{25}$ and has been summarised by Dr E L C White.

'The first [patent] describes a coherent integration method for improving the signal/noise ratio, by strobing a radio frequency or intermediate frequency stage and then integrating by successively narrowing the pass-band down to, say $\pm 200\,Hz$. If the strobe pulse is made narrow compared with the echo, e.g. $0.1\,\mu s$, and its timing relative to successive echoes is slowly varied so as to pass through the echo, then the integrated output reproduces the shape of the echo on a much slowed-down time scale. To align the three echoes it is then necessary to switch to each antenna/receiver in turn at the same rate as the strobe scan, typically $25\,Hz$, so that the echo from each individual receiver repeats at $8.3\,Hz$ — just about fast enough for persistence of vision to allow all three to be seen together. [British patent] $585\,909$ describes the method used to distinguish the three echoes, by modulating the [cathode ray oscilloscope] beam differently for each one, or by displacing them vertically.

'The figure of $\pm 200\,Hz$ bandwidth suggested for the coherent integration allows for the Doppler shift of a target travelling radially at $500\,ms$, adequate for aircraft of the period. It is suggested in $585\,907$ that the receiver can be tuned to the Doppler, thus allowing a still narrower pass-band to be used, and providing a measure of radial velocity. It is pointed out that this helps to distinguish fixed from moving targets. There is also a suggestion that interference from other signals, deliberate or not, can be reduced by slowly varying the carrier frequency.'

Patent $585\,908$, which relates to 'methods of and apparatus for modifying recurrent signals and applications thereof', contains a profusion of ideas, as evidenced by the 29 claims of the patent. Apart from the principal topics of waveform sampling, the influence of moving targets, and the effects of deliberate and random interfering signals, the patent deals with several mundane but important practical matters such as the setting-up procedure; using a test cable for the antennas and receivers; the equalisation of the receiver gains; the bandwidths of the pre-detector and post-detector passbands; the effect of non-target signals; the elimination of an ambiguity by the replacement of phase measurement by time measurement; the reduction of any deleterious effect due to drift of the local oscillator frequency; and signal integration.

The patent is too complex and extended to be described and discussed here; suffice it to mention that it illustrates once again not only Blumlein's great creative engineering ability, but also his prowess as a practising engineer. Innovative ideas and practical considerations are described side by side.

Technical note 3

The Blumlein system of repetitive observations[26,27] (British patent no. 585 908) is a system of pulse transmission and reception for locating reflecting objects. It is designed to enhance the signal-to-noise ratio for a given transmitter power, and pulse length. The improvement is obtained by the integration of the effects from a succession of radiated pulses.

Consider the '60 MHz job' in which a wave having a frequency of actually, 66 MHz, is modulated, by a train of rectangular 0.5 µs pulses having a pulse recurrence frequency of 5 kHz, so that the carrier signal in the pulses is coherent If now the received pulse modulated carrier signal is frequency translated, using a superheterodyne receiver, the new carrier signal in successive pulses is still coherent. The process of heterodyning may be continued several times. In the system of the '60 MHz job' the input tuned circuit of the 66 MHz receiver was followed by mixer stages which produced signals having intermediate frequencies of 10 MHz, 300 kHz, and 20 kHz, i.e. frequencies equal to integer multiples of the pulse repetition frequency.

Now if a pulse modulated carrier signal is applied to a single tuned circuit the circuit will continue to 'ring' after the pulse has ceased. The decay of the 'ringing' (a damped oscillation) for such a circuit, varies as $\exp(-t/\tau)$, where $\tau = 2L/r = 2CR$ and where L, C, and r are respectively the inductance, capacitance, and series resistance of the tuned circuit and R is the dynamic resistance Thus τ is the time taken for the amplitude of the carrier oscillation to fall to $1/e$ of the value it had at the end of a pulse. In terms of the bandwidth, B, of the tuned circuit, τ is given by $1/\tau B$.

In Blumlein's system of repetitive observations the bandwidth of the input 66 MHz tuned circuit was 2 MHz, and the bandwidths of the succeeding intermediate frequency stages were 200 kHz, 4 kHz, and 200 Hz. Consequently, since τ is proportional to $1/B$, the duration of the 'ringing' extends for an increasing length of time as the received signal passes through the various stages of the system. Indeed the possibility arises that the damped oscillation set up by a given pulse, P1, might not have decayed away to noise level by the time the next pulse, P2, is received. Hence, because the carrier signal within the pulses is coherent, the output from P1 will add directly (i.e. be integrated) to the output from P2 and so on until an equilibrium value is reached which may correspond to the effects of several pulses. The limit to the number of pulses which can be used is set by the time which can be allowed for an observation.

If the time constant of the output circuit is $NT/2$, where N is an integer and T is the pulse recurrence period, then the output corresponds to the combined effect of N pulses all adding in phase. The gain in the signal-to-noise ratio is the same as if a single pulse N times as long as the actual pulse were observed, or as if the transmitted power were increased N times. The output signal voltage is increased N times and the output noise voltage is increased $\sqrt{N}$ times because the noise components associated with the pulses are in random relative phases and so the net gain in signal-to-noise ratio is $\sqrt{N}$. In the '60 MHz job', N was 25.

A further limit to the number of pulses which can be used is imposed by the Doppler effect associated with reflection from a moving object. An aeroplane which has a radial component of velocity towards an observer of v_R, which reflects radiation having a wavelength λ, produces a Doppler frequency shift of $2v_R/\lambda$. For v_R equal to 300 m.p.h. and λ equal to 4.5 m the shift is 60 Hz. If, then, the bandwidth of the final receiver stages were much less than 100 Hz, the output would be much reduced.

Alternatively, the transmitter output power may be reduced, with receiver integration, to give the same signal-to-noise ratio as would be given by an increased transmitter power and no receiver integration. Or a greater resolving power in range may be obtained for a given signal-to-noise ratio and a given transmitter power.

Technical note 4

The discussion of pre-detector and post-detector integration given below follows that of A D Blumlein[28] (1940) (see Figure 11.5).

Consider a signal s, with noise n, and a number of successive observations which are either integrated or are subject to a reduction of

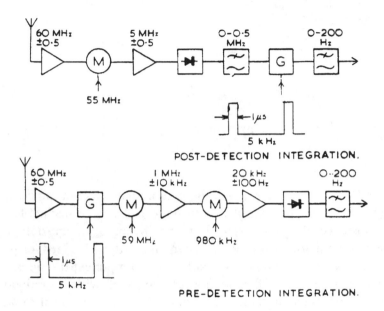

POST-DETECTION INTEGRATION.

PRE-DETECTION INTEGRATION.

NOTE:– DOPPLER SHIFT FOR MACH 0·5 - 60 Hz

Figure 11.5   Improvement of signal/noise ratio in radar

Source: *BKSTJ*, July 1968

bandwidth by a factor m. If integration or bandwidth reduction is applied prior to the detector, the effective noise arriving at the detector will be reduced to $n/\sqrt{m}$. Consequently with a square law dectector the output components from the detector will be proportional to s^2, $sn/\sqrt{m}$, and n^2/m. However, if integration is applied after the detector, the output components from it will be proportional to s^2, sn, and n^2. The subsequent integration or bandwidth reduction will be, effectively, to modify these terms to s^2, $sn/\sqrt{m}$, and $n^2/\sqrt{m}$. In addition there is also a steady component in n^2, which expresses the average direct current (d.c.) due to noise, which cannot be reduced by bandwidth reduction. Table 11.1 shows the relative terms for the two types of integration or bandwidth reduction.

Table 11.1   Integration

	Pre-detector		Post-detector	
Signal	s^2		s^2	
Fluctuating components	$\dfrac{sn}{\sqrt{m}}$	$\dfrac{n^2}{m}$	$\dfrac{sn}{\sqrt{m}}$	$\dfrac{n^2}{m}$
Unwanted d.c.		$\dfrac{n^2}{m}$	n^2	

Case 1. $s \gg n$

Then

$$\frac{sn}{\sqrt{m}} > \frac{n^2}{m}$$

and $n^2/\sqrt{m}$ and the steady components are negligible.

Case 2. $n \gg s$

Here the n^2/m and $n^2/\sqrt{m}$ terms are the predominant fluctuating components and the steady d.c. output due to noise will not be negligible. 'In these circumstances the pre-detector band narrowing or integration gives a reduction of the noise term which is the square of that obtained for post-detection and as regards the steady rectified d.c. component pre-detection band narrowing gives its full reduction as compared with no reduction obtained by post-detector band narrowing.' Hence 'the signal-to-noise ratio at the detector can be raised to a value at which it becomes advantageous to use further post-detector band narrowing or integration'.

References

1 SAYER, A. P.: 'Improvements in sound locators', October 1938, AVIA 7/2763, PRO, Kew, UK
2 TUCKER, W. S., and COSTELLOE, W. H. G.: memorandum to RESB, 28th April 1939, AVIA 7/2763, PRO, Kew, UK
3 TUCKER, W. S., and COSTELLOE, W. H. G.: memorandum to RESB, 14th April 1939, AVIA 7/2763, PRO, Kew, UK
4 WHITE, E. L. C.: memorandum, 9th June, THORN EMI Central Research Laboratories archives, Hayes, UK
5 BLUMLEIN, A. D., and WHITE, E. L. C.: 'Improvements in or relating to methods and apparatus for determining the direction or position of sources of radiant energy', British patent no. 581 920, 20th July 1939
6 Director of Communications Development (Air Ministry): 'EMI radio locator (RDF)', 16th May 1940, WO 195/151, PRO, Kew, UK
7 WILKINS, A. F.: memorandum, 15th April 1940, AVIA 7/2011, PRO, Kew, UK
8 LEWIS, W. B.: letter to I. Shoenberg, 24th April 1940, AVIA 7/2011, PRO, Kew, UK
9 LEWIS, W. B.: letter to Secretary, MAP, 29th June 1940, AVIA 7/2011, PRO, Kew, UK
10 Superintendent, AMRE.: letter to Secretary, MAP, 9th June 1940, PRO, Kew, UK

11 BLUMLEIN, A. B.: 'Note on development contract No. 8 B 9678/C.31 (a), Reception of pulses from moving objects only', 19th September 1940, AVIA 7/2011, PRO, Kew, UK
12 CHURCHILL, W. S.: 'Their finest hour' (Cassell, London, 1949), p. 126
13 Ref. 12, p. 278
14 CLARK, R. W.: 'Tizard' (Methuen, London, 1965), chapter 11
15 BURNS, R. W.: 'Factors affecting the development of the radio proximity fuse 1940–1944', *IEE Proc.— Sci. Meas. Technol.*, January 1996, **143**, (1), pp. 1–9
16 ROWE, A. P.: letter to Secretary MAP, 28th June 1940, AVIA 7/2011, PRO, Kew, UK
17 BROWN, R. H.: 'Boffin' (Adam Hilger, Bristol, 1991)
18 WILLIAMS, F. C.: taped interview with the author, personal collection
19 WATSON-WATT, R. A.: letter to Secretary, Air Ministry, 12th January 1937, Acc. 9343, No. 2, 1937, National Library of Scotland
20 TIZARD, H. T.: letter to R. A. Watson-Watt, September 1936, Acc. 9343, No. 1, 1936, National Library of Scotland
21 ROWE, A. P.: 'One story of radar' (Cambridge University Press, 1948), p. 43
22 CRIPPS, S.: Address, 31st August 1945, Acc. 9343, No. 2, 1937, National Library of Scotland
23 WHITE, E. L. C.: 'A review of the "60 MHz job"', 9th October 1991, personal collection
24 BLUMLEIN, A. D., and WHITE, E. L. C.: 'Improvements in or relating to methods and apparatus for reducing the effects of interference in the observation of recurrent oscillating signals', British patent no. 585 907, 1st September 1939
25 BLUMLEIN, A. D.: 'Improvements in or relating to methods of and apparatus for modifying recurrent signals and applications thereof', British patent no. 585 908, 4th December 1939
26 BLUMLEIN, A. D.: 'Repetitive observation', 1940, AVIA 7/2011, PRO, Kew, UK
27 LEWIS, W. B.: 'The EMI system of repetitive observation', 11th April 1940, AVIA 7/2011, PRO, Kew, UK
28 Ref. 26

Chapter 12

The battle against the night raiders and AI Mark IV

In the autumn of 1940 when the German night air offensive on London and other British towns and cities commenced, the means available to protect the country were quite inadequate and ineffective. The position was exemplified in a memorandum sent by Sir Archibald Sinclair, the Air Minister, to Mr W S Churchill, the Prime Minister. On 15th November 1940 Sinclair wrote[1].

> '[Last night] 300 German aircraft converged upon a previously known target. Round that target were five times as many guns per head of population as there are round London. 100 British fighter aircraft were airborne. Yet the only German casualty is claimed neither by the fighters nor by the guns.'

Clearly, this was an alarming situation. Germany, from 7th March 1936, had reoccupied the demilitarised zone of the Rhineland; had annexed Austria, Bohemia, Moravia and the Sudentenland (September 1938); had conquered Poland (September 1939), Denmark (April 1940) and Norway (April 1940); and, in just a few weeks, from 10th May 1940 to 26th June 1940, had subjugated the Netherlands, Belgium, Luxembourg and France.

The German High Command's strategy in the summer of 1940 seemed to be to extend these conquests by an invasion of Great Britain. For this purpose operation SEELOEWE had to be implemented. Following the fall of France, preparations were made to acquire the necessary invasion vessels. By the beginning of September 168 transports, 1910 barges, 419 tugs, and 1600 motor boats had been requisitioned[2]. However, before the German High Command could launch SEELOEWE, the UK's air defences had to be rendered ineffective.

From 10th July to 31st October 1940 extensive daylight air assaults on English fighter airfields, radar installations and command posts were part of the German tactical plan. During the battle, known as the Battle of Britain, 1492 German aircraft and 892 British fighters were destroyed[3]. The daylight attacks which had proved so efficacious in Poland and the Low Countries were unsuccessful and from 7th September these attacks were augmented by night-time bombing raids on London and other British towns and cities. An average of 200 German aircraft assailed the country every night from 7th September to 3rd November 1940.

The new onslaught caused much loss of life and destruction to the civilian population. For their purpose the German night bombers had excellent means to find their targets. As Air Marshal Sir Hugh Dowding, the Commander-in-Chief, Fighter Command, noted on 11th October 1940: '[The] enemy's navigational aids are so effective that he will be able, if necessary, to bomb this country with sufficient accuracy for his purposes without ever emerging from clouds'. Furthermore: '[The] Germans can fly and bomb with considerable accuracy in weather in which our fighters cannot leave the ground'[4].

The position mentioned by Sinclair to Churchill poses several questions. First, why were Anti-Aircraft Command's guns so ineffective at night? Secondly, why were the RAF's fighters incapable of destroying German night-flying aircraft? And thirdly, what was being done, and what could be done, to strengthen the air defence of Great Britain?

Several reasons can be adduced to account for AA Command's poor performance.

1 The shortage of heavy and light AA guns and searchlights was grave, On 15th December 1940 ChurchilL informed President Roosevelt, of the USA, that 4000 AA guns were required for the defence of the UK alone (apart from those needed overseas) but less than 1500 were available[5].

2 The majority of the guns in service depended on visual sighting and so could not operate at night without the aid of searchlights.

3 Only 11 gun-laying (GL) radar sets existed in the London area and these were of a primitive nature. Just four GL sets had been calibrated by October 1940 and by beginning of this month not a single round had been fired by AA-assisted guns[6].

4 Searchlight crews, unaided, had great difficulty finding targets at night[7].

5 Enemy aircraft flew at altitudes higher than 25 000 ft and were beyond the range of searchlights.

6 Layers of mist and clouds scattered the searchlight beams and caused difficult, if not impossible, sighting[8].

Again, several reasons can be advanced to explain the inadequacies of the country's night fighter defence system.

1 Fighter pilots had extreme difficulty in seeing hostile targets on dark nights. Trials demonstrated that under full moon conditions a fighter had to approach to within 1000 ft of an enemy aircraft before a sighting and a consequential attack could be initiated; with a moonless sky this distance was a mere 400 ft[9].
2 The tolerances of the country's early warning radar system (CH) were too large to enable ground controllers to vector fighters to within attacking distances of enemy planes. These tolerances were[10]:

Range	$+/-1\%$
Bearing	$2°$ to $10°$
Height	$+/-10\%$

3 Ground Controlled Interception radar (GCI) which would have allowed ground controllers to position accurately fighter aircraft was not generally operational in 1940[11].
4 Single-seat fighters, such as the Hurricanes and the Spitfires, were without air interception radar.
5 The existing, operational, AI Mark III radar's minimum detection range of about 1000 ft was too large for most interceptions.
6 The Blenheim night fighters, which were equipped with AI Mark III, were unable to overtake and engage enemy bombers since many of them could outrun the Blenheims[12].

An illustration of the difficulties experienced at night by aircrews was given in a report, dated 28th March 1940, written by Air Vice-Marshal R E Saul, of No. 13 Group. He wrote that after six months of war experience in his group's area there was no instance of a hostile aircraft engaged in night attacks on coastal shipping, being engaged by the RAF's fighters.

'Single aircraft have been sent on patrols in conditions of blackout, in front of the sound locators, and on occasion have been vectored to intercept hostile aircraft when plots have appeared. [This has been] a waste of effort. . . . Without the aid of searchlights on even a clear night and with good visibility and no moon, pilots have said that it is impossible to see another aircraft even if it were within a few spans of their own. . . . [A] pilot in these conditions is dependent entirely on his instruments for keeping his balance and can only take momentary glances out of the cockpit.'[13]

The consequences of all the above-mentioned deficiencies was the almost total failure of the night fighters and the AA guns to shoot down any raiders. The figures of expenditure of AA shells were extremely large (260 000 rounds fired in September, for example), and the 'claimed results of the [AA gun] barrage infinitesimal'.[14] During the autumn of 1940 an average of 18 000 shells had had to be fired to bring down one enemy plane, and at one stage in the Blitz offensive the figure was as high as 30 000. On this period General F A Pile, the General Officer Commanding-in-Chief, Anti-Aircraft Command, in his book *Ack-Ack* wrote:

> 'Our London [gun] barrage was a policy of despair. Every available gun had been told to fire every available round on an approximate bearing and elevation. Morale had been saved, if there had been no other result.'[15]

These disappointing results were, in the opinion of Sir Henry Tizard, 'very largely due to a lack of calibration, maintenance and training, and not to a lack of science'.[16] Tizard had discussed the position with Blackett and he had calculated that if the GL sets were fully calibrated, and if the organisation and training were such as could be reasonably expected in course of time, the number of rounds fired for a kill would be 3000 instead of 11 000.

And on the poor performance of the night fighters, Tizard was similarly of the view that lack of training, maintenance and good engineering, rather than a lack of science, were the factors which inhibited success. Tizard's views were not supported by future events.

As stated in Chapter 10, several possible methods of detecting aircraft at night were extensively investigated prior to the commencement of hostilities in September 1939. These depended on the emission of infrared[17,18] or acoustic radiations from an aircraft's engines; visual observation aided by ground-based or airborne searchlights; methods to silhouette an aircraft against an illuminated background such as the ground, or clouds, or the night sky; and the reflection of electromagnetic radiation from an aircraft.

Further work on infrared methods was undertaken in 1937 by R V Jones and J Anderson, at the Royal Aircraft Establishment, Farnborough[19]. They proved, first, that the exhaust gases from an aircraft engine do not produce sufficient infrared radiation for detection purposes except within a few centimetres of the exhaust manifold; secondly, that only the infrared radiation from the exhaust manifold can be detected at a distance; and thirdly, that a screened exhaust manifold negated the observation of this radiation[20]. These important results, and the discovery early in the war that a crashed German Ju88 aeroplane had a screened exhaust system led to a cessation of activity on this method of detecting aircraft.

In his book *Three Steps to Victory*, Watson-Watt, wrote:

> 'During raids, searchlight beams waved about the sky but rarely found and held a target. . . . SLC [a type of radar] grew out of the exasperation we experienced at the futile efforts of our searchlights to pick up the enemy bombers who flew over us [at the Air Defence Research and Development Establishment] night after night during the summer months of 1940.'[21]

Ground-based searchlights were essential components in the UK's air defence system of the early war years since, it was hoped, they would illuminate night-flying aircraft and enable visually sighted guns to fire at the raiders. Moreover, during the 1914–18 war, searchlights had given great service in assisting fighters to destroy enemy planes. 'Three beams accurately pointed at the enemy [was in 1940] still far the best way of indicating his position to our fighters. [It had] the advantage over all scientific methods in that any day fighter [without AI radar] could be used.'[22]

To aid the searchlight crews, sound locators were positioned at the searchlight sites to provide elevation and bearing data. These locators comprised two orthogonally mounted pairs of sound collecting horns which, prior to 1939, required the operators to point the horns in the direction of maximum sound intensity by listening means (usually stethoscopes) only. From 1939 Blumlein's Visual Indicating Equipment allowed the bearing and elevation of a sound source to be determined more reliably than by purely acoustic means. Tests showed that the VIE was more accurate than aural detectors in the presence of random noises, and also was safer. Many thousands of VIE detectors were constructed[23]. Until gun-laying radars able to give accurate range, bearing and elevation information, and AI radars having a minimum range of 400 ft became available in adequate numbers, the VIE played a significant role in the air defence of Great Britain.

In 1940 Squadron Leader (later Air Commodore) Helmore conceived the idea of mounting a searchlight in the nose of an AI radar-fitted aircraft, such as the Havoc, to permit single-seat fighters, which were unequipped with AI radar, to attack night raiders. Helmore envisaged that each searchlight plane would be supported by two 'satellite' fighters.

The searchlight, called 'Turbinlite', was developed by GEC and became operational in May 1942 following a maiden trial on 9th April 1942. After a great deal of training ten flights of aircraft were fitted with Turbinlites. In operations, Helmore's conception was unsuccessful. By May, enemy action over the UK had dwindled to a small scale; the use by the enemy of its beam navigation systems had ceased; and enemy planes had resorted to taking evasive action by 'weaving' $\pm 30°$ about a mean course. Turbinlite teams found much difficulty in dealing with such targets.

> 'It is found that even if the Turbinlite does expose successfully the enemy immediately takes the most violent evasive action and although the satellite fighter might see the enemy aircraft it is extremely difficult to get into a position from which to deliver a lethal attack. In fact this has never been done.'[24]

This lack of achievement is highlighted by the data in Table 12.1 which compares, for the period 1st April to 31st August 1942, the results obtained by aircraft fitted with AI Mark IV radar (described below) and aircraft fitted with Turbinlites.

Nocturnal observation of hostile aircraft is enhanced if they can be seen silhouetted against a bright background[25]. For ground observers the background may be illuminated clouds (if these are at a height greater than that of the aircraft), or an illuminated cloudless night sky. Similarly for airborne observers flying higher than an enemy plane, the background may be either illuminated clouds, or an illuminated ground region.

Weather records showed that on a large number of nights in a year low clouds rendered searchlights ineffective to ground observers. Late in 1935 the suggestion was made that if a low cloud layer were illuminated by means of lights on the ground, aircraft flying above the cloud would be visible as silhouettes to other aircraft flying above them. Various experiments were carried out to test this proposal. During the autumn of 1936 a small-scale trial was conducted at Farnborough[26]. An area of cloud, four square miles in extent, was illuminated and observed from the air at night. The experiment was successful and established the necessary intensity and distribution of lights. With suitable illumination it was found that small aircraft flying above a cloud layer, having a thickness of up to 2500 ft, could be seen from other aircraft flying at up to 5000 ft and at ranges of up to two miles[27].

From these conclusions the Air Staff estimated the minimum size of the illuminated area necessary to provide an effective contribution towards

Table 12.1   Results for aircraft with AI Mark IV radar or Turbinlite

	Detections	Visuals	Combats	Destroyed	Prob. Des.	Damaged	No claims
AI Mark IV	200	91	56	27	11	12	6
Turbinlite	37	11	5	1*	1	2	1

* light not exposed

the defence of London as 40 miles by 26 miles. Implementation of the scheme would cost about 2.5 million[28].

At this time, the summer of 1937, much money was being expended on rearmaments and the development of the RDF (radar) early warning system. With this in mind, the Air Ministry felt it prudent to seek expenditure for illuminating, as a first step, an area, north of the Thames Estuary, of 200 square miles (20 miles by 10 miles), at a cost of £400 000. Fifty lighting units would be required. Initially the Chancellor of the Exchequer was reluctant to agree the expenditure on an experiment which might prove to be valueless[29], but on 26th July 1937 Treasury approval was given. It was anticipated that the scheme would be completed by the end of July 1939[30].

Actually, means to silhouette aircraft against clouds were not a feature of the UK's air defence system, primarily because after May 1941, following the successes mentioned later, and the invasion by Germany of the USSR (Operation BARBAROSSA), there was little air activity by the enemy over the country. Over Germany much use was made of silhouette methods. Silhouetting was accomplished by star shells, by flares dropped from aircraft, and by groups of searchlights illuminating suitable cloud bases[31]. Allied bombers could also be seen silhouetted against the background of extensive fires caused by their bombing. In the Kammhuber plan an extensive region of northwest Europe was provided with illuminants for this purpose[32].

Of the various systems which were implemented to detect hostile aircraft at night, none were more effective than that which comprised a fast night fighter, fitted with air interception (AI) radar, operating under the instructions of a ground controller aided by Ground Controlled Interception (GCI) radar.

The progress achieved in 1935 and 1936 by Watson-Watt's radar development group at Orfordness and later at Bawdsey led in September 1936 to the formation of an airborne radar group. It consisted of Dr E G Bowen, Mr A P Hibberd, and Dr A G Touch. Initially, their work was handicapped by the lack of a small v.h.f. (very high frequency) transmitter valve suitable for use in a 200 MHz airborne transmitter. Nevertheless, by illuminating a target plane from a 6 m ground transmitter, the group in June 1937 demonstrated that reflections from the target aircraft could be detected by an antenna-receiver system in another aircraft. The system was known as RDF 1.5[33].

During 1937 the Samuel triode, type 4316 (sometimes known as the 'door knob') became available; it could operate at 200 MHz. With this valve the group constructed the first British ASV (Air-to-Surface Vessel) radar and showed that signals from a 2000 t freighter at a distance of four to five miles could be detected. ASV radar work continued from July 1937 to May 1939 when activity commenced on AI radar. The circuit techniques which had been developed for the ASV radar were applied to AI radar: they included the use of the 'squegging' 200 MHz transmitter, based on TY-150 valves, and the mixer which utilised an RCA 954 tube.

The AI antenna system comprised a radiating dipole and reflector, and four receiving dipoles, each having a reflector (Figure 12.1). The two pairs of dipoles were lobe switched to enable signals to be obtained from which the bearing and elevation of a target could be determined (Figure 12.2). Experiments conducted with very crude equipment mounted in a Battle aircraft, no. K.9208, during the period May to June 1939 showed that the AI radar had a maximum range of 12 000 ft and a minimum range of 1500 ft.

Although this prototype was not entirely satisfactory, the ominous political situation in Europe led to hasty contracts being placed in July 1939 with Metropolitan Vickers Ltd, Pye Radio Ltd and Bawdsey Research Station, for the supply of the transmitters, indicators and receivers respectively of the Mark I version of the AI radar. Six Blenheims were fitted out. The radar failed operationally, and there were many failures due to poor manufacture. There was a complete failure of the power pack made by Metropolitan Vickers. Also the radar was not without some design defects: the transmitter valves glowed in the darkened cockpit of a fighter aircraft and hence showed its position to an enemy; also the antennas gave rise to ambiguities.

Further contracts for a Mark II radar set having an improved TY-150 transmitter were given, in October 1939, to E K Cole and to Pye Radio for the transmitter and power pack, and receiver units respectively, but again the sets were deficient in several important respects. They were very unreliable, and the receiver suppression means was not functional. Twenty-one sets were installed in short-nose Blenheims[34]. They were soon taken out of service. By March 1940 a Mark III AI set had been evolved which had maximum and minimum ranges of 18 000 ft and 1000 ft; the transmitter power output was 15 to 20 kW at the operating frequency of 200 MHz. Twenty sets were fitted into Blenheims, but again the standards of workmanship, by two very well known companies, left much to be desired[35]. The radars previously known as AI and RDF 1.5 were now referred to as AIH and AIL respectively[36].

In May 1940 the FIU (Fighter Interception Unit) was formed to investigate AI radar problems and equipments. It found that the horizontal polarisation used gave rise to anomalous bearings: this led to the adoption of vertical polarisation for all later metric AI radars. Furthermore the value of R_{MIN} was still much too large for operational use; urgent steps had to be taken to reduce this range. Tizard's prognosis after some recent visits

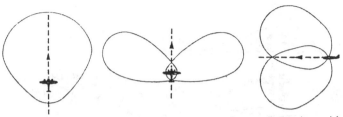

(A) Transmitter aerial   (B) Azimuth receiver aerial   (C) Elevation receiver aerial

AI Mk. IV., typical aerial polar diagrams

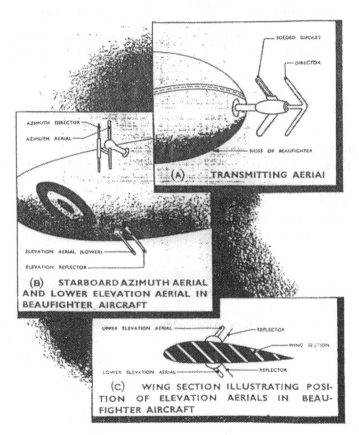

AI Mk. IV., aerials

Figure 12.1   Air Interception (AI) principles

Source: Public Record Office

in May to defence establishments at Tangmere, Bawdsey, Swanage and Christchurch, was quite gloomy. 'I [felt] convinced that no form of instrument [was] going to give good results for interception on dark nights within the next 18 months or two years. The problem [was] soluble, but we [were] a long way from practical success.'[37]

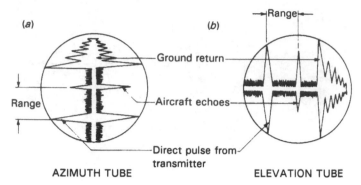

Figure 12.2   *The 200 MHz AI display patterns in azimuth and elevation respectively. In (a) the target aircraft is shown to starboard, and in (b) it is above the fighter aircraft*

Source: Public Record Office

The solution of the R_{MIN} problem was of such crucial importance that it was considered by three organisations. From April 1940 Dr W B Lewis, the Deputy Superintendent of AMRE (which was now stationed in Dundee), was personally engaged on the problem, as were staff at the Royal Aircraft Establishment, Farnborough. Additionally, EMI was given a contract to devise a solution. Blumlein and White quickly solved the problem.

EMI's involvement in the minimum range problem dates from early April 1940. As noted in Chapter 11, Wilkins, Lewis and AD/RDCl (an individual known only by job title) had visited the Hayes research laboratories to view apparatus, which employed several ingenious devices, which might find application in radar. It seems that on this occasion Shoenberg was told that AMRE would probably seek his assistance in the problem of shortening the pulse effect in its existing AI apparatus[38].

A few days later Blumlein and White visited AMRE at Dundee and on 23rd April they returned to Hayes with part of the AI system. White has recalled that the equipment travelled in the first-class railway sleeper with Blumlein and himself since it was too secret to be left in the goods van. The earliest reference in the Public Record Office files to a contract for this work is a letter dated 15th April 1940 from the Director Communications Directorate to the Superintendent, AMRE, Dundee, that EMI wanted a contract for developing a high-voltage anode modulation scheme[39]. A similar system had been used in EMI's 60 MHz private venture gun-laying radar.

Blumlein and his colleagues, White and Harker, worked rapidly and by 27th May Lewis was able to inform Group Captain de Burgh, who was responsible for the coordination of all developments in connection with AI radar, of 'the AI modifications which we hear have been successfully carried out at EMI. Their apparatus should approach the scientific limit of minimum range. The apparatus does not differ very markedly from that at present being produced (Mark III) except in the [transmitter] pack which is altogether different'. Two days later de Burgh saw the new modulator: '... I like the look of it very much, particularly as it seems capable of being introduced more or less anywhere along the production line'[40]. The receiver contained minor modifications[41].

Coded AI Mark IV, the modified Mark III radar was taken to Christchurch aerodrome and installed in a Blenheim IV. The first flight took place on 14th June 1940. Later tests by the FIU showed the maximum and minimum ranges to be 19 000 ft and 450 ft respectively. The latter figure was a marked improvement on the minimum range of 1000 ft for AI Mark III and satisfied the Air Ministry's requirement. Following AI Mark IV's first flight in a Beaufighter, on 15th September 1940, the radar was fitted, from 18th September 1940, to the RAF's Beaufighter night fighters[42]. By 4th November 1940, 29 of these aircraft had had the new AI radar fitted (Figure 12.3). These and later AI Mark IV equipped Beaufighters operating under the control of GCI radar were to transform the battle against the night raiders and cause Watson-Watt to exclaim: 'At last we [had] a successful airborne interception radar, AI Mark IV.'[43]

Much to the surprise of Leedham[44], of the Ministry of Aircraft Production, the contract for 900 modulators and power packs was completed by EMI by the forecast delivery date (see Figure 12.4).

The speed at which Blumlein found the solution to the R_{MIN} problem and the rapidity by which EMI manufactured the new equipment lends weight to the hypothesis that the early evolution of radar systems would have been greatly improved and accelerated if EMI had been contracted before the 1939–45 war to engage in its development. It would seem possible to argue that the devastation wrought by the German Blitz attacks of autumn 1940 would have been much reduced and that thousands of lives would have been saved.

Both the Blenheim and Beaufighter night fighters suffered from either inherent limitations or post-production defects. The fighters were initially either too slow or too unreliable to be effective against fast German bombers. On the night of 19/20th August, for example, a German aircraft was exposed by searchlights, was detected by AI radar, and was followed for 20 minutes by a British fighter to Liverpool. The Blenheim could not overtake the enemy and eventually a shortage of petrol compelled it to return to base. Another Blenheim, during the same night, pursued an enemy aircraft for some time between Portland and the Isle of Wight but could not close[45]. These factors led to the comparatively few interceptions

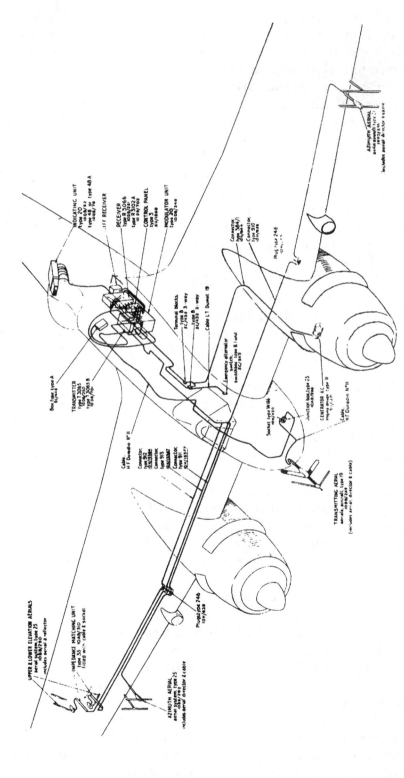

Figure 12.3   The layout of the AI Mark IV radar in a Beaufighter

Source: Public Record Office

MINISTRY OF AIRCRAFT PRODUCTION, MILLBANK,
S.W.1.

22nd November,1940.

Dear Mr. Shoenberg,

      I would like to congratulate and
thank you and your staff for the magnificent
achievement in delivering the whole of the
A.I. Mark IV Modulator Units by the original
date forecast by you.

      In view of the fact that the
delivery date and final completion date
forecast were considered to be an almost
impossible target I look upon this as a
particularly creditable achievement.

        Yours sincerely,

        (Sgd.) Lee-lham

I. Shoenberg, Esq.,
Electrical Musical Industries Ltd.,
Bayes,
Middlesex.

Figure 12.4   *(a) Copy of a letter from the Ministry of Aircraft production to EMI*

Source: EMI Archives

made during the night of 14/15th November 1940 when 300 German
bombers converged upon a previously known target. An investigation
highlighted the serious difficulties of the night interception problem[46]
(Table 12.2).

Thus 11 AI 'blips' led to just one sighting, because of the 'well known
poor view from these aircraft'. The difficulty of maintaining the serviceability of the Beaufighter was illustrated by the fact that only ten sorties
could be flown from the 43 aircraft available of this 'very undeveloped
type which [were] allotted [then] to Service squadrons'. It was not until
November 1940 that the first 'kill' was effected by a FIU Blenheim. By
using 100 octane fuel, by removing the mid-upper gun turret, and the
cockpit heater, and by generally tidying the streamlining of the fuselage, it
was hoped an extra 15 m.p.h. could be added to the Beaufighter's top
speed. Post-production problems associated with these fighters included
faulty attachment of the carburettors and difficulties with the elevators.
Furthermore, there was reason to suspect that one of this aircraft's
handicaps was its glowing exhaust rings in flight, which made the plane
visible to the enemy[47].

Figure 12.4   (b) Units of AI Mark IV, and an AI fitted Beaufighter

Source: D J Martin

Apart from these troubles little success was accomplished by the night-fighters, in the autumn of 1940, because they could not be vectored, by ground controllers, to positions from which interceptions could take place. Professor F A Lindemann[48], Churchill's scientific adviser, in several memoranda, written in September and October 1940[49,50], argued strongly

Table 12.2   *Air fighting at night*

Aircraft	No. of sorties	No. of AI blips	No. of sightings	No. of unassisted sightings	Combats	Results
Beaufighter	10	5	1	2	0	—
Blenheim	39	6	0	2	2	1 enemy a/c dam.
Defiant	22	saw 1 illuminated and 4 unilluminated targets				—
Hurricane	45	saw 1 illuminated target				—
Gladiator	4	saw nothing				—
Spitfire	1	saw nothing				—
Total	121					1 damaged

for GCI. He stated that Air Marshal Sir Hugh Dowding, Commander-in-Chief, Fighter Command, had studied the technique but mentioned that 'apparently [it is] not being used'. Lindemann urged that it was most desirable at once to try out the GCI technique since it would help the AI fighters enormously. Finally on 2nd October 1940, after the German night offensive had been in operation for about four weeks, Lindemann[51] asked Churchill: 'Are you satisfied that everything is being done to cope with night bombers? . . . [It is] only within the last fortnight that the principle of ground control has been accepted in night interception'.

Churchill sought the views of the Chief of Air Staff, and, on 1st October, Marshal of the RAF, Sir John Salmond, informed the Prime Minister of the need for a change in the holder of the important position of Commander-in-Chief, Fighter Command. '[The] change, in my opinion, [is] imperative. . . . [This] opinion is also very strongly held by most, if not all, Service members of Air Council'.[52]

The Admiralty, too, was critical of the apparent inaction in combatting enemy night activity. Admiral Sir Vivian Phillips, on 10th October, wrote to Churchill and noted: 'With regard to AI, of course as this is perfected it should provide the ideal means of interception by night but the hard fact is that it is does not deliver the goods today. At the beginning of the war, AI was stated to be a month or two ahead. After more than a year we still hear that in a month or so it may really achieve results'.[53]

Two days after Salmond sent his memorandum, Churchill chaired a meeting to consider 'Night Air Defence'. GCI was accepted and it was hoped that six GCI radar sets would be ready by 31st December 1940 and that 12 sets would be ready by 31st March 1941[54]. At the end of the second week in December 85 Blenheims and 57 Beaufighters had been fitted with

AI radar and were attached to squadrons. Twelve more were being wired at maintenance units and a further seven were being installed by No. 43 Group[55].

Much training by both aircrews and groundcrews was necessary before the combination of Beaufighter, AI Mark IV radar, and GCI radar could be integrated into an effective fighting unit. Fortunately, in January and February 1941, the bad weather over northwest Europe severely constrained enemy air activities over the UK, but in March the raids recommenced. Now, however, everything was in place to counter the German night attacks. By this month the weeks of preparation had produced well trained crews and in March 27 German aircraft were destroyed, 9 were probably destroyed and 12 were damaged. In April the German losses doubled to 53 aircraft destroyed, 18 probably destroyed, and 25 damaged. A further doubling occurred in May when 111 aircraft were destroyed, 25 were probably destroyed and 54 were damaged[56]. These were losses which the German Luftwaffe was unable or unwilling to sustain and, by the end of May 1941, apart from occasional flurries, such as the Baedeker raids, there was no large-scale aerial bombardment of the UK of the type suffered by the country during the Blitz of 1940. This state persisted until the start of the Vergeltungswaffe campaign in June 1944. SEELOEWE was abandoned; the German High Command looked to the East for further conquests and Operation BARBAROSSA was implemented.

An analysis for the period January to May 1941 showed that during these months the number of German night bomber sorties per German bomber destroyed, and the number of British night fighter sorties per German night bomber destroyed steadily decreased[57] (Table 12.3).

Further analysis of the effectiveness of the GCI and AI radars, for the six-week period 6/7th March to 12/13th April 1941 gave additional figures[58] (Table 12.4). Thus, GCI resulted in AI aircraft being approximately 5 times more potent in initiating night combats.

Table 12.3   Air sorties

Month	German night bomber sorties per German bomber destroyed	British night fighter sorties per German night bomber destroyed
January	143	198
February	107	150
March	88	47
April	57	38
May	33	25

Table 12.4   Radar effectiveness

Combats (with AI aircraft)	Aircraft destroyed/ probably destroyed	Total aircraft destroyed/ probably destroyed/damaged
24	16 per 100 sorties under GCI control	21
5	2.5 per 100 sorties not under GCI control	3.5

The analysis also determined that, during this period, 'Catseye' aircraft (i.e. aircraft without AI radar) operating under ground control did succeed in encountering enemy planes; but that one fighter aided by AI and GCI radars was roughly as efficacious as six fighters operating unaided.

The early marks of both AI and GCI radars operated at metric wavelengths (usually about 1.5 m). Later versions worked at centimetric wavelengths (e.g. 9.1 cm for AI Marks VII and VIII). The use of ultra high frequencies permitted a reduction of the beam width, for a given antenna aperture, of the radar set and hence an increase in its target resolution. For air interception purposes the narrower radiated beams of AI Marks VII and VIII, compared to those of AI Marks IV/V/VI, enabled fighters to pursue and engage hostile, evasive targets at much lower altitudes than were possible with metric radars. The figures in Table 12.5 show the improved performance of centimetric versus metric AI radars under 'no-moon' conditions[59].

Meanwhile during the period of 1940 to 1941 determined efforts were being made to improve the performance of AA Command's gunnery. A radar, known as 'Elsie, which could be fitted to a ground-based searchlight was rapidly designed and from September 1940 to February 1941 100 SLC (Search Light Control) radars (Figure 12.5) were manufactured; followed by 8796 during the period April 1941 to December 1943. The radar was

Table 12.5   Performance of AI radars

AI	Dusk or moon		No moon	
	Attacks	Visuals	Attacks	Visuals
IV and V	70	16	37	1
VII	43	12	25	8

Figure 12.5   An early example of a searchlight directed by radar equipment for spotting enemy aircraft at night and revealing them to AA gunners and RAF fighters. The Yagi antennas serve for transmitting and receiving purposes

Source: Imperial War Museum

introduced into service in Great Britain in 1941 and enabled visually sighted guns to fire at night.

The existing gun-laying radar, GL Mark I, could give accurate range and approximate bearing information, but it could not provide elevation data. This limitation was ameliorated by L H Bedford, of Cossor Ltd, who designed an elevation finder for GL Mark I. The finder improved the effectiveness of gun sites by a factor of approximately four.

A Mark II version of GL radar (see Figure 12.6) was generally deployed by mid-1942: it led to a further decrease in the number of rounds which had to be fired to destroy an enemy aircraft. The figures given in Table 12.6 are from a January 1944 War Office report and illustrate the improvements which radar effected[60]. (Because Blumlein was killed in

Table 12.6   Ground radar improvements

Radar type	Period	Rounds fired/aircraft destroyed
GL Mark I	Sept.–Oct. 1940	18 500
GL Mark I with elevation finder	1941	4 100
GL Mark II	From mid-1942	2 750
GL Mark III	Expected to be 2×3× more accurate than GL Mark II	

June 1942, a detailed consideration of air defence post 1942 is beyond the scope of this book.)

Two quotations will suffice for an appraisal of the technical resources employed during the German night offensive on Great Britain. In his book *One story of radar* A P Rowe wrote[61]:

'During this period the casualty rate suffered by night bombers rose from less than half of 1% to more than 7% and enemy attacks substantially ceased. . . . The night war of the air was the second major British victory of the war and, as with the first, victory depended upon radar. This second victory was achieved with the aid of a few GCI sets and perhaps not more than 100 AI radars.'

General Sir Frederick Pile, in his account *Ack-Ack* observed[62]:

'We were at long last reaping the benefits of all our training and of all our technical devices. The initial difficulties had largely been smoothed away, and on May 11 – 12 [1941], when the raids were so widespread that we were given greater scope, we obtained 9 victims, with one probable and no fewer than 17 others damaged. This was a gratifying "climax" to the battle against the night bomber. I say "climax" because the Blitz virtually ended that night. The next night raids were "light", the night after that "smaller", and the night after that "very small".

'And soon the German bombers were engaged upon the newly opened Russian front.

'By the end of the Blitz we had destroyed 170 night raiders, probably destroyed another 58, and damaged, in varying degree, 118 more.'

AI Mark IV radar was the mainstay air interception radar used by the RAF throughout 1941 and 1942 (Figure 12.4b). From the beginning of 1943 centimetric AI radar, Mark VIII, began to replace the metric Mark IV version. Even so, five squadrons of night fighters were still using AI Mark IV radar as late in the war as February 1944. Blumlein's genius in solving the crucially important R_{MIN} problem of AI Mark III and EMI's engineering and manufacturing prowess most certainly played a vitally important role in the defeat of the German night raiders.

Figure 12.6   Radar equipment of a mobile heavy AA Regiment of AA Command. The receiver and generator are shown in the foreground, and the two transmitters can be seen in the background. The Mark II version of the gun-laying equipment is illustrated

Source: Imperial War Museum

Technical note 1

A brief description of AI Mark IV radar

The radar worked at a frequency of approximately 193 MHz and was observer operated. Radar returns were received by a pair of azimuth antennas and a pair of elevation antennas and their outputs, after signal processing, were displayed as a function of range on the screens of two cathode ray tubes, see Figure 12.2. The sharpness of direction finding at dead ahead was ±5°.

In operation the transmitting antenna produced a very broad beam of radiation. Since some of this radiation was reflected from the earth's surface and gave rise to ground echoes, the maximum range of the radar set could not exceed the altitude of the aircraft above ground. The minimum and maximum ranges were approximately 300 ft to 400 ft, and 15 000 ft to 20 000 ft.

Figure 12.3 shows the installation layout in a Beaufighter, and Figure 12.4b illustrates the individual sub-systems which made up the overall system. Facilities were provided in later models for IFF, beacon homing, and beam approach.

AI Mark IV-A and AI Mark V were similar to the Mark IV version but had a TRE modified receiver and indicator so that target information could be presented, by the observer, on the pilot's spot indicator. Again two cathode ray tubes were provided for the observer but now, instead of showing range–azimuth and range–elevation information as in the Mark IV, the tubes displayed range, and elevation–azimuth data respectively.

The pilot's cathode ray tube spot indicator also gave elevation–azimuth information but differed from the corresponding observer's tube in that, when the target range was less than 2500 ft the tube's spot indication 'grew wings', the width of which gave a direct estimate of range. This was achieved by the observer positioning a manually operated strobe pulse in coincidence with the selected echo pulse shown on the range tube. When this had been achieved the pilot's spot indicator was automatically brought into operation. Only the strobed echo was displayed on the azimuth–elevation tube.

The maximum range of AI Mark V (which was the engineered version of AI Mark IV-A) was of the order of 15 000 ft. and the minimum range was between 300 ft to 400 ft.

Technical note 2

Some radar parameters

AI Mark IV

Wavelength, 1.5 m; c. 193 MHz
Pulse recurrence frequency, 750 Hz
Pulse width, 0.8 μs
Peak pulse power, 10 kW
Antenna system, half-wave dipoles, vertical polarisation
Maximum range, over 18 000 ft
Minimum range, 400 ft
Sharpness of DF at dead ahead, $\pm 5°$
Facilities for IFF, homing beacon, and beam approach

GCI Type 7

Frequency, 209 MHz
Pulse recurrence frequency, 400 Hz to 600 Hz
Pulse width, 4 μs to 3 μs
Peak pulse power, 100 kW
Antenna system, half-wave dipoles, horizontal polarisation
Beam width, 15°
Antenna rotation speed, 2 r.p.m. to 4 r.p.m.

GL Mark I

Frequency, 54.5 MHz to 85.7 MHz
Pulse recurrence frequency, 1 kHz to 3 kHz
Pulse width, 3 μs
Peak pulse power, 50 kW
Antenna system, half-wave dipoles
Accuracies, range ± 50 yd, bearing ± 1°

GL Mark II

Frequency, 54.5 MHz to 85.7 MHz
Pulse recurrence frequency, 1.0 kHz to 2.5 kHz
Pulse width, 1.0 μs to 1.2 μs
Peak pulse power, 150 kW
Antenna system, half-wave dipoles
Accuracies, range ± 50 yd, bearing ± 0.5°, elevation ± 1°

SLC types I to VII

Frequency, 204 MHz
Pulse recurrence frequency, 1.5 kHz to 2.5 kHz
Pulse width, 3 μs
Peak pulse power, 10 kW
Antenna system, half-wave dipoles, conical scanning achieved by
    switching
Accuracies, range ± 1000 yd, bearing ± 1°, elevation ± 1°

References

1 SINCLAIR, A.: memorandum to Mr W. S. Churchill, 15th November 1940, PREM 3/22/3, PRO, Kew, UK
2 CHURCHILL, W. S.: 'Their finest hour', Vol. 2 of 'The Second World War' (The Reprint Society, London, 1951), p. 253

3 DEIGHTON, L.: 'Battle of Britain' (Book Club Associates, London, 1980), pp. 115, 149, 167–169

4 DOWDING, Air Marshal Sir H.: paper on 'Night interception', 11th October 1940, PREM 3/57/1, PRO, Kew, UK

5 CHURCHILL, W. S.: draft message to President F. D. Roosevelt, on 'Supply of heavy AA guns', 15th December 1940, PREM 3/57/1, PRO, Kew, UK

6 ANON.: report of a meeting held on 7th October 1940 to consider 'Night air defence', PREM 3/22/1, PRO, Kew, UK

7 Ref. 4

8 ANON.: 'Air fighting at night', 20th March 1940, H/Q No. 11, RAF, Uxbridge, AIR 16/428, PRO, Kew, UK

9 Minutes of a meeting, held on 18th September 1940, to discuss 'Operational use of 10 cm wave AI', AVIA 7/137, PRO, Kew, UK

10 Air defence, Pamphlet No. 2, April 1942, AIR 10/3758, PRO, Kew, UK

11 ANON.: 'Notes on discussion on AI', 8th October 1940, AVIA 13/1024, PRO, Kew, UK

12 DOWDING, Air Marshal Sir H.: memorandum on 'Night interception', 30th September 1940, PREM 3/22/2, PRO, Kew, UK

13 SAUL, Air Vice-Marshal R. E.: report, 28th March 1940, AIR 16/428, PRO, Kew, UK

14 PILE, General Sir F.: 'Ack-Ack' (Harrap, London, 1949), pp. 167–168

15 Ref 14, p. 172

16 TIZARD, H.: memorandum, AIR 20/1489, PRO, Kew, UK

17 HOFFMAN, S. O.: 'The detection of invisible objects by heat radiation', *Phys. Rev.*, 1919, **14**, (2), pp. 163–166

18 BURNS, R. W.: 'Aspects of UK air defence from 1914 to 1935; some unpublished Admiralty contributions', November 1989, *IEE Proc.*, **136**, Pt. A, (6), pp. 267–278

19 JONES, R. V.: 'Infrared detection in British air defence, 1935–1938', *Infrared Physics*, 1961, **1**, pp. 153–162

20 ANDERSON, J. S., and JONES, R. V.: 'Infrared radiation from aeroplanes', 18th November 1935, AIR 20/145, PRO, Kew, UK

21 WATSON-WATT, R. A.: 'Three steps to victory' (Odhams Press, London, 1957), p. 381

22 BURNS, R. W.: 'A.D. Blumlein—Engineer extraordinary', *Engineering Science and Education Journal*, Feb 1992, **1**, No. 1, pp. 19–33.

23 WHITE, E. L. C.: 'Blumlein's contributions to television and radar', 9th June 1947, THORN-EMI Central Research Laboratories' Archives

24 ANON.: 'Turbinlites', paper dated 24th September 1942, and an appendix by the Operational Research Section, F. C., dated 16th September 1942, AIR 20/1495, PRO, Kew, UK

25 ANON.: paper on 'Silhouette scheme'; 1938, Air Defence Sub-committee, CAB 21/636, PRO, Kew, UK

26 ANON.: CSD Paper No. 153, 'Silhouette detection of aircraft. Preliminary consideration of trials', 23rd November 1938, AIR 20/81, PRO, Kew, UK

27 ANON.: paper on 'Air defence research', 14th July 1939, CAB 21/636, PRO, Kew, UK

28 PEIRSE, Air Marshal R. E. C.: letter to Wing Commander Elliott, 8th July 1937, CAB 21/629, PRO, Kew, UK

29 ANON.: note on 'Silhouette detection of aircraft', 1937/38, CAB 21/636, PRO, Kew, UK

30 Minute 243 of CSSAD, 32nd meeting, 26th July 1937, AIR 20/80, PRO, Kew, UK

31 STREETLEY, M.: 'Confound and destroy: 100 Group and the Bomber Support Group' (Jane's Publishing, London, 1978)

32 GOSSAGE, E. L.: letter to Air Marshal Sir Charles Portal, 3rd October 1940, AIR 16/428, PRO, Kew, UK

33 TOUCH, A. G.: 'Chronological history of airborne radar (1936–1941)', 13th April 1945, AIR 20/1464, PRO, Kew, UK

34 Minutes of 50th meeting of CSSAD, 16th August 1939, AVIA 8/515, PRO, Kew, UK

35 'AI apparatus for operational use', 19th May 1940, AVIA 13/1024, PRO, Kew, UK

36 DIXON, E.: note on AI, c. March 1940, AVIA 7/88, PRO, Kew, UK

37 TIZARD, H.: note on night interception, 22nd May 1940, AIR 20/4316, PRO, Kew, UK

38 AVIA 7/2011, PRO, Kew, UK

39 AVIA 7/2011, PRO, Kew, UK

40 LEWIS, W. B.: letter to Captain D. H. de Burgh, 27th May 1940, AVIA 13/1024, PRO, Kew, UK

41 BURGH, Captain de: letter to Dr W. B. Lewis, 30th May 1940, AVIA 13/1024, PRO, Kew, UK

42 Progress report on AI to 4th November 1940, AVIA 13/1024, PRO, Kew, UK

43 Ref 21, p. 255

44 LEEDHAM, Group Captain: letter to I. Shoenberg, 22nd November 1940, EMI Central Research Laboratories' archives

45 PREM 3/22/5, PRO, Kew, UK

46 ANON.: 'Air fighting at night', 20th March 1940, AIR 16/428, PRO, Kew, UK

47 AVIA 7/831, PRO, Kew, UK

48 BIRKENHEAD, The Earl of: 'The Prof in two worlds' (Collins, London, 1961), chapters 6 and 7

49 LINDEMANN, F. A.: memorandum to W. S. Churchill, 8th September 1940, PREM 3/22/5, PRO, Kew, UK

50 LINDEMANN, F. A.: memorandum to W. S. Churchill, 19th September 1940, PREM 3/22/5, PRO, Kew, UK

51 LINDEMANN, F. A.: memorandum to W. S. Churchill, 2nd October 1940, PREM 3/22/5, PRO, Kew, UK

52 SALMOND, Marshal of the Royal Air Force Sir J.: letter to W. S. Churchill, 5th October 1940, PREM 3/22/5, PRO, Kew, UK

53 PHILLIPS, Admiral Sir V.: memorandum to W. S. Churchill, 10th October, PREM, 3/22/3, PRO, Kew, UK

54 Report of the 4th meeting of the War Cabinet, 9th December 1940, CAB 81/22, PRO, Kew, UK

55 PREM 3/22/5, PRO, Kew, UK

56 Progress report by Air Officer Commander-in-Chief, Fighter Command, on 'Developments and results obtained in night interceptions for the period 1st August to 10th November 1941', 12th November 1941, CAB 81/22, PRO, Kew, UK

57 Chief of Air Staff to W. S. Churchill, enclosing a paper on 'Night fighter combat reports', 28th March 1941, PREM 3/22/3, PRO, Kew, UK

58 WATSON-WATT, R. A.: 'Notes on night fighter operations' 30th April 1941, AIR 26/3469, PRO, Kew, UK

59 WATSON-WATT, R. A.: 'Notes on AI by SAT', c. January 1943, AIR 26/3469, PRO, Kew, UK

60 SCHONLAND, Brigadier B. J.: 'The operational performance of army radar equipment, 1939–1943', 25th January 1944, WO 195/5060, PRO, Kew, UK

61 ROWE, A. P.: 'One story of radar' (Cambridge University Press, 1948), p. 74

62 Ref 14, p. 206

Chapter 13

The Blitz and AI Mark VI radar

AI Mark IV radar had two fundamental limitations. First, since it was observer operated, it was unsuitable for single-seat fighters such as the Hurricane and the Spitfire, and secondly, the need for an observer introduced a time lag between the observation of an echo and the initiation of the pursuit. The observer had to determine the direction of the hostile target from the relative amplitudes of the two pairs of echoes and then had to communicate his judgement to the pilot.

Mathematical analysis indicated that a ten-second delay in communicating the tube readings to the pilot 'rendered observation useless, and a delay of only two seconds might give serious error'.[1]

The lag gave rise to an oscillation in the approach of the fighter to an enemy plane which, at ranges of less than 2000–3000 ft, became so marked that it was extremely difficult to bring the fighter within visual range of the enemy[2]. The effect was, of course, exacerbated by any target which was continually altering course.

In the Mark V equipment, information (free from unwanted signals due to the transmitter pulse, noise and ground clutter) selected by the operator, concerning the range and direction of a single enemy aircraft, was presented visually to the pilot on an indicator of the spot and line type. This method eliminated the communication link and therefore reduced the lag in correcting the course which contributed largely to the oscillation phenomena mentioned earlier. It was hoped that with good training and practice the pilot's indicator would allow estimates of 'off-bearing', in elevation and azimuth, to be determined to an accuracy of three to five degrees, and would indicate the range up to 7500 ft. Target selection was achieved by the observer using a strobe pulse and manually tracking the chosen echo (Figure 13.1).

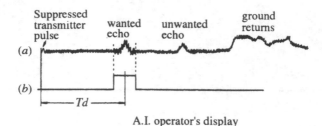

A.I. operator's display

Figure 13.1   An AI operator was able to select a wanted echo, from his display, for viewing by the fighter pilot

Source: *J.IEE*, **93**, Part IIIA, (1), 1946

The use of a strobe pulse for manually selecting radar echoes having a particular range was probably due to Blumlein. He devised the technique in connection with EMI's '60 MHz job'[3].

The advantages of the strobe technique were, first, that it selected a particular echo and allowed the pilot's attention to be concentrated on it; and secondly, that by the subsequent use of suitable integration circuits it could considerably improve the signal-to-noise ratio[3].

It would appear that the proposal for the manually operated strobe for AI Mark V was made, independently, by R Hanbury Brown of TRE, as the result of his flying experience with AI Mark IV[4]. Since the first flight with AI Mark IV took place in 1940, Blumlein's use of the technique predated that of Hanbury Brown.

The development of an automatic strobe-following circuit led to AI Mark VI, which was suitable for single-seat fighters. The earliest reference in the documents of the PRO to such an air interception radar appears to be a letter dated 2nd June 1940, from the Air Ministry to Rowe. 'The idea [was] to have equipment of a small weight to rule out the cathode ray tube as an indicating device, substituting other visible apparatus in the form, say, of a meter or some aural means of interpretation.'[5] Lewis, Rowe's deputy, noted on 20th June that the problem was 'considerably more complicated' than the memorandum implied: 'For example, discrimination between an aircraft echo and the ground return [was] not mentioned and [was] not simple; also it [was] difficult to make such a system work to a short minimum range.'[6] A month later Rowe informed the Ministry of Aircraft Production: 'The early provision of a satisfactory form of AI on a single-seat fighter aircraft [called] for a striking advance in some direction as yet unknown. The most probable source of a new factor in the situation [was] felt to be wavelengths of c. 10 cm and probably less.'[7] He expressed the view that if GEC, EMI, and TRE worked together on the problem, a solution within a few months was likely.

A similar opinion had been given, on 17th July 1940, by Air Marshal Sir Hugh Dowding[8]. He, too, was keen that the information presented to the pilot should either be non-visual or, if visual, should not involve his looking at anything but a very dim light.

Rowe's letter followed a visit, on 21st July 1940 during the Battle of Britain, to AMRE by the Assistant Chief of the Air Staff (R) when he had outlined the war situation in relation to AI radar. A logical conclusion of ACAS (R)'s statement was that 'the outcome of the war might depend upon the early provision of a large number of aircraft of adequate performance, fitted with AI'. This view had previously been expressed by Sir Henry Tizard[9].

The earliest involvement of Blumlein in the design of AI Mark VI dates from 20th July 1940 when he attended, with Shoenberg and Condliffe, a meeting at EMI to discuss 'proposed contracts for a single-seat fighter AI and [a] repetitive observation (Doppler) ground station'.[10] Five days later, after further discussion with Blumlein and White at the AMRE, Worth Matravers, the establishment requested the MAP to issue a development contract, to EMI, for a new type of AI radar. Thirteen points were listed in the specification, which was for apparatus 'giving a direct indication to the pilot in a simple form of range, azimuth, and elevation of all aircraft within the field of view' (Figure 13.2). The indication had to be such that it would not require 'close and prolonged scrutiny' and had to be 'suitable for observation without serious distraction from the other observations the pilot [had] to perform'. It was suggested that 'all ranges might be progressively explored by a running strobe pulse'.[11] Instructions to proceed with this work were given by the Director of Contracts on 7th October 1940[12].

In the meantime grave events were unfolding. The Battle of Britain had commenced on 10th July 1940 with the first of many aerial bombardments, and on 15th August Hitler had issued Directive No. 17 which authorised the intensification of the air war against England. For this purpose the Luftwaffe, by August, had assembled 2699 operational aircraft, comprising 1015 bombers, 346 dive-bombers, 933 fighters, and 375 heavy fighters[13]. Their objective was initially to destroy the UK's defensive air power system and extensive attacks were made against the RAF's forward airfields, sector stations, and radar installations. Heavy losses were suffered by the two combatant air forces. From 10th July to 31st August, 826 enemy aircraft were destroyed (according to German records) for the loss of 458 British fighter aircraft[14].

Then on 7th September Goering publically assumed responsibility for the bomber onslaught. He changed the tactics of the offensive from daylight to day and night attacks; and from assaults on the fighter airfields

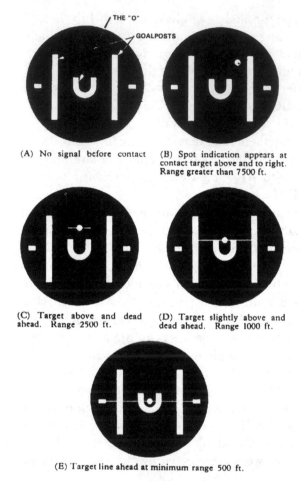

(A) No signal before contact

(B) Spot indication appears at contact target above and to right. Range greater than 7500 ft.

(C) Target above and dead ahead. Range 2500 ft.

(D) Target slightly above and dead ahead. Range 1000 ft.

(E) Target line ahead at minimum range 500 ft.

Figure 13.2   *The figures illustrate the cathode ray tube displays for various positions of a target relative to the AI fitted aircraft*

Source: Public Record Office

of Kent and Sussex to the indiscriminate bombing of the vast conurbations of London and other cities. It was a fortunate change for the Allies. As Churchill has written: 'Goering should certainly have persevered against the airfields, on whose organisation and combination the whole fighting power of our Air Force at this moment depended. By departing from the classical principles of war, as well as from the hitherto accepted dictates of humanity, he made a foolish mistake.'[15]

As previously noted for 57 nights, from 7th September to 3rd November 1940, an average of two hundred German bombers attacked London every night[16]. The statistics of the Blitz make grim reading. In the three months September to November, 12 696 civilians in the London region were killed and about 20 000 seriously injured, by about 36 000 bombs weighing perhaps 6600 tons. The main weight fell upon Stepney, Poplar, Bermondsey, Southwark, Lambeth, Deptford, Shoreditch, Bethnal Green, Holborn and the City itself, but some riverside boroughs further west as far as Fulham also suffered.

During this period Blumlein was living in the Hanger Hill, Park Royal area of London. His friend J B Kaye, and his wife Millicent, lived in a flat in 'a fairly biggish building, brick built', in Eaton Avenue, London NW3. It was soon determined that such buildings had 'a nasty habit of collapsing' if they were bombed. Knowing this, Blumlein, in an act of generosity, phoned his friend, on 15th September, and suggested that he and his wife should reside with him. Kaye has recalled the occasion[17]:

'Blumlein said: "Look here, things have been a bit hot and they seem a bit lively now. Doreen has gone with the boys to Cornwall, which was the wisest thing to do. Would Millicent and [you] like to use [my] house?" The upshot was we said yes, very kind. It was typical of him because [his] house was just a normal size house and could not be compared with the place, which was composed largely of flats, we were living in. He immediately came over that Sunday morning, with machine gun fire overhead and all sorts of things. They were really hard at it the RAF and the Luftwaffe then — the fighters and I suppose the German bombers. And there was hell's delight going on up aloft. I think the car he brought was the eight-cylinder Ford — the V8.

'So there we were [at Blumlein's house]. He was having to spend/work funny hours ... at EMI and [used to] come over in the evening at dusk or just before dusk to get to the [anti-aircraft] gun pits at Gunnersbury Park, where [with Nind and Clark] he was testing out the acoustic direction finding system [VIE] for directing anti-aircraft fire.... Millicent was looking after things with the cooking and generally trotting around and looking after us. Then in the mornings back they would come looking pretty unshaven, bleary eyed and ready for breakfast; and then off they went to EMI. I [headed] for Shaftesbury Avenue every morning and [returned] in the evenings.'

Great community spirit was shown by the people of London and the other great cities which were blitzed. Churchill told the nation that Hitler expected to 'terrorise and cow the people of this mighty imperial city [London].... Little does he know the spirit of the British nation, or the tough fibre of the Londoners'[18]. Fortunately much had been done to establish an effective Civil Defence Force.

In 1937 Local Authorities were ordered by the Home Office to organise air raid precautions and, during the weeks following the Munich crisis of 1938, various measures were introduced, including the distribution of 38 million gas masks. Sir John Anderson was made Lord Privy Seal, in November 1938, and became responsible for ARP (Air Raid Precautions) and Civil Defence measures. Government expenditure for these services increased from £9.5 million in 1938 to £51 million in 1939 and when war was declared on 3rd September 1939 approximately 1.5 million men and women had enrolled in the ARP force. About 400 000 of them were full-time members, the rest being volunteers. All were mobilised when hostilities commenced. Initially their task was to encourage the populace to take sensible precautions to safeguard their properties. Stirrup pumps, buckets, spades and sand became part of the domestic scene, and since much damage and injury could be caused by broken flying glass, householders were persuaded to take steps to reduce the impact of such fragments. Some windows were boarded up, others had netting glued to the glass panes, or were criss-crossed with strips of gummed brown paper. In addition, the structural integrity of buildings was improved by the judicious use of sand-bags and internal scaffolding.

Incendiary bombs and fires were a great hazard. The public were told to 'leave a bucket of sand outside [their] front door, give the key to [the ARP] warden and leave him a plan showing where in the house the [family slept]. Thousands [could] be rescued from bomb and fire that way'[19].

Blumlein and Kaye duly made some arrangements for fighting fires caused by incendiary bombs[20].

'First of all, in the porch there was a bucket of sand, a stirrup pump and a spade [for use against] an incendiary [bomb]. We were mainly thinking about incendiaries because if you got an HE [a high explosive bomb] you couldn't muck about with it with a bucket of sand. So we arranged that Blumlein would carry the stirrup pump, and Millicent and I would carry the bucket between us, and one of us ... would carry the spade.

'We didn't sleep because the nights were very noisy: it was bombing practically all the night [in] Park Royal and around us. I think I'm right in saying that none of us slept at all [soundly] from that September until the night blitz had more or less eased off'.

Blumlein devised a number of devices to safeguard his property and its occupants. At the back of the house, in the garden, a blast wall protected the ground floor room which the Kayes used as a bedroom. The wall extended in height to the top of the window of this room and its purpose was to prevent blast, from exploding bombs, from shattering the glass and causing injury and damage. Flying glass shards could cause much injury, even at some distance from an explosion.

One of the effects of blast was suction, which would tend to force the glass of the windows of a house to break and fly backwards. Blumlein reasoned that if the opening window of a window frame were allowed to open quickly and freely under the action of blast suction the pressures on either side of the frame would be equalised and less damage would result. For this purpose a retaining pin was required which would shear and allow the opening frame to move under the action of a differential pressure. So[20]:

'We got a nail, an ordinary nail used in woodwork, bent it at right angles and, instead of latching the latch lever into a slot in the woodwork of the window frame, we latched it behind this nail. [Thus] the head of the nail, which was obviously a thin bit of metal, would hold the latch in position, [but] if the window were pulled open or struck from one side it would shear the head off the nail and open. . . . To prevent the [window] being flung open or pulled open by winds we used elastic bands from the office. That was the only precaution we had downstairs.'

Thousands of incendiary bombs were showered on London and other British cities during the Blitz. If the bombs were located sufficiently early, before they created a conflagration, a bucket of sand tipped on to the burning incendiary material would suffice to prevent substantial damage. It was, of course, of crucial importance to detect whether such a bomb had penetrated the roof of a house and was burning in the confined space of a house loft. Blumlein's incendiary bomb detector was a very Heath Robinson affair, but it probably would have worked. With Kaye, Blumlein placed lengths of sewing cotton across the loft of his house. One end was attached to a tobacco tin, filled with sand, and then the thread was run over a small pulley (of the type used for hanging curtains from a curtain rail) and taken across the loft. At the far end the thread was fixed to the open leg of a very large Woolworth's safety pin. The other leg of the pin was rigidly screwed to one of the rafters of the room. Sufficient sand was placed in the tobacco tin to maintain the thread under tension and to keep the open leg of the safety pin away from the body of the pin. Between these two parts a piece of thin brass strip was mounted. From this strip and from the body of the safety pin two wires were led to a battery and bell. Other tobacco tin – thread – safety pin switch detectors were connected in parallel to the battery and bell. Hence, if an incendiary bomb crashed through the roof of the house it might break, or burn through, a thread and cause the bell to ring. Fortunately Blumlein's ingenious system was never tested. However, he was involved in tackling an incendiary bomb. Kaye has described the occasion[20].

'The road [in which we lived] was pretty well deserted apart from ourselves and someone who was doing a voluntary fireman's job. . . . What happened was there was a very nasty night — a Saturday night — and Mrs H's husband [a neighbour] who was an officer in the artillery had walked all the way from

Marble Arch. Stuff [bombs and gunfire] was rattling round the place and we were sitting in the backroom waiting for the action — Blumlein, Millicent and I — and there was an almighty clatter and I said: "I wonder what that was?". We thought for a while and I said: "I thought it sounded like a nose-cap or a piece of shell or something". Blumlein said: "I know what it was, it was an incendiary — those were slates and things". We rushed upstairs, found that we were intact and said: "It's the H's." So we charged in. There was a figure prone on the floor — it was Lieutenant H — fast asleep, whacked out completely after his long walk. We gave them a kick and said: "You've got an incendiary somewhere we think". He said: "Oh, let the b — y thing burn", and turned over and went to sleep again. Blumlein and I went tearing up the stairs, opening bedroom doors and rushing around generally, [but saw] no sign of anything burning. We said it must be another house but it sounded terribly like this one. Just as we were about to come downstairs one of us opened the only door [to the room] we hadn't looked in. We were leaving it, and that was the lavatory door at the top of the stairs. And lo and behold there was the incendiary in the lavatory pan. [It] hadn't burst, [and] hadn't gone off [and its] fins stuck out, all looking very symmetrical and decorative.'

Compulsory fire watching was introduced by the government. Every man between 16 and 60 years of age had to engage in this activity for 48 hours each month. As Ernest Bevin, the Minister of Labour, said in a broadcast to Australia: 'We have adopted the great democratic principle that everybody must become a fire watcher, either at their works, their office or on the street'.[21]

Not everyone was skilled in dealing with the many fires which were started by incendiary bombs, but all were enthusiastic and eager to help. There was an urgent need to tackle the fire quickly, not just to prevent damage to property but to deny the enemy bombers beacons on to which more bombs could be dropped. Many incendiaries fell on Park Royal — 'we were getting showered with them' (Kaye) — and a well rehearsed drill was essential if they were to be extinguished expeditiously. At first, before practice led to perfection, amateur enthusiasm sometimes led to humorous situations. On one occasion Blumlein, Kaye and his wife went out to tackle a fire[22]:

'We were normally dressed in flannel trousers, Wellingtons standing by, and a sweater or something. [When the fire started] we went leaping out and acting according to the prescribed routine. Off [went] Blumlein up the road with the stirrup pump: Millicent and I then set off with a bucket of sand and the spade charging up the road. . . . We passed a figure standing in the middle of the road looking like a very celebrated piece of sculpture, I can't remember who did it, (it was somebody wrestling with a serpent), but this one was Blumlein wrestling with the stirrup pump [hose]. He had put it over his shoulder and then the coils [of the hose] did the rest. They descended and he was in one heck of a mess; so we left him to it really, saying: "Come on sort yourself out, we must deal with this quickly."

'We got up to the far end [of the road], and [at] just about the same time Blumlein had detached himself and joined us. . . . Somebody had a bucket of water and [Blumlein put] the stirrup pump in it. Meanwhile Millicent and I somehow parted company — she was very cross about it. I had the spade and was accused of having taken the easy course and leaving her, with the heavy load of the bucket of sand, and charging up [the road] with this spade.

'There were one or two of these [incendiaries] cooking away because they had a nasty habit of clinging to [the] sticks they were attached to. They were supposed to burst in the air and scatter, [but] sometimes they burst on the ground and sometimes they had explosive charges in them just to brighten the party up.

'On this [occasion] I seized the shovel [but] no sand. So I hastily shovelled some earth and flung it across, as I thought, at the bonfire. It [the earth] passed straight through and hit the Civil Defence boys on the far side who were trying to do that. . . . Blumlein was working with the stirrup pump; [and] Millicent arrived playing merry hell for having been left by me. I was accused of desertion under fire.

'Blumlein got the stirrup pump going but, not having [had] any practice with it before, he managed to get the jet right across the flaming things, and saturated some of the people on the far side. By [this] time, with this panto-mine going on, the other characters, the Civil Defence boys, had put it out: they had had some practice.'

On another night Blumlein decided he wanted a souvenir of the bombing — a fragment of a bomb, or a nose-cone, or whatever. So when a row of flats not far from Park Royal tube station caught fire Blumlein dashed across some undeveloped ground between his house and the flats. These were two-storey buildings containing ground-floor and first-floor apartments. It seems that the fire brigade's policy at that time was to ensure the evacuation of a building and then take immediate action to extinguish the fire. No household contents could be saved since the prime action was to prevent further bombs being dropped on the burning target. During the fire Kaye suddenly missed Blumlein[22]:

'We were looking down at the bottom [of the hillside towards the flats and decided that it was] no good going down there because the squad down there was the one officially dealing with it. I couldn't see Blumlein anywhere. I said: "That's very funny". And then way down in the fire light and glow there was a figure looking rather as if he were enjoying a Guy Fawkes Day. These darn things [incendiaries] were popping away and fizzing about and this figure was dancing around the fire. [The] fire was put out and back came Blumlein with a broad grin on his face carrying an incendiary bomb. I think I'm right in saying he kept it on the table in the hall. That was the sort of thing that went on.'

When Churchill became Prime Minister on 10th May 1940 he faced a desperate situation. On the day he took office the German army invaded the Netherlands and Belgium; and in a very short time the French army was routed, France fell and the British Expeditionary Force was evacuated from Dunkirk. Nevertheless Churchill embarked on his task with immense confidence. His oratory stirred and inspired the nation, his human touches endeared him to the common man and woman, and his great intellect, prescience and sound common sense ensured that practical policies were immediately put in hand which were vital to the well-being of Great Britain.

The threat from invasion, Operation SEELOEWE, caused much anxiety. There was an imperative need to strengthen the country's defences, to mobilise the male civilian population in a Home Guard, to maintain morale, and to improve the means by which people could be protected during air raids.

On 4th July 1940, a few days after Dunkirk and a few days before the Battle of Britain, Churchill asked General Ismay, Head of the Military Wing of the War Cabinet secretariat: 'What [was] being done to encourage and assist the people living in threatened sea-ports to make suitable shelters for themselves in which they could remain during an invasion?' Churchill wanted active measures taken forthwith. Local Authority officers or representatives had to go round explaining to families that 'if they [decided] not to leave in accordance with our general advice, they should remain in the cellars, and arrangements should be made to prop up the building overhead', he insisted. Moreover: 'They should be assisted in this with advice and materials.'[23]

Blumlein heeded this advice. In the ground-floor back room of his house he had installed some builder's scaffolding poles, of the type known as Acrow props, which have a jacking action. These were used to hold stout wooden planks against the ceiling of the room so that, in the event that a collapse or partial collapse of the house occurred, he and the Kayes would have some protection and 'some chance of not having a lot of stuff on [their] heads'. Blumlein appreciated that it was not sufficient to support the props on the wooden floor of the room and arranged for them to bear on the solid ground beneath.

Kaye has recalled Blumlein's lifestyle during the hazardous days of the Blitz. After a day's work at EMI, he would return home for a meal and then either go to Gunnersbury Park to watch the Visual Indicating

Equipment in action, or start work again 'pushing a slide-rule and calcu-lating'[24].

'He never showed any sign of nerves at all. He seemed quite happy. Millicent and I would be dozing — we had Parker Knoll settees in this place — but not sleeping. We couldn't sleep with what was going on, and when you heard a whistling bomb you knew it was somewhere in the vicinity. And if it sounded rather near Blumlein would come hurtling into this back room.... And if a bomb fell over the garden it [might] or [might] not blow the windows in. Really it was quite fun at times.'

Blumlein had a passion at this time for playing classical music on a 'great big radiogram'. According to Kaye, he had a theory that to appreciate classical music the music had to be played over and over again at full volume.

'And there he [would be] sitting at the table with his slide-rule and this radio-gram going full blast — I think it was Caesar Franck and things of that type which I didn't appreciate at all because my ears were sort of pinned back listen-ing for whistling bombs.

'On one occasion it was obvious something fairly big was whistling down and that really got him off his feet. He shot through the door into [the back room], jumped into the air, hung on one of the horizontal [members of the scaffolding] just below the ceiling and managed to turn upside down and land with his feet. [With] all his money and the contents of his pockets scattered all over the floor [Blumlein] hooted with laughter. That was a typical evening at the Blumlein house in Hangar Lane in the Blitz.'

When the Kayes joined the Blumlein household in Park Royal the Blitz was in full operation and during their first night a total of 26 factories were on fire. Slowly the Blitz subsided. One night there was little activity when 'suddenly we heard the noise — the drumming — of a bomber overhead and suddenly a whistle and a crash and then another crash and then another whistle'[24].

'It sounded absolutely on us.... We moved while the whistle was on from the front room to the room at the back. There was an almighty crash sufficient to press ... one's ear drums in to an extent that prevented them from vibrating. In other words [one] felt the crash rather than heard it. And also [one] felt the suction of the air being drawn out of [one's] lungs. It didn't hurt either of us. We [Kaye and his wife] were perfectly alright. [Blumlein was not at home.]

'The windows were flung open due to suction — air going upwards — the nail heads sheared beautifully according to plan, and the curtains, hanging down by the windows, went straight up in the air outside. They were not wrenched off the rails; it looked so funny to see the windows open and the cur-tains going upwards ... '

The Kayes were uninjured, although J B Kaye thought the house was 'a bit cracked afterwards'. On looking out of the back window Kaye

'said to Millicent: "What the hell is that great lump in the garden? Has it always been there? And look where the fence is." The garden fence in the middle was lifted up in the air and appeared to be resting on . . . what looked like a miniature volcano in the middle of the garden.

'Then I saw white helmets bobbing about and these were the ARP people. [The distance between where the Kayes were standing in the room and the bomb's impact crater was later measured to be about 12 yards.] I let out a shout and the ARP man turned, and looking very pale said: "Good God, are you alive?" . . . They were completely staggered. Fortunately I [had] a bottle of whisky handy so they all came in and we had a drink to celebrate the occasion.'

Blumlein's house had been built just before the commencement of hostilities. The subsoil was clay and it seems that the bomb had 'drilled' into the clay for several feet before exploding. Kaye likened the situation to a charge exploding at the bottom of a vertical gun barrel. 'The force of the explosion blew everything out at the top end, forcing the ground up around it, and the blast was projected upwards creating a vacuum that opened the windows, took the curtains out, and more or less drew our breath out.'

One feature of the bombing which Kaye remembered well was the smell—exactly like Guy Fawkes Night and the squibs which he used to set off as a boy. Another effect was the ringing of the electric bell of the Blumlein incendiary bomb detector. The explosion had rocked the house and caused the tobacco tin weights to swing to and fro, thereby producing intermittent contacts at the safety-pin switches. There were at least half a dozen of these cotton thread indicators in Blumlein's roof loft.

About a week or so after this incident, the Kaye's left the Blumlein house in Park Royal. Kaye's boss, at Phillips, was concerned that the constant night attacks on the Park Royal area and Kaye's consequential lack of sleep would have an adverse effect on him. One of the managers at Phillips had a relative living in Radlett who kindly agreed to accommodate the Kayes. J B Kaye then travelled each day between Radlett and St Pancras Station. As he has mentioned: ' . . . things were a bit quieter'. He never saw Blumlein again.

EMI, too, was concerned about the possible effect on its researchers of continuous sleepless nights, and in October 1940 it bought some iron bedsteads and blankets for use by the laboratory staff. Blumlein and White would work until midnight and then bed-down in the laboratory. They set an example and others followed. If the staff stayed late a local pub, the

Blue Anchor, would provide a meal. Blumlein negotiated with the landlord for breakfasts to be provided[25].

Sleeping in the laboratory, though having some advantages, was not without its hazards. On one occasion, when Blumlein arranged to sleep in the laboratory, he asked the commissionaire to wake him at 7.00 a.m. During the night the watchman shone his torch into the lab. Blumlein, thinking it was the commissionaire, got up and shaved: it was 3.00 a.m.[25]

Life in the autumn of 1940 was hectic for Blumlein and his senior colleagues. Following the design conference held on 20th July on AI Mark VI, the Director of Contracts (MAP) on 7th October gave instructions to EMI to proceed with the development and engineering of the new single-seat fighter air interception radar. Both TRE and RAE took a keen interest in EMI's latest AI project. TRE had to ensure that its specification, based on Air Staff requirements, was being implemented; and RAE had to be assured that the engineered equipment was satisfactory and could be installed, in the restricted spaces available in fighter aircraft, without adversely affecting the handling of the aircraft. All of this required meetings, conferences and visits between EMI, and TRE and RAE.

A brief chronolgy for the period October to November (inclusive) will highlight the intense pressure under which Blumlein and, usually, White and Browne worked. In addition it will demonstrate the importance, at the time, of the AI Mark VI effort.[26,27]

9th October, liaison visit to EMI by RAE staff
12th October, liaison visit to EMI by RAE staff
16th October, liaison visit to EMI by RAE staff
17th October, meeting at EMI with RAE staff
21st October, liaison visit to EMI by RAE staff
24th October, conference between EMI, RAE and TRE
30th October, liaison visit to EMI by RAE staff
1st November, meeting at EMI with TRE, RAE, SRS and RP Tech
7th November, meeting at EMI with RAE staff
11th November, meeting at EMI with RAE staff
13th November, meeting at EMI with RAE staff
26th November, conference at RAE
27th November, liaison visit at EMI by RAE staff

Apart from his key role in these events, Blumlein had to provide overall project direction, write reports, supervise junior staff, personally engage in design work, visit the ack-ack batteries at Gunnersbury Park, participate in fire-watching duties, look after his home and contend with the nightly disruptions caused by enemy aircraft.

At first, swift progress on the AI Mark VI radar was made. Two days after EMI was awarded the contract, two members of RAE were able to say: 'EMI seem to be pushing the development work of the AI Mark VI hard.'[28] Two weeks later TRE's ideas, as exemplified by Dr F C Williams, on the circuitry needed for the automatic strobe had been examined, and rejected, and before the end of November EMI had substituted its own version of the strobe circuit to TRE. Since EMI and TRE had a close association in the development of the new AI radar, the issue of who realised the practical form of this vitally important circuit might appear difficult to resolve. Fortunately Condliffe, of EMI, related some details of the division of effort in a letter, dated 4th November 1940, to RAE. An extract from this is quoted below because it highlights an important difference, in the approach to a problem, taken by a government research establishment, and an industrial research and manufacturing company[29].

> 'The critical parts of their [TRE's] circuits were bread-boarded and could not be made to work satisfactorily when modified to operate with lower slope valves on the approved list and kept within the manufacturer's ratings.
>
> 'Concurrent with the above [EMI] designed circuits, still basically along the TRE lines, but more adapted to lower slope valves. The critical part of these circuits were also bread-board and trouble experienced in that components had to be critically adjusted to obtain satisfactory operation. At the same time as this work was going on, [EMI] made an examination of the sets from the point of view of AVC [automatic volume control] time constants and strobe drift, and came to the conclusion that [EMI] should have great difficulty in meeting the inter-related numbers in the circuit constants while still leaving margins sufficient to make the operation of low slope valve tolerances.'[30]

As a consequence, EMI departed 'almost completely' from the TRE circuits. EMI's circuits were functionally the same, but they met the requirements of the signals likely to be encountered under working conditions, i.e. fading and off-bearing signals. Moreover, the circuits were designed to allow for probable variations in the parameters of commercially available valves.

As noted previously, the essential difference between the AI Mark V and AI Mark VI radars was the change from manual to automatic strobing of the target. In the automatic stroboscopic system of utilising the output of a radio receiver, the time period between successive transmitter pulses is scanned by a strobe pulse which has a width of the same order as the transmitted pulse. Observation of the receiver's output took place only during the time of the strobe pulse. Consequently when searching, the time interval between the transmitted and the 'observation' or strobe pulses had to be automatically increased at each successive transmitter pulse until the strobe pulse was coincident in time with an echo signal (see Figure 13.3). At this instant the automatic scanning motion had to be stopped and the strobe pulse held in synchronism with the echo signal. In this way observations could be made on a single target echo. The scanning action of the strobe pulses was achieved by arranging that the repetition frequency of these pulses was slightly less than that of the transmitter pulses. When the strobe and echo pulses were 'locked' together, amplitude measurements from the elevation and azimuth antennas could be made.

A simplified functional block diagram of an AI Mark VI radar set is shown in Figure 13.4. The master oscillator provides timing pulses for the modulating circuits of the transmitter, and, via the strobe pulse and drift timing circuit, for the strobe pulse generator. The strobe pulse drift timing circuit enables the strobe pulses to drift relative to the transmitted pulses.

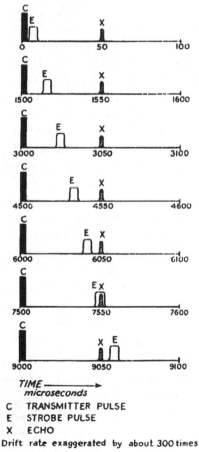

Figure 13.3   Stroboscopic observation pulses

Source: Public Record Office

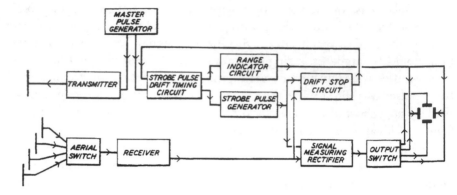

Figure 13.4   Simplified functional diagram of AI Mark VI

Source: Public Record Office

At the receiver, the output from each of the four antennas is connected sequentially to the receiver's input by a four-way antenna switch which is operated synchronously, with the output switch, at about 10 Hz. The output from the receiver is applied to a signal-measuring rectifier and also to the drift stop circuit. When the strobe pulse is coincident with an echo, an output is produced by the drift stop circuit which is fed back to the strobe timing circuit. This output is arranged to stop the drift of the strobe relative to the transmitter pulse and thereby ensure the strobe is locked to the echo to permit measurement by the signal-measuring rectifier. The output from the rectifier is applied via the synchronous switch to the cathode ray tube indicator. No output appears from the rectifier until the strobe pulses overlap the echo. Controls are provided so that the strobe may be 'unlocked' from an undesired signal and allowed to drift along to the next signal. When the drift stops and the strobe is synchronised with an echo, the time delay between the strobe and the associated transmitted pulse is equal to the time taken by the radiated pulse to travel to the target and back to the receiver. This time (which is measured by the range indicator circuit) is directly proportional to the range of the target since the transmitting and receiving antennas are effectively spatially coincident. The cathode ray tube indicator is biased off by means of a voltage which is produced by differentiating that which causes the strobe pulse drift. When the strobe 'locks' onto a target echo, the differentiated voltage becomes zero and the tube scanning spot brightens to show the target image. In this way no tube display is provided until a target is highlighted.

EMI's version of the automatic strobe circuit was demonstrated to representatives of TRE and RAE on 27th November 1940 and the first flight test of the AI Mark VI radar took place, at Christchurch, on 19th December 1940[31]. The bench and flight tests were considered to be satisfactory except for the in-flight minimum range of 700 ft and the maximum range of only 10 000 ft which were outside the specified limits[32]. Modifications to the transmitter and modulator and the Pye receiver to decrease the pulse width led to a minimum holding range of 150 ft when measured in the laboratory. However, further in-flight testing again gave rise to poor minimum and maximum ranges. The difference between the observed in-flight range of 500 ft and the laboratory value of 150 ft was ascribed, by Blumlein[33], to repeated pulse reflections in the transmitter-antenna feeder caused by mismatching. 'Unless these [antennas were] better matched', noted Blumlein, 'shorter minimum ranges will probably not be obtained'.

The transmitter and receiver feeder and antenna installations were the responsibility of the Royal Aircraft Establishment and until these were properly matched no improvements could be expected. A particularly annoying problem concerned the maximum range of just 7000 to 8000 feet. Tests showed that the transmitter output and receiver sensitivity of the AI Mark VI were 'well up to the best performance of AI Mark IV', but the latter type gave twice the range of the Mark VI. Much laboratory and flight testing was required before it was found that the range limitation was due to sparking occurring, at altitudes greater than 10 000 ft, in a cable which connected the modulator and transmitter[34].

Blumlein participated in some of these flights, and on 7th April 1941 he flew with Houchin (of EMI) in Blenheim IV, No. L. 4839. The flight, in the Christchurch area, was very successful: range indications to within 150 ft of a target were demonstrated, and the automatic strobe held consistently to the echo and was 'only defeated temporarily by fading'. This caused the strobe to return to the ground echo before rescanning. In a report on this flight the FIU (Fighter Interception Unit) observed: 'It is apparent EMI Ltd have solved the problem of automatic strobe control'[35].

The speed at which Blumlein and his colleagues worked on the AI Mark VI radar was remarkable. In just over two months from the date of the award of the contract, 7th October 1940, to the first flight, on 9th December, they had developed and engineered the prototype system and circuitry of a new air interception radar. Their accomplishment is all the more praiseworthy given Rowe's July statement on the need for 'a striking advance in some direction as yet unknown'. EMI's brilliant R&D team and the company's manufacturing skill and resources lend further weight to the hypothesis advanced in Chapter 12 that a pre-war involvement of EMI in the radar programme would have been greatly beneficial.

After the fatal accident of 7th June 1942 Blumlein was not, of course, responsible for any developments of auto-strobing. However, his *modus operandi*, in circuit design had been adopted by Williams. When he applied, in 1947, to the Inventions Awards Committee, Ministry of Supply, for an award in respect of his contributions to radar, a supporting paper noted[36]:

'There was a great tendency among radar workers to "cut and try" when it came to designing radar circuits. This succeeded in the laboratory, but caused trouble in production when commercial tolerances were used. Dr Williams, continuing along lines originally stated by the late A D Blumlein of EMI, was concerned with finding circuits which were designable.'

Blumlein was held in high regard by the staff of TRE, Swanage. He was often referred to as Dr Blumlein in the discussions, and in the minutes of meetings, with non-EMI staff and his eminence in the field of electronics was well known.

One former member of TRE, writing in 1991, has recorded: 'There is no doubt that the circuits [for the AI Mark VI] developed by Mr Blumlein for this equipment caused something of a sensation at Swanage when the plans appeared and I am sure influenced a lot of engineers at TRE!'[37]

Dr F C Williams, who later became Professor and Head of the Department of Electrical Engineering, University of Manchester, was one of the engineers who was influenced. He recorded his 'great indebtedness' to Blumlein in a paper on circuit techniques presented at the IEE's 1946 Conference on Radiolocation.

Apart from engineering AI Mark VI, EMI also developed the test equipment for the radar set. This comprised Signal Generator type 50; Resistance Unit type 126 (a dummy load); Aerial type 11 (for testing Transmitter type T.3074); Monitor type 22; and Impedance Bridge type 2 (a bridge for the *in situ* measurement of resistance and capacitance). Moreover, it was quickly realised that the original contract did not include the provision of beacon homing and IFF (Identification of Friend or Foe) facilities. These were urgently required for AI Mark VI and were developed by Blumlein's team. Of course, these necessary additions delayed the introduction into service of the AI Mark VI A radar and led to Blumlein in early August 1941 expressing some concern about 'the amount of equipment [which had] to be squeezed into the Mark VI A'[38].

Nevertheless, by this date the AI Mark VI had progressed to the stage where orders could be placed with EMI for 35 hand-made sets. Fifteen sets were delivered by 11th August 1941 and the balance was delivered by the end of that month[39]. The main production contract for 1500 sets, later reduced to 1125, commenced in December 1941 with an initial production rate of 50 per month, and then of 40 per week in 1942[40]. By 19th October 1942, 969 sets had been delivered[41].

Curiously, the rate of manufacture was not matched by a comparable aircraft installation schedule. AI Mark VI was fitted into one squadron of Defiants in the autumn of 1941 but was abandoned due to the slow speed of this particular aircraft. The single-seat fighter AI radar was later intended to be fitted into 12 Mark II C Hurricanes, but by November 1942 only six had been equipped and had gone to Fighter Command for service trials. The single-engine Mark II C Hurricane was, in July 1942, considered to be 'still very unpleasant to fly' and there was a view it might not be capable of operational use[42]. It would seem that this uncertainty precluded the early utilization of AI Mark VI radar.

The radar itself was 'very definitely' favourably commented upon by the TFU pilots who had flown Hurricane Mark II C No. BN 288 fitted with AI Mark VI radar. Its minimum and maximum ranges were 350 ft and 12 000 ft and a report, dated 7th June 1942—the day Blumlein was killed in an air accident—concluded:

> 'If the AI continues to work as accurately and reliably as it has during the 30 hours flying at TFU and the 27 hours flying at FIU [Fighter Interception Unit] it will be worth its weight [in the aircraft], although inexperienced pilots are not recommended to fly an AI Hurricane in bad weather at night.'

The main problem with AI Mark VI concerned the antennas. Ideally AI aircraft for night interception required an antenna system (whether metric or centimetric) which could be accommodated in the front of the fuselage. This constraint meant that AI aircraft had to be twin-engine without guns mounted in the nose of the fuselage. Single-engine aircraft were far from ideal for AI radar purposes, since both the transmitting and receiving antennas had to be mounted behind the engine and in front of the forward part of the aircraft frame. The consequences were that the polar diagrams of the radiating and receiving antennas were far from being operationally perfect and the antennas had to be specially designed for each type of single-engine fighter. This limitation, the proven effectiveness of the twin-engine Beaufighter as a night fighter, the impending introduction into service of the twin-engine Mosquito fighter, and the development of centimetric AI radar, led to the almost complete abandonment of AI for single-seat fighters.

The Air Officer, Commander-in-Chief, Fighter Command, in reviewing, in early February 1943, the requirements for night defence concluded that Hurricanes were 'unlikely to have any operational value in the UK against the more modern enemy bombers, and where flying conditions [were] unsuitable for single-engine night fighters'. The 12 Hurricanes fitted with the AI Mark VI were sent to India.

Though AI Mark VI saw only very limited service, the major value of its development lay in providing what was to become one of the standard 'tools of the trade', the automatic strobe[43].

Williams's contribution to this system were the ideas of, first, causing the position of the strobe pulse to move automatically so as to follow range variations of the echo, and causing the strobe pulse to 'search' between defined limits of range and, secondly, when it encountered an echo of sufficient amplitude, to 'lock' onto the echo[44]. Since the efforts of Blumlein, White and Williams in devising automatic strobing circuits were interrelated, the patent which was taken out in October 1943 for 'Radar system for searching (in range) and locking onto any echo found' bears all three names.

Automatic strobing was an outstanding advance in the art of radar. Apart from its value in AI Mark VI, the automatic strobe was later employed in both 1.5 m and 10 cm ASV blind bombing equipment, in Oboe, and in various distance measuring equipment (e.g. Rebecca). Furthermore, it was widely used in US wartime equipment after it had been communicated to them in 1940. It was also used in H_2S Mark IV and later Marks.

Auto-following was of central importance for successful gun-laying (GL) equipments and was utilized in practically all S and X band GL radars, whether land, sea, or air-based. Its value lay not only in the obvious saving of human effort, but also in the fact that 'smoother' and more reliable information was available for the gun-laying predictor, with a consequential improvement in accuracy of aiming. In addition, the automatic strobe was the basis of nearly all the distance indicators for post-war civil aviation aircraft (British, Canadian, Australian and USA).

References

1 Notes of conclusion reached at the 4th meeting held by the Secretary of State, 28th June 1940, AIR 20/3470, PRO, Kew UK

2 'Technical aids to night fighting', MAP, April 1941, AIR 20/4316, PRO, Kew, UK

3 WHITE, E. L. C.: letter to the Secretary, Inventions Awards Committee, Ministry of Supply, 21st October 1947, AVIA 53/630, PRO, Kew, UK

4 BROWN, R. H.: letter to Sir Robert Watson-Watt, 8th December 1951, Acc. 9343, No. 19, Nuffield College

5 Air Ministry: letter to AMRE, 2nd June 1940, AVIA 7/106, PRO, Kew, UK

6 LEWIS, W. B.: letter to the Secretary, MAP. 20th June 1940, AVIA 7/106, PRO, Kew, UK

7 ROWE, A. P.: letter to Secretary MAP, 25th July 1940, AVIA 7/106, PRO, Kew, UK

8 DOWDING, Air Marshal Sir H.: letter to Under Secretary of State for Air, 17th July 1940, AVIA 13/1024, PRO, Kew, UK

9 Ref. 7

10 'Notes on meeting at EMI 20th July 1940 to discuss proposed contracts for a single seat fighter AI and repetitive observation (Doppler) ground station', AVIA 7/2011, PRO, Kew, UK

11 Superintendent, AMRE: letter to Secretary MAP, 25th July 1940, AVIA 7/2011, PRO, Kew, UK

12 Director of Contracts: letter to EMI, 7th October 1940, AVIA 13/1047, PRO, Kew, UK

13 CHURCHILL, W. S.: 'Their finest hour', Vol. 2 of 'The Second World War' (The Reprint Society, London, 1951), p. 266

14 Ref 13, p. 278

15 Ref 13, p. 272

16 Ref 13, p. 281

17 KAYE, J. B.: taped interview, National Sound Archives

18 CHURCHILL, W. S.: radio broadcast, 11th September 1940

19 ANON.: report in the *Daily Express*, 1st October 1940, British Library, Colindale, London, UK

20 Ref. 17

21 BEVIN, E.: radio broadcast to Australia, 17th February 1941

22 Ref. 17

23 Ref. 13, p. 222

24 Ref. 17

25 JAMES, I. J. P.: letter to the author, personal collection

26 See AVIA 13/1047, and AVIA 7/775, PRO, Kew, UK

27 See AVIA 7/106, PRO, Kew, UK

28 Report of visit to EMI by Paddon and Harwood, 9th October 1940, AVIA 13/1047, PRO, Kew, UK

29 CONDLIFFE, G.: letter to the Chief Superintendent RAE, 4th November 1940, AVIA 7/114, PRO, Kew, UK

30 EMI: letter to RAE, 23rd October 1940, PRO, Kew, UK

31 AI production progress report to 23rd December 1940, AVIA 13/1024, PRO, Kew, UK

32 Report on AI Mark VI flight test, 23rd December 1940, AVIA 7/106, PRO, Kew, UK

33 BLUMLEIN, A. D.: EMI report, 17th January 1941, EMI Central Research Laboratories archives

34 Notes of meeting at RAE to discuss AI programme, 3rd April 1941, AVIA 7/106, PRO, Kew, UK

35 Report: AI Mark VI—result of flight trials by FIU crew, 7th April 1941, AVIA 7/106, PRO, Kew, UK

36 Report on further investigation relating to Dr F. C. Williams's application for an award in respect of his contributions to radar development, AVIA 53/630, PRO, Kew, UK
37 WATERS, L.: letter to the author, personal collection
38 BLUMLEIN, A. D.: letter to the Superintendent, TRE, 5th August 1941, AVIA 7/775, PRO, Kew, UK
39 Notes of meeting 'To consider the development and production of items of scientific equipment', 11th August 1941, AVIA 20/3471, PRO, Kew, UK
40 Memorandum, 'Production of and orders for AI equipment. As known at present', 26th November 1941, AVIA 13/1024, PRO, Kew, UK
41 Notes of meeting 'To consider the development and production of items of scientific equipment', 19th October 1942, AIR 20/2355, PRO, Kew, UK
42 Notes of conclusions reached at 82nd meeting held by Secretary of State, 14th July 1942, AIR 20/3472, PRO, Kew, UK
43 WILLIAMS, F. C.: memorandum, 1947, in file on 'Dr Williams (late TRE). Radar inventions, question of award', AVIA 53/630, PRO, Kew, UK
44 Ref. 3

Chapter 14
Miscellaneous wartime activities

By the summer of 1941, metric radars, operating at wavelengths of about 1.5 m, had been developed for air interception, gun laying, searchlight control, coastal defence, ground-controlled interception and early warning of low-flying aircraft. The radiating structures of some of these radars produced broad antenna beams which were not wholly suitable for the applications for which the radars were designed. The beams of the AI (Air Interception) Marks I, II, III, IV, V and VI radars, for example, were such that the effective maximum range of the sets was equal to the height of the AI radar-fitted aircraft above the ground. Hence a night fighter flying at 25 000 ft could obtain a radar echo from a hostile target at ranges of up to about 4.5 miles, but when flying at just 5000 ft the night fighter's target acquisition range was only about one mile. This limitation of AI metric radar follows from diffraction theory[1].

If a plane electromagnetic wave of wavelength λ is incident on an infinite conducting sheet having an elongated aperture of small dimension a, Figure 14.1b shows how the electric field strength, on the far side of the sheet, varies as a function of θ. Since, from the definition of a parabola, a suitably illuminated paraboloid antenna produces a plane electromagnetic wave, the radiation pattern of such an antenna (Figure 14.1c) is similar to that given in Figure 14.1b. From radiation theory, the paraboloid's beamwidth in radians (at the half-power points of the pattern) is given approximately by λ/d, and the gain of the antenna, relative to an isotropic radiator, is equal to $(\pi d/\lambda)^2$ where d is the diameter of the paraboloid. Thus, if a metric radar, operating at a wavelength of 1.5 m, uses a paraboloidal antenna having a diameter of 1.0 m (the maximum size, approximately, of the antenna which could be fitted into a wartime night fighter aircraft), the beamwidth and gain of the antenna would be 86° and 4.4 respectively. But if the wavelength of operation is 10 cm the corresponding quantities have values equal to 5.8° and 1000.

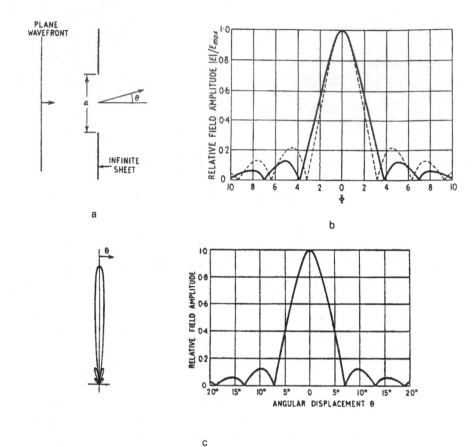

Figure 14.1   (a) *A uniformly illuminated aperture in an infinite plane conducting sheet.*
(b) *The variation of electric field strength with* θ *on the far side of the sheet
for a circular aperture of diameter D (full line); and a rectangular aperture
of side a (dashed line), where* $\phi = (\pi D/\lambda)\sin\theta$, *and* $(\pi a/\lambda)\sin\theta$ *respectively.* (c) *Radiation pattern of a paraboloid of diameter D = 10λ, shown in
polar coordinates and in Cartesian coordinates*

Source: the author

   This fundamental property of a radiating structure confers several significant advantages on centimetric antennas. These were listed by Watson-
Watt, in 1940, as[2]:

1 their ability to illuminate certain objects to the exclusion of others at
   relatively small angular separations, in particular to exclude the
   profound influence of the ground-reflected rays;

2 their ability to obtain high discrimination in measurement of angle by the use of very steep-sided polar diagrams;

3 the economy of power concentration of most of the transmitted energy within the pencil beam; and

4 the reduction of unwanted signals, in particular naturally occurring radio noise and deliberate jamming signals, by confining the reception substantially to a very small angular sector.

Subsidiary, but still important, advantages of such antennas are:

1 their ability to radiate and to receive, efficiently, signals modulated over a wide band of frequencies;

2. the intrinsically low level of atmospheric noise at centimetric wavelengths;

3 the exclusion of effects due to ionospheric bending;

4 the improved transportability of the equipment; and

5 the increased effective mobility, for gunfire control apparatus, conferred by an independence of site conditions.

Since the physics of diffraction was well known when radar research and development commenced in the UK in 1935, the beam-forming property of radiators operating at centimetric wavelengths was, of course, similarly well known. Indeed, the Standard Telephone and Cable Company had in March 1931 utilised parabolic reflectors, of about 3 m diameter, in its demonstration of microwave communications, at a wavelength of 17.6 cm, across the English Channel[3]. Later, from 26th January 1934, the company had established a service between the aerodromes at Lympne, Kent, and St Inglevert, Pas de Calais, France[4]. However, the use of such large antennas was quite impractical for AI radar purposes, and the one-way (transmitter–receiver) transmission path was much less demanding in its power requirements than the two-way (radar–target–radar) transmission path of radiolocation apparatus. In free space the received power density decreases as $1/r^2$ for the former system, but as $1/r^4$ for the latter system.

Notwithstanding STC's success and the known dependence of antenna beamwidth on wavelength, the earliest work, in the UK, on the development of centimetric radar dates from 1940. Essentially, progress before then was hindered because the valves needed to generate and to detect u.h.f. (ultra high frequency) signals were either unsuitable for high-power radar applications, or had not yet been developed to a satisfactory operational state.

The history of the Services' interest in valves for v.h.f. and u.h.f. purposes effectively commenced sometime in about 1932 when the Air

Ministry found difficulty in obtaining from the commercial valve manufacturers the special valves which were needed for the distinctive needs of
the Service[5]. During that year a meeting was held at the Admiralty's HM
Signal School, which was attended by representatives of the three Services,
to discuss the possibility of the establishment of a central valve development organisation. At that time the only Service valve laboratory in
existence was that at the Signal School, so the meeting agreed that the Air
Ministry should put to the Signal School problems, in valve development,
which required solution. The War Office representative declined these
facilities, since they had found the commercial valve companies could cope
quite adequately with their more restricted needs.

A year later the valve section at HM Signal School was slightly
increased to permit valve development and fabrication work to be undertaken for the Radio Research Board.

In May 1936 C S Wright, the Director of Scientific Research (DSR),
Admiralty, represented to the Board of Admiralty his concern with the
state of R&D on very short wavelength techniques[5]. Wright's unhappiness
was possibly based on four facts. First, in 1921 Brigadier General W
Mitchell, Director of Military Aviation and Assistant Chief of Air Staff
(USA), had demonstrated[6] in practical sea tests that the mightiest capital
ships were highly vulnerable to bombing attacks from hostile aircraft. (His
conclusions were vindicated many times in the Second World War.)
Secondly, it was manifest in 1936 that Germany was rearming and had recreated its Luftwaffe (on 1st March 1935). Thirdly, Germany was dis-
playing aggressive military tendencies. Fourthly, the only method in the
mid-1930s of detecting aircraft at sea was by means of lookouts equipped
with 7×50 binoculars.

There was thus an urgent need for a ship-borne early warning surveillance system. The CH radar warning system[7] being developed at the
Bawdsey Research Station worked at frequencies in the 20–30 MHz band
and utilised 350 ft transmitting antenna masts and 240 ft receiving
antenna masts and was totally unsuitable for adaptation for use on board
ships. There had to be a move towards the employment of very high frequencies (30–300 MHz) and ultra high frequencies (300–3000 MHz) to
permit practical naval radar antennas to be effected.

The Board of Admiralty was persuaded of the validity of its DSR's
arguments and gave its approval for the DSR to enter into negotiations
with a commercial firm. Discussions were held with GEC, STC, and the
Metropolitan Vickers Company and subsequently an Admiralty contract
was placed with GEC for the development of valves and components
suitable for communications at wavelengths below one metre[8].

Twelve sets of equipment for operation in the range 40 to 60 cm were manufactured by GEC Telephone Works, Coventry: most of them had been fitted to HM ships by the commencement of hostilities in 1939. The power generated was 20 W. Modulation of the carrier wave was carried out by square wave switching of the anode voltage of the output valve[9].

The contract led to very close cooperation between the Admiralty and GEC and, on 30th November 1938, J F Coales of the Admiralty and E C S Megaw of GEC showed by tests on a glass magnetron (E 821), designed for 150 W c.w. operation at 300 MHz, that such a valve could be operated under pulsed conditions. The pulse output of 1.5 kW at 37 cm was limited principally by the emission of the tungsten filament which was overrun for the experiment[10].

The following year the Sub-committee on Air Defence Research invited Sir Frank Smith and Mr Pye to examine the question of the manufacturing resources necessary for the production of the special valves required for RDF (Radio Direction Finding). Sir Frank Smith's committee, in October 1938, decided there was an urgent demand for R&D on an increased scale to fabricate valves of greater output at the very high frequencies which had to be utilised for RDF[11].

The urgency arose from the ominous political situation in Europe in 1938. The *anschluss* of Austria, by Germany, had taken place on 15th March 1938, and, following the Munich crisis, the Sudetenland had been annexed on 29th September 1938. These offensive moves by the Third Reich heralded a situation which demanded a British rearmament policy. There was an imperative need for the Services to be equipped with the most modern supplies which could be engineered and manufactured in the short term.

Smith's committee recommended that the Committee of Imperial Defence should ask the Board of Admiralty to sanction a substantial enlargement of HM Signal School's Valve Section, under the direction of H G Hughes, and also that R&D, subject to government control, should proceed in close cooperation with one or more industrial organisations. It was thought that the cost of such enlargement and the subsequent programme of research and development should be divided between the Services. Their views were to be borne in mind when the programme was drafted. These deliberations caused an appreciable expansion of the Valve Section at HM Signal School, in both staff and facilities, and prompted the Admiralty to expand its contract with GEC to cover the R&D of special valves for RDF[12].

Curiously, although de Forest had invented the audion (triode) valve in 1906, the complete theory of the operation of the three-electrode valve had not been delineated until 1938. Triode valves are limited in their performance by three effects due to[13]:

1. the anode–grid inter-electrode capacitance;
2   the cathode lead inductance; and
3   the cathode–grid transit time of the thermonically emitted electrons.

The first of these constraints was described in 1919 by Miller and was well known in the 1930s: circuit techniques existed which could neutralise the adverse effect of the capacitance at high frequencies. Any valve, whether triode, tetrode or pentode, is limited in its utility by the damping effects produced in its grid–cathode circuit by the effects of the cathode lead inductance, and the finite transit times of the electrons travelling between the cathode and the grid. These limitations were only extensively investigated in 1936 and 1938 respectively[14,15]. Prior to these years, valves had been developed for wireless telegraphy, wireless telephony, sound broadcasting, and high-definition television broadcasting, but in all these communication systems the valves had operated at frequencies below 50 MHz. Now, in the second half of the decade 1930–40, there was an immediate demand for valves which would function satisfactorily at metric and centimetric wavelengths.

Further thought on valves for radar was given at a meeting, held on 27th January 1939, which was attended by the Directors of Scientific Research of the three Services, and which was presided over by Sir Frank Smith. It was decided that the national interest necessitated still further development, for communication purposes, of apparatus employing ever decreasing wavelengths. The DSR, War Office, undertook to examine certain aspects of the problem, and as a result a contract was placed by the Ministry of Supply with STC for the evolution of specific communication systems at very short wavelengths[16].

Contracts were also allocated to the British Thompson Houston Company and to Electric and Musical Industries Limited. The work was coordinated by the CVD (Communication: Valve Development committee) which acted on behalf of the Army, the Royal Navy and the Royal Air Force. It operated through seven sub-committees, and shared out the work appropriately between them.

From this brief historical background it is apparent that DSR, Admiralty, who chaired the CVD, had considerable responsibility for the integration of the improvement of transmitting valves for RDF.

In connection with his responsibility, C S Wright, DSR, Admiralty, often visited the Bawdsey Research Station. On one of these occasions, in 1939, he asked Dr E G Bowen what wavelengths would be most desirable for airborne applications. Bowen's assessment was that to minimise ground return a beamwidth of 10° was wanted and to obtain this from an antenna of 75 cm diameter — the largest that could be fitted into the nose of an existing fighter — a wavelength of 10 cm would be required. This agreed with Wright's own view of what was needed for naval applications[17].

During the autumn of 1939 the CVD committee placed contracts with the Universities of Birmingham, Oxford and Bristol for the development of transmitting and receiving valves able to operate at 10 cm.

The failure of the Bawdsey Research Station to develop an effective operational AI radar by the outbreak of the Second World War has been ascribed by Rowe to 'a failure to recruit an adequate scientific staff for the work to be done'[18]. Before 3rd September 1939 'only a handful of staff had been engaged on the night-defence problem'. Furthermore, as mentioned earlier, 'with three or four exceptions, the Bawdsey scientists were of average ability'[19].

Early in 1938, at about the time of the Austrian *anschluss*, Sir Henry Tizard, the Chairman of the CSSAD, recognised that in the event of war with Germany there would be a demand for a large number of physicists. They would be needed not only to man the expanding early warning radar chain (CH), but also to devise new types of radar equipment. A possible source of these physicists would be the universities.

In the spring of 1938 H Tizard conferred with Dr J D Cockcroft, of the Cavendish Laboratory, University of Cambridge, on this matter[20]. It was the first move in a plan that enlarged during the following months. By the autumn Sir Lawrence Bragg, the Head of the Cavendish Laboratory, was approaching the Secretary of the Department of Scientific and Industrial Research, Sir Frank Smith, and stressing the need for more university men to learn the nation's defence secrets. Events now seem to have moved slowly, possibly due to the Prime Minister's pact with Hitler, which was signed on 30th September 1938.

In the summer of 1939, because of the foreboding situation in Europe, it was agreed that about 80 physicists should be introduced to RDF so that if war was declared their skills could be deployed to aid the war effort. Each of them would spend a month at a CH station. Subsequently groups of nine to ten physicists[21], sworn to secrecy, from the Cavendish Laboratory, the Clarendon Laboratory (University of Oxford) and the Universities of Birmingham, London, Manchester and Bristol were stationed, from 1st September 1939, at several CH radar sites, all of which had been placed on continuous alert from Good Friday, 1939. Among the physicists who operated the radars were Dr J T Randall and Mr H A H Boot of the University of Birmingham. Later the physicists were transferred to government and other research establishments. On their worth, Rowe has opined[22]:

> 'I am certain of this; without these specially recruited university men, the few scientists of equal calibre, recruited under pre-war Civil Service conditions, would not have come within sight of achieving radar as it is known today [1946].'

When war was declared, Tizard became, officially, Scientific Adviser to the Chief of the Air Staff, Air Chief Marshal Sir Cyril Newall. During the early months of the conflict Tizard travelled extensively and worked tirelessly to strengthen the radar coverage of the United Kingdom. He did all he could to improve the premature, poorly engineered AI radar, but was of the view that additional industrial resources had to be applied to the problem. Moreover, the severe limitation of metric AI radar necessitated a move towards centimetric AI radar.

On 6th November 1939 Tizard visited the Wembley site of the Research Laboratories of GEC and addressed the senior staff[23]. He stressed the vital importance of 'radio' as a war weapon and the desire for much more of the work to be allocated to industry for quick progress. Later, on 11th December, Watson-Watt (who was now the Director of Communication Development) had a meeting at the Research Laboratories: he highlighted the importance of R&D work proceeding at wavelengths not greater than 20–30 cm. Some preliminary information of what was required was given, together with the initial terms of reference for the work.

The first detailed discussion between DCD and GEC on the AI project took place on 22nd December 1939. A pulse radar system suitable for operation in an aircraft and functioning at wavelengths of 20–30 cm with a minimum transmitted power of 100–250 W (but 1 kW if possible) was the design aim. It was hoped the range would be five to ten miles. One week after this discussion DCD asked GEC to accept an official development contract on AI radar. EMI was asked to collaborate in the same contract and subsequently, from 16th January 1940, a series of joint meetings with Air Ministry officials and others where held alternately at Wembley and at Hayes at approximately monthly intervals. Dr C C Paterson acted as Chairman at Wembley and Mr I Shoenberg at Hayes; Dr E G Bowen was the coordinator of the meetings. Prominent members of the committee were Gosling, Megaw and Jesty of GEC, and Shoenberg. Blumlein and White of EMI. The objective of the committee was the design and implementation of a prototype centimetric AI radar. In his book *Radar Days* Bowen states (apropos this committee): 'Blumlein soon proved that he was a master of circuit design and produced a flow of innovative ideas which transformed circuit design during the next few years.' The GEC AI research and development group comprised G C Marris, D C Espley, G W Edwards, R J Clayton, and E C Cherry[24].

Good progress was made and by 17th February 1940 a complete schematic and circuit arrangement had been drawn up, based on the use of a scanning horn antenna system, and construction of the various units was beginning to take place. Espley suggested the possibility of having one antenna for both transmission and reception, (using two diodes on the transmission lines), and circuit work on the common transmitter and receiver antenna problem began about 18th February 1940. Contemporaneously, continuous progress by the Research Laboratories was being achieved on the engineering of the E 1130 'milli-micropup' transmitter valve. Elsewhere, R&D contracts were in progress in March 1940 for the production of 10 cm AI units[25].

A 'lash-up' of the 25 cm AI radar was shown at the AI meeting held on 3rd March 1940, and during the month preliminary work on crystal mixers was initiated. The 25 cm system was demonstrated on 14th March, and, following Dr E G Bowen's opinion that the wavelength should be between 10 and 15 cm, it was decided on the 28th that AI systems operating at 25 cm and at '12 cm or thereabouts' should be GEC's double objectives. The decision to press forward with the additional 12 cm AI radar was based on the limitation of the size of radiator which could be accommodated in a night fighter. Sir George Lee, the new DCD, on 15th April confirmed this policy and agreed to GEC developing an AI radar for operation at 10 cm. He also suggested the 25 cm work should be terminated at a later date if the former proved promising. The contract for the AIS (10 cm) radar was received by GEC on 3rd May 1940[26].

In the first week of April the first 'real comparison on a pulse system receiving reflected echoes' was made between a crystal mixer and a mixer using a concentric diode. An appreciable improvement was observed with the crystal mixer. From this time the unit 'steadily improved its lead over competitive mixer arrangements', many of which, including various diode designs, autodyne triode circuits and velocity modulation tubes, were tried.

An important benchmark was reached in the middle of April 1940 when a pair of pressurised E 1130 transmitter valves gave an output of 1 kW at 20 cm: at the AI meeting held on 26th April an output of 2 kW at 25 cm was reported. Thus in just four months GEC's research staff had met, and indeed had exceeded, the tentative power specification put forward by Watson-Watt in December 1939. This success led to a trial of the complete 25 cm radar set, using crystal mixers. DCD was informed on 8th June 1940 that ranges of six miles on ground targets had been obtained. GEC's achievement was unique: at that time no other industrial or government laboratory/establishment was actively engaged on all aspects of the development of a complete centimetric AI radar set operating at 25 cm[26].

In the first week of September 1939, after the invasion of Poland, the staff at the Bawdsey Research Station, with the exception of Bowen's AI group, moved to premises in Dundee and the establishment was renamed the Air Ministry Research Establishment (AMRE). The AI group was based in Perth, but in November 1939 it moved to St Athan in South Wales[27]. A further transfer of the main establishment took place on 5th May 1940 when AMRE was relocated at Worth Matravers (near Swanage), Dorset. On this period and AMRE's interest in centimetric radar, Rowe has stated[28]:

'We need not dwell on this phase of centimetric history because no facilities existed at Dundee or St. Athan for work in the centimetric field. It is, of course, equally true that no facilities for the work existed at Worth Matravers in May 1940.'

AMRE's experimental work on centimetric radar dates from June 1940 when a klystron, based on a design produced in Professor M Oliphant's Department, University of Birmingham, was constructed by the Mond Laboratory, Cambridge. The Oliphant klystron, with its power supplies, was a large piece of laboratory equipment which 'not even Dee and his team [the centimetric radar group] could conceive being installed in an aircraft. . . . The look of it did nothing to encourage hopes that the installation of centimetric wavelengths in the aircraft was within sight; it even had its own pumping plant for creating the near vacuum needed in the working space of the valve.'[29]

Meanwhile, in June, at the Research Laboratories, GEC staff had patented the sleeve coupling, and the contactless switch, for use on rotating scanners; arrangements were being made for the requisition of aircraft for range trials at Wembley; and on the 25th of the month the first useful trials on common antenna working with resonant transmission and receiving feeders had been carried out, although the problem of protecting the receiver detector from the transmitter power had still to be solved. Units and parts for both the 25 cm and 10 cm systems were demonstrated at the AI meeting held on 8th July 1940, with Watson-Watt present. Strangely, Watson-Watt, in his book *Three steps to victory* ignores GEC's contributions[30].

The invention of the resonant cavity magnetron by Randall and Boot was the crucial invention which enabled high-power centimetric radars to be implemented. According to Tizard: 'The cavity magnetron had a more decisive influence on the war than any other single weapon.'[31] And after the war the Royal Commission on Awards to Inventors observed: 'the inventions [of the cavity magnetron and strapping] were of outstanding brilliance and exceptional utility for offensive and defensive purposes'.

Randall and Boot continuously pumped prototype magnetron operated for the first time in the laboratory on 21st February 1940 and was first seen by GEC staff on 10th April 1940[32]. It was agreed that GEC would fabricate sealed-off magnetrons for the Birmingham group and would assist in every way possible with techniques and materials. The company had vast experience and expertise on valve technology and manufacture since the Research Laboratories undertook research on valves on behalf of the MO Valve Company which was jointly owned by GEC and EMI.

Blumlein was not directly concerned with valve design, but A F Pearce, of the Electron Tube Division, EMI Electronics, has recalled two contributions which Blumlein made.

'In order to have a source of micro-wave power for measurement purposes, in 1940 we made replicas of a c.w. split-anode magnetron. . . .

'These were glass envelope tubes and the output power was taken from the anode tank circuit by means of a Lecher pair [a short length of balanced transmission line], the wires being sealed through the glass. On seeing this in operation, Blumlein suggested that the seal through the glass and losses associated therewith could be avoided by coupling a half-wave dipole to the tank circuit inside the envelope, and placing the whole valve inside a circular wave guide resonator. This was tried and found to work, but not very well, possibly because of a mismatch between the impedance of the dipole and the impedance of the guide. The idea was not pursued, but [it] shows that Blumlein had a firm grasp of the nature of microwaves and of the fields existing in a hollow resonator at an early stage of development.

'His second contribution concerned reflex klystrons of the disc seal variety which we were developing in 1940 and subsequent years. . . . Blumlein suggested the use of a cavity working in a harmonic mode. . .which enabled the need [for a 3 cm local oscillator] to be satisfied with an oscillator of proven principle and manufacturability.' (This is considered later in this chapter.)

GEC's achievements led to a directive, dated 19th July 1940, that all the company's developments had to be passed to AMRE, Swanage. A 'marriage meeting' was convened on 29th July at Worth Matravers, to combine the parallel efforts of AMRE and GEC 'in order to obtain a solution [to the AI problem] in the minimum possible time'. However, according to Dr C C Paterson, the Director of the Research Laboratories, AMRE was 'considerably behind' GEC on AI[33].

The 'marriage' caused Paterson much frustration and dismay. P I Dee, a nuclear physicist, without any specialist knowledge of radio, was in charge of AMRE's centimetric work[34]. His group's inexperience was noted by Paterson and he recorded in his wartime diary: 'No background of radio engineering to establish their operations, measurements, etc on a sound reliable basis.'[35]

'What a day!', he wrote on 12th August. 'Lewis, Dee, Skinner, Bartlett, Atkinson, Ward from Swanage to talk AI and means of co-operation. Except for Lewis [the Deputy Superintendent] there was no articulate idea on the subject from their end. They appear to be a group of individualists with little experience in teamwork—except for Lewis.'[36]

Six days later Paterson wrote to Rowe[37]:

'You are asking us, and being supplied for *your* 10 cm AI with specimens of the components which we develop and put in hand for *our* 10 cm AI. Your and our 10 cm AI appear more or less identical. We have a commission from the Air Ministry to develop our system and have made considerable financial commitments therefor. We now have the feeling of being sucked dry of our work and ideas as they materialise, and then being left out of the picture! Ought you not really to bring us into the effort in a formal way?

And after a meeting, with Rowe, about four months after the marriage, Paterson recorded[38]:

'I had to be frank with Rowe and told him that I thought Dee seriously intolerant and entirely satisfied with his own ideas which are often narrow and unsound. Skinner may be better, but has no conscience in appropriating credit to himself and Swanage which should go to Wembley.'

And:

'It appears that [some people at Swanage] have sold themselves and their institutions very effectively to their chiefs, and whatever they say, however immature, goes in MAP [Ministry of Aircraft Production] circles.'[39]

On political manoeuvrings and intrigues, Paterson's opinions were likewise trenchant and down to earth:

'I have the impression from various sources that our academic ridden radio effort is poisoned with intrigue, jealousy and uncharitableness. Industrial and commercial life seems clean in comparison. . . .'[40]

'Truly the professor type needs to learn how to give honour where honour is due.'[41]

Notwithstanding this unease[42], substantial progress was being made by the GEC Research Laboratories. The 10 cm and 25 cm equipments were shown, working on the roof and taking shape for air trials, to Sir Frank Smith and Sir George Lee; and at the AI meeting held on 5th August the two systems were demonstrated and engineered units were displayed. On the same day GEC received a memorandum from RAE (the Royal Aircraft Establishment) 'laying down that GEC were to provide the mock-ups for [the] initial fitting into [a] plane'. A few days later Professor Oliphant in a report for DCD on crystal mixers referred to the 'GEC lead on silicon/tungsten arrangements[43]. Puzzlingly, the official *History of the Development and Production of Radio and Radar* states: 'The early development of the crystal mixer was done mainly at TRE; it was subsequently [*sic*] improved as a result of fundamental work by the GEC and the BTH.' And on 10 cm AI radar the official history does not mention GEC's pioneer work. 'Ultimately [*sic*] the GEC AI work took the form of development to TRE's directions rather than independent research.'[44]

GEC continued to develop the common T and R antenna method of working and circulated, on 18th August 1940, the first report on 'A single radiator for both transmission and reception'. The method gave about 40 dB protection of the crystal detector against the transmitter's output and was used in all the flight trials in the early months of 1941 (until May) and was in the system when the first signals were received in the air from a submarine target (April 1941). The system was patented in 1941.

During September 1940 work on AI was 'dragging' (Paterson). 'Swanage...wants to run everything as though they are cock of the walk. I see no special signs of real partnership.'[45] Even a government scientist, F S Barton of the Royal Aircraft Establishment, felt 'the new arrangement [i.e. the 'marriage'] seemed to have the only effect of slowing Wembley down to the pace of Swanage!'[46]

There is no doubt that Rowe felt his establishment should have overall charge of the 10 cm AI development, notwithstanding GEC's official written approvals for both 25 cm and 10 cm AI work; the excellent R&D facilities at Wembley; the experience and ability of the GEC staff engaged on the AI project; and the good progress being made by the AI group. Rowe stated his view in a note dated 21st July 1940:

'Dr Paterson thinks that he is developing the whole of 10 cm AI whereas we feel that firms should be doing specific parts. All this has been said to the Air Ministry and we are awaiting for a reply ...'[47]

To resolve this issue a new committee was set up. It was called the 'Committee on 10 cm wave AI', and was chaired by Dr T Walmsley of the Ministry of Aircraft Production (MAP), and comprised representatives from the MAPRE (formerly AMRE), GEC, EMI, Fighter Command, RAE, and MAP. Blumlein was one of EMI's representatives on the committee[48].

At the first meeting, on 3rd September 1940, the Chairman pointed out that the 'purpose of [the] committee was to make sure that results were obtained quickly and that the resources of the various organisations engaged on this work were being used to the full'. He sought the state of developments at MAPRE, Birmingham University, GEC, and EMI. Marris of GEC mentioned that the 'GEC equipments were drawn up and blueprinted, and ready for discussion with MAPRE [Swanage]'; but Dee 'contended that better progress would be made if MAPRE formulated the technical character of the complete AI system'.

> 'To this contention Marris replied that the work which was the basis of the present degree of advancement was not done by MAPRE and only recently had they been in a position to make any real contribution towards the solution of the AI centimetric problem. In deciding future action the historical background should be borne in mind.
>
> 'The chairman sympathised with Mr Marris's point of view. At the same time he reminded him that supervision and control was naturally vested in the Directorate through their authorised officers.
>
> 'It would be better in the early stages of development for the contracting firms to submit their technical proposals to Dr Dee for approval . . .'[49]

Dr Paterson agreed not to press any objection to this minute but 'said he would like the idea of reciprocity to obtain as far as possible'.

Among the matters discussed, it was noted that EMI and GEC had exchanged information on pulse modulators; that EMI had provided GEC with a modulator which Blumlein had designed for the 1.5 m AI equipment; and that EMI was considering as part of its contract[50] the application of its modulator to the GEC magnetron[51].

Later at a conference on 10 cm measurements, held on 1st October 1940, EMI's 'very substantial efforts in this field' were noted; and at the third meeting of the committee, held on 23rd October 1940, Blumlein outlined the results which EMI had obtained in tackling the problem of modulating GEC's type E 1189 magnetron.

Apart from its work on centimetric valve modulation techniques and 10 cm measurements, EMI initially also provided designs for 10 cm cables, plugs and sockets. A 'half-wave connector' for a cable was described by the DCD as 'a very useful achievement'[51]; and in a paper on 'ultra short wave cable measurements', EMI–Siemens cable type 1 was listed as having an attenuation constant at 8 cm of just 0.6 dB/m[52]. The cable used 'crinkled' conductors similar to those employed in Blumlein and Cork's 1937 television cable design.

On 18th November 1940 GEC dispatched its AIS 'mock-ups' to Swanage for the first fitting into a Blenheim (N 3522). The GEC helical scanner was sent out on 5th December 1940 and in February 1941 GEC staff participated in the first AIS flight trial at Christchurch. Apart from

the scanner, which had been developed by Nash and Thomson in cooperation with AMRE, the system was mainly the engineered version of the
GEC experimental system, basically as reported on 15th August 1940. The
first flight trial was satisfactory, except that the range of the set left
something to be desired. GEC's new type 2 helical scanner was demonstrated on 12th July 1941 and passed RAE's tests on 31st July. According to
Bowen, the scanner was superior to the American type fitted in a Boeing
aircraft. Installation and trials of the scanner began at Christchurch on 1st
August 1941. Later, a new comprehensive AI scheme (AIS II) was
produced and approved by Swanage, and was delivered to Christchurch
on 18th November 1941.

EMI's R&D expertise was certainly of considerable importance in the
progress of centimetric radar. On 18th January 1941 the Chairman of the
'Committee on 10 cm wave AI' wrote to A P Rowe, the Superintendent of
TRE (Telecommunications Research Establishment) and mentioned[53]:

> 'Messrs EMI have pointed out that frequent requests are made upon them to
> supply information and apparatus at the request of members of TRE. . . .They
> might have some enquiries from TRE about circuit improvements, which
> means that some of their staff would have to donate time to considering the
> matter and probably to testing out circuits. Again, [they] might be asked to
> make certain measurements on behalf of TRE utilising known technique.
> They might be asked to make a test which is not covered by an existing con-
> tract . . .'[54]

On 1st January 1941 Shoenberg, Blumlein and Cork had discussed the
question of a new contract with Dr Walmsley of MAP and Dr Lewis of
TRE and it was agreed that the contract should cover miscellaneous items
such as the development of centimetric equipment capable of measuring
voltage, wavelength, power, impedance and attenuation; the design of
components (for example, cables and connectors for cables) essential to the
development and use of the measuring equipment; and the development of
methods of measurement at a wavelength of 3 cm[54]. Lewis's view[55] was
that the contract should cover the development of local oscillators, velocity
modulation mixers, low-loss cables, and components such as line length-
eners, connectors and resonant chambers. 'We would find it convenient',
he wrote[56], 'to call on EMI for any such components which they develop in
the course of their work towards establishing a 3 cm technique'. Two
months later the Ministry of Aircraft Production awarded a miscellaneous
contract to EMI for £250 per month in the first instance, for a period of
three months[57].

The detailed assistance which EMI and Blumlein gave to others on a
non-contractual or *ad hoc* basis is not known in detail, for few records exist.

Life was hectic for Blumlein and his colleagues following the 1940 fall of France. Precious time could not be spent on the administrative nicety of recording what was undertaken in the common endeavour of preserving the integrity of the United Kingdom. Solutions had to be found to problems of central importance to the war effort and they had to be found quickly. Fortunately two letters exist which gave an indication of Blumlein's genius in identifying and analysing engineering situations in entirely disparate fields of enquiry.

Dr (later Professor) J Sayers was, in the early 1940s, a member of Professor Oliphant's group in the Physics Department at the University of Birmingham. He worked with Randall and Boot, in 1940 and 1941, on the development of the cavity magnetron and in September 1941 invented 'strapping' to limit mode switching in the valve. This work entailed the measurement of power and frequency/wavelength at centimetric wavelengths, and the practice of coupling power from the valve to a waveguide. After the war £36 000 was given to Randall, Boot and Sayers for their fundamental work on cavity magnetrons. (About 1950 a good honours degree physics graduate, on leaving university and entering industry, could expect to receive an annual remuneration of £350 to £400.) Clearly, in 1941, Sayers was as great an expert in the field of centimetric wave technology as existed at the time. Yet Blumlein was able to inform him of a point on 'the amount of energy in the unwanted modes' of a waveguide which he (Sayers) had not previously appreciated (Figure 14.2)[58].

Dr (later Professor) E B Moullin was an acknowledged authority on electrical measurements. He had had many learned society papers published, from 1924, on aspects of measurements. In January 1942 he was the Chairman of an 'Informal Conference on the behaviour of magnetic materials under conditions of pulse magnetization'. At the Conference Blumlein presented some thoughts on the relationship between the losses in and the changing permeability of such materials. His experience and knowledge of magnetic materials was probably much inferior to those of Moullin—after all, Blumlein was a circuit and systems engineer. Nevertheless, such was his ability to appreciate new aspects of a problem, which others had not perceived, that he was able to make 'a very important point'. Moullin's admiration[59] for Blumlein is readily apparent from his letter of 15th February 1942 (Figure 14.3).

The delay in the introduction of 10 cm AI radar into the RAF's fighter squadrons stemmed from the 'tremendous pressure' put on the Air Ministry to improve the UK's centimetric gun-laying radar. The Blitz attacks on London and other British cities were causing great damage and grievous loss of life. There was a desperate need to improve the nation's defences against German night bombers.

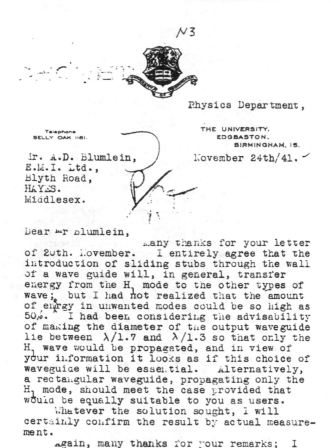

N3

Physics Department,

Mr. A.D. Blumlein,
E.M.I. Ltd.,
Blyth Road,
HAYES.
Middlesex.

THE UNIVERSITY.
EDGBASTON.
BIRMINGHAM. 15.

November 24th/41.

Dear Mr Blumlein,

            Many thanks for your letter
of 20th. November.   I entirely agree that the
introduction of sliding stubs through the wall
of a wave guide will, in general, transfer
energy from the H_1 mode to the other types of
wave; but I had not realized that the amount
of energy in unwanted modes could be so high as
50%.   I had been considering the advisability
of making the diameter of the output waveguide
lie between $\lambda/1.7$ and $\lambda/1.3$ so that only the
H_1 wave would be propagated, and in view of
your information it looks as if this choice of
waveguide will be essential.   Alternatively,
a rectangular waveguide, propagating only the
H_1 mode, should meet the case provided that
would be equally suitable to you as users.

      Whatever the solution sought, I will
certainly confirm the result by actual measurement.

     Again, many thanks for your remarks;   I
shall be glad to hear of any further results
affecting magnetron outputs that may emerge
from your experiments.

         Yours sincerely,

                   J Sayers.

Figure 14.2   Copy of a letter from Dr J Sayers to A D Blumlein

Source: EMI Archives

Fortunately in 1940 some action had been taken to strengthen the
RAF's night-fighter capability and by the end of 1940 the twin-engine
Beaufighters, fitted with AI Mark IV radar, were entering service. GCI
(Ground Controlled Interception) radar had been approved as an
essential aid to night-fighter interceptions and soon the combination of
these three components of the air defence system would be wholly operational.

On the ground the AA gunnery defences had not been successful
during the night bombing raids. Much had to be done to improve Anti-

7364

From:- Admiralty Signal   To:- A.B. Blumlein,Esq.,
      Establishment Extension,   Electric & Musical Industries,
      Hambrook House,   Blythe Road,
      FUNTINGTON,   Hayes,
      Nr Chichester,   MIDDLESEX.
      SUSSEX.

RECEIVED
- 6 FEB 1942
ANSWERED

Feb 5th. 1942.

Dear Blumlein,

     I now send you the minutes of the meeting which we
held at the I.E.E. on Jan 2nd. They have been very nicely and
kindly prepared by Mr. Mc Fadyen of the G.E.C., and I am sure we are
all very grateful to him for his help. One of the points most
striking to me which arose at the meeting, was your discussion of
the relation between loss and changing permeability. Although I was
familiar enough with the idea, it had not occurred to me that it is
theoretically possible to make the loss occur elsewhere than in the
iron, and I think this is a very important point. Mc Fadyen submitted the minutes to me and I wrote considerable additions much in
the form you will see, but in fact put the words into your mouth.
For I said that though you had'nt actually said them, you certainly
would have said them if you had more opportunity.

     In his final draught, most of these remarks seemed to be
from my mouth after all, but I think it is clear from the top of
page 4 that it was you who made the essential contribution, and that
I have only amplified it.

     I saw Redfearn yesterday, who told me his report on eddy
currents is now typed, and copies will reach me in a few days. When
they come, I'll go through his solution and compare it with Mr.
Manifold's report later. I am very glad you came to the meeting,
which was, I think well worth while.

     Please may I draw your attention to a slight slip in the
notice calling the Modulation Committee on Feb 19th. I should read:-
Members must enter by the main door in **Whitehall** ; apparently if you
go to the entry in Horse Guards Parade you may very easily get lost
and never be found. Will you and Mr. Collard note carefully if you
intend to come.

                 With Kind Regards,

                     Yours sincerely,

                     E.B. MOULLIN.

Figure 14.3   Copy of a letter from Professor E B Moullin to A D Blumlein

Source: EMI Archives

Aircraft Command's figure of rounds fired per aircraft destroyed. The
average figure of 18 500 for September 1940 was quite unacceptable.

As a consequence of the pressure put on the Air Ministry, Sir Frank
Smith, the Controller of Telecommunications Equipment, devoted consid-
erable effort to the production of 10 cm gun-laying radar (GL Mark III).
He formed a team at BTH (the British Thompson Houston Company)

under the direction of Dr T Walmsley (the Deputy Director of Communications Development, Air Ministry), to expedite the production of this radar. So successful were the efforts of this team that by June 1941 GL Mark III existed in prototype form.

Meanwhile 10 cm AI was 'progressing very slowly'. In an appraisal written in June 1941 the Assistant Chief of the Air Staff (R) noted[59] that:

'there were at the most [just] three equipments available in [the] country, still in rather primitive form. There [was] no prospect at the present scale of effort of any improvement in the situation till next winter. No work of any importance [was being] done on shorter wavelengths, such as 3.5 cm, which [was] without question the real centimetric AI that [RAF Fighter Command] required. . . . GL Mark III manufacture [was] so complicated it [could] not get into production before the early summer of next year [1942], and any sets that [were] made now [would] be made at the expense of the production of GL Mark II and of Bofors predictors.'

In view of these points the ACAS (R) recommended most strongly that effort behind GL Mark III be 'slowed up and that CTE be instructed to proceed with the manufacture of 10 cm AI as a matter of the highest priority'.

An alternative explanation for the slow rate of progress of centimetric AI was given, also in June 1941, by Dee. He wrote[60]:

'Owing to the inadequacy of the General Electric Company Limited model shops, less than two complete sets of equipment have been available to us for aeroplane fitting and for the essential ground trials which should be carried out before and in parallel with flight experiments. Much delay in the production of a final scheme has resulted from this shortage of experimental units. . . . From [15th June 1941] production is expected to be about one off per week.'

On 24th December 1941 Rowe wrote[61] to Paterson and mentioned that an aircraft fitted with AIS (10 cm AI) had had its first combat. The result of the combat was unknown, but 'the the AIS worked perfectly and gave a long range at the low height of 2000 ft'. This was a vindication of the decision taken towards the end of 1939 to move from metric to centimetric wavelengths. In his letter Rowe also noted: 'Although it must seem to your people that we get the fun at the end, I would like to assure you that we always link your people with ours in referring to this common effort.' (Regrettably, in his book *One Story of Radar* written just a few years later in 1948, Rowe gave no credit to GEC for their pioneering work on centimetric radar.)

In the same month AI Mark VII was used in a particularly important trial to determine whether airborne centimetric radar could aid navigators to find their targets over enemy territory. The successful nature of these trials led to a contract being placed with EMI, in January 1942, for H_2S bombing/navigation radar. A Coastal Command variant of this equipment was coded ASV Mark III. Together these radars transformed the bombing offensive over Germany and led to the defeat of the U-boat in the Battle of the Atlantic. The development of these radars owed a great deal to Blumlein's genius and is considered in the following chapter.

Further action with AIS aircraft during the first Thursday night in May 1942 led to seven enemy planes being destroyed. Six of these were 'caught and held by Mark VII AIS which was responsible for their destruction. Even when [the enemy aircraft] came down quite low they were followed and held'.[62]

Dee, in June 1942, informed Paterson of further successes: 'You will be interested to know that one night last week there were three certain [kills] and one probable achieved by AI Mark VII. The 100 sets of Mark VII has clearly been a tremendous success and I am fully aware of the GEC's contribution.'[63]

After the commencement of the German night-bombing offensive in September 1940 and the demonstrated inadequate performance of the Blenheim and (initially) Beaufighter night fighters, there was an imperative need to equip the RAF's faster single-engine fighters with AI radar. Fortunately during the Blitz and the first few months of 1941, German bombers flew at 20 000 ft and so the AI Mark IV radar's maximum range of 4 miles was adequate. There was some apprehension that hostile night attacks undertaken at 5000 ft would severely limit the efficacy of metric airborne radar.

With a twin-engine fighter, such as a Blenheim, Beaufighter and Havoc, it is possible to accommodate a 10 cm AI radar in the nose of the aircraft. With the single-engine Hurricane and Spitfire fighters, which proved so successful in the daylight Battle of Britain, AI antennas can only be mounted on the wings, possibly inside streamlined nascelle housings. Thus 3 cm radar was desirable for these fighters.

The major operational advantage of such a radar *vis-à-vis* 10 cm radar lies in the improved directivity of a 3 cm airborne antenna; primarily in the availability of a narrower beam, and secondly in the ease of movement of the beam. Alternatively, since an antenna's beamwidth is given approximately by λ/d it follows that the diameter of a 3 cm paraboloid antenna is 0.3 times that of a 10 cm antenna, for the same beamwidth in the two cases.

At first it was hoped that 3 cm transmitter and local oscillator (receiver) valves could be manufactured by scaling down suitable 10 cm working valves. EMI had developed, as part of its contract on centimetric techniques, a range of 10 cm local oscillator valves (CV35, CV36 and

CV67) based on the work of Sutton's groups at ASE on velocity modulation valves. Theory showed that it would be impracticable to scale down the CV35, since the u.h.f. losses in the glass of the resonator would be prohibitive. At the same time it was considered desirable to retain the copper discseal technique which was a feature of the CV35, since it had proved easy to handle, and in 1941 the need for 3 cm radar was too compelling to permit a new technique being developed. The solution[64] to the problem was put forward by Blumlein, who suggested using a resonator which operated in a harmonic mode. At about 3 cm the size of the first harmonic E_0 resonance, i.e. E_{02}, is such that a voltage node occurs at about 18 mm diameter: and so the glass-to-copper seal was made at this point (Figure 14.4a). Tuning was carried out by flexing one of the copper discs outside the envelope. The discs had V-shaped portions at the voltage antinode, and movement of one of them changed the capacitance at this point, and therefore the resonant frequency. Hence, although the glass envelope was placed at the nodal surface between the central gap and the capacitance rim and was within the resonator, it was not in an electric field and caused no loss in the u.h.f. region.

Samples of the new klystron (known as KRN2) were available for experimental use in March 1941, and during the next 12 months about 100 tubes were constructed. Many of them were supplied to TRE for application in their X band work. The first pre-production samples were ready in September 1942.

The valve proved to be very satisfactory in operation and was used principally as a beat oscillator, and for measurement work. Operating at 1500 V, 10 W input, the power output was about 100 mW over a tunable range of 3.05 cm to 3.45 cm. It was used by the Services with two different types of tuner (CV87 and CV129) in the early X band AI and H_2S radar sets[65], Figure 14.4b.

The development for airborne radar of X band technology, including the design of high-gain, high-resolution antennas, led to the shorter wavelengths being considered for ship-borne radar. There was a view that 3 cm electromagnetic waves might be more effective in illuminating small surface targets than 10 cm radiation and that consequently the echoes from such targets might be more easily seen on a p.p.i. display tube. However, the absence of data on the propagation characteristics and the reflection coefficients of targets at 3 cm necessitated a sea trial of an experimental ship-borne radar. A development contract was placed, in the autumn of 1941, with EMI for this purpose and, as usual, Blumlein made a major contribution.

The requirements of the radar, which was coded N3 (later Type 261), were:

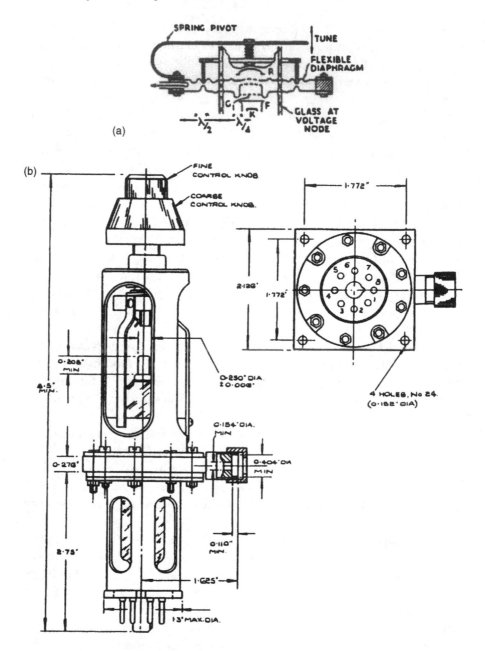

(a)

(b)

Figure 14.4   *(a) Using Blumlein's solution the metal-to-glass seal was positioned at a nodal surface of the resonator electric field. Tuning was varied by flexing the diaphragm which formed part of the resonator. (b) The CV87 (KRN2) klystron*

Source: 'Metres to microwaves', by E B Callick (Peter Peregrinus Ltd, London, 1990)

1 to detect reflecting objects, particularly of small dimensions;
2 to measure their distance, and direction to about $\pm 0.5°$, without the possibility of giving false indications;
3 to be capable of continuous rotation at mast height;
4 to be of small size, weight and wind resistance;
5 to be unaffected by the roll of the ship, which implied that the radiant beam must have a vertical width of about $\pm 12.5°$ to half voltage amplitude;
6 to allow the transmitting and receiving apparatus to be located at a distance from the aerial system;
7 to permit rapid installation; and
8 to be unaffected by weather and extremes of temperature.

Such a specification would make N3/Type 261 an appropriate replacement for the S band Type 272 radar in the event that X band radar working had demonstrated advantages over S band radar for general surface vessel warning.

Of the several sets of equipment which were constructed, one set was delivered, in May 1942, to the Eastney Laboratories of the Admiralty Signals Establishment, and another, modified and improved, was installed in October 1942 in the trial ship HMS *Saltburn*. Figure 14.5, which is self-explanatory, shows the arrangement of the Type 261 radar.

Only a very few letters (nearly all of which bear Blumlein's signature or initials) on the N3 project are still extant in EMI's archives. Fortunately, a most illuminating letter[66], Figure 14.6, written by Blumlein but sent out under Condliffe's signature, has survived. It describes the background to Blumlein's work on the problem of modulating the resonant cavity magnetron which led to the evolution of a very important hard valve modulator, using two artificial delay lines, which was employed in several types of centimetric radar.

In centimetric, pulsed, radar systems the modulator has to provide very short pulses (c. 0.5 µs to c. 5.0 µs), of considerable power, to the cavity magnetron. To achieve this, the modulator must incorporate a component able to switch power on and off at power levels which, in the later stages of the war, were as high as 1 MW. Hard valve modulators are characterised by charging and pulse-forming circuits, and may utilise hydrogen thyratrons, mercury thyratrons, rotary spark gaps, or triggered spark gaps as switches. With high-power modulators the function of the switch is limited to that of 'make'; the termination of the pulse, or 'break', being effected by the circuit and not by the switch.

Figure 14.7 shows the Blumlein circuit with a resistive load and a unidirectional switch (a hydrogen thyratron)[67]. The pulse-forming networks

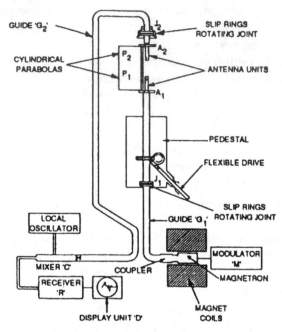

Figure 14.5   Type 261 radar installed in HMS Saltburn for trials of 3 cm radar

Source: DERA, Malvern

have a characteristic impedance which is normally made equal to half the load impedance, and each line has a one-way delay time which is half the required pulse length (T).

In operation, the two lines are charged in parallel, by any suitable method, to a voltage (V). When the switch is closed at $t = 0$, thus short-circuiting the left-hand end of line L1, a voltage wave travels from left to right of the line discharging the line capacitance to zero voltage. At $t = T/2$ this wave reaches an impedance discontinuity and is partly reflected back along L1—recharging the capacitance to a voltage of $-V/2$, and partly transmitted down L2—partially discharging its capacitance to a voltage of $+V/2$. At the same time a voltage of $-V$ appears across the load impedance. Subsequently the voltage wave travelling in L1 is again reflected, with a change of sign, at the short-circuited end of this line, and returns discharging the capacitance of L1 to zero voltage. After this reflection the switch passes no more current. The voltage wave travelling in L2 is reflected without change of sign at its open-circuited end and then discharges the capacitance of L2 to zero voltage. When the two waves return simultaneously to the junction of L1 and L2 no further reflections occur and the cycle of operations can be repeated. Thus the properties of the circuit are:

Copies to C.O.B.
    E.Nind.
    White – Cork to see

SECRET

Ref. RD1/ADB/GB.

14 th October 1941

The Officer-in-Charge,
H.M.Signal School Extension,
Eastney Fort East,
Royal Marine Barracks,
PORTMOUTH. [sic]

Contract CP. Br.4F/54212/41/F.9286.

Modulator.

Dear Sir,

    We have been examining the design of the modulator for the above job and have come to certain conclusions as to how to proceed, and we would be glad to have your remarks on these conclusions.

    Following our visit to your station, two of our engineers visited T.R.E. and there learnt that the magnetron was expected to work at 10 to 14 KV and 8 Amps as against the 8 KV 10 Amps anticipated. We therefore feel that the construction of a modulator using series valves limited to 12 KV maximum anode potential would be unwise. Of course a transformer might be used, but owing to the short wavelength it was thought undesirable to risk any pulse shape distortion which would lead to frequency modulation.

    We got in touch with the G.E.Co. Wembley, and confirmed the maximum voltage rating of NT.100; also we obtained information on the operation of their gas filled relays. We also visited B.T.H. and obtained information on the operation of their thyratrons and also information on the construction of artificial line condensers, pulse transformer and their synchronous charging circuit.

Figure 14.6 Copy of a letter, drafted by A D Blumlein, which mentions some of the factors that influenced his design of the voltage doubler modulator

Source: EMI Archives

1 a delay equal to half the pulse length between the firing of the switch and the output pulse (the delay may be used to trigger other subsystems of the radar set);

2 an output voltage that equals the voltage to which the lines are initially charged;

It would appear that gas filled relays are
limited in voltage to 15 to 20 KV, so that a plain
series discharging circuit would not give us the
possible 10 to 14 KV required by the magnetron. We
therefore designed a voltage doubler modulator which
we tried out in model form, and which appeared to
work quite successfully. The arrangement of the
modulator and the test results obtained are given in
the attached report

We therefore propose to construct a
modulator as follows:-

Maximum output voltage - circa 18 K

Maximum output current - 12 Amps

Matched load impedance   - 1500 Ω

Circuit   - Voltage doubler

Discharge Device   - B.T.H. Thyratron BT.45

Charging Circuit   - B.T.H. Choke system.

In view of the fact that the charging circuit produces a very
rough curent waveform, which might upset the voltage regulator,
we proposed to "split" the machine into two effective machines,
one for modulation output and one for steady load, by means of
a tapped choke, the machine feeding into the tapping and the
mutual between the two choke windings being equal to the machine
reactance. The voltage regulator would control from the "steady
load" volts.

We should be very glad if you would inform us of the
following machine characteristics:-

Voltage.

K.V.A. (1500 ?)

Range of frequency to be expected

Percentage reactance.

We should be glad later to have an opportunity of running our
equipment on one of your machines with one of your regulators
before bringing it to Signal School.

Figure 14.6   (continued)

3 an output voltage opposite in sign to the voltage at which the lines are
   initially charged.

Although Blumlein's modulator was mainly utilised during the Second
World War to modulate cavity magnetrons, it found an application after
the war as a modulator of klystron and travelling-wave valves as used in
some pulsed microwave power amplifiers. Since these valves function
according to the well known three-halves law of the diode ($I = kV^{1.5}$

We have been considering the matter of pulse length,
and as there is no shortage of power, we wonder whether we might
not be well advised to make the pulse length 2 μsecs. instead
of 1 μsec., or possibly make the pulse length in the experimental
modulator adjustable. It appears to us if the pulse were
very flat topped to avoid frequency modulation, weak echoes
might be more easily observed on the trace with the longer
pulse. We should be very glad to have your opinions on this
point.

As the voltage doubler modulator may be of general
interest, we are forwarding copies of our report to the
Secretary of the C.V.D. Sub-committee on modulation technique,
whom we are also asking to forward us copies of B.T.H. reports
on charging circuits and transformers.

Yours faithfully,

G.E.Condliffe,
RESEARCH LABORATORIES.

ENC.

Figure 14.6   (continued)

where k is a constant), they present to the modulator a changing load
impedance (V/I) which varies inversely as $V^{0.5}$. If this impedance matches
the characteristic impedance of the pulse-forming network at the normal
voltage, the valve will have an impedance of four to five times the
impedance of the network. With a conventional modulator, a series of
voltage steps of decreasing amplitude will travel up and down the network
and the hydrogen thyratron will conduct for many pulse periods. There
will be no negative voltage on the anode of the thyratron to allow it to
de-ionise and it will 'arc through' and short-circuit the supply[68].

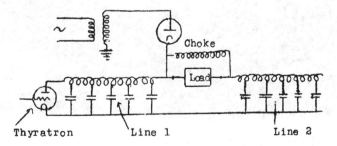

Figure 14.7   The Blumlein modulator circuit

Source: EMI Archives

This difficulty is avoided by the use of the Blumlein circuit because a negative voltage appears on the thyratron switch for all values of load impedance (Z) greater or less than a perfect match. By making Z at full voltage slightly greater than that for an exact match, a negative voltage is always applied to the thyratron to allow it to de-ionise.

Blumlein's modulator was patented in October 1941[69] and was utilised in the N3 and Type 261 radars and in all GL (gun-laying) Mark III S band radars of UK design. A total of 876 of these radar sets was delivered between December 1942 and April 1945. GL Mark III radar gave range, elevation and azimuth data relating to hostile aircraft and was the principal radar of British design used by Anti-Aircraft Command towards the end of the war. It was distributed to the Dominions and to the UK's Allies, including the USSR.

The GL mark III modulator (Figure 14.8) gave a peak output of 40 A at 25 kV to drive the Type CV76 magnetron and replaced a twin-line unit using a pair of thyratrons. The change was introduced to overcome problems with the short life of the mercury tubes caused by overvoltages, and to provide rapid starting without the need to keep the modulator on 'stand-by'.

Figure 14.8   GL Mark III spark gap modulator fitted to all equipment after No. 249 and as a modification to all earlier sets

Source: Mr D H Tomlin

Blumlein type modulators were also fitted to the Naval Type 274 and Type 275 S band gunnery radar systems, which were in operational use from mid-1943. The modulator provided a peak power output of 120 A at 9 kV in 0.5 μs pulses.

Sea trials of Type 261 (N3), in November 1942, were good but not outstanding. It was determined that ranges on various surface targets were comparable to those obtained with a Type 271 S band radar with its antenna mounted 35 ft above the sea. More precisely a Type 261 set installed in a corvette would have had a 10% to 20% range advantage over a Type 271. Since this radar was already obsolescent and was about to be replaced by the more powerful Type 271Q, the formal ASE conclusion was that there seemed to be no naval application for which Type 261 could compete with the S band equipments.

The history of centimetric radar is an examplar of the need for a measure of circumspection when unreferenced historical accounts are read. Since both Watson-Watt and Rowe held senior positions of authority in wartime government departments dealing with the evolution of radar systems, it might be expected that their writings on this subject would be definitive. But neither of their books, written in 1957 and 1948 respectively, gives GEC's Research Laboratories and EMI's Central Research Laboratories any credit for their pioneering centimetric radar work. The impressions given are that the Telecommunications Research Establishment (the replacement name for MAPRE) developed AI centimetric radar.

Watson-Watt quotes Cockcroft's description of 'the little centimetric group at TRE', of which 'Dee and Skinner were the very distinguished "ring leaders"', as 'always furiously engaged in argument, breaking the first ground in centimetres ...'. Watson-Watt further wrote that'.Skinner himself went "back to the fork in the road", resuscitated and re-vivified the ancient "cats whisker", and guided to commercial production "crystal mixers" ...'.

Rowe, notwithstanding his 24th December 1941 letter to Paterson, opined that 'The moving spirit behind the drive for an early application of centimetric radar was Dee', and that 'The beginning was in May 1940 when Dee visited Oliphant at Birmingham'.

A recent (1991) historical account[70] of radar development written by a former member of AMRE, Swanage, states: '...GEC Wembley had already been working on a 25 cm system previously initiated by Bowen [*sic*] and they had no desire (nor indeed intention) to drop the promising venture in favour of some hair-brained 10 cm set-up concoted by a "bunch of academic nuts" who knew nothing about radar.'

Two pages further on, after mentioning a Swanage 10 cm experiment carried out in July 1941, the author wrote: "Grudgingly the firm [GEC] agreed there might be some future in the 10 cm system ...'. And: 'While the basic centimetre technology was being perfected Lovell and Hodgkin set about devising a possible AI system to be known as AI Mark VII. Once a prototype had been built and successfully tested, production could be put in hand by GEC.'

These accounts are historical travesties. Paterson seems to have had good cause to write, in his diary, the comments which have been quoted previously. The Research Laboratories' Director's note that there was a 'noticeable inclination to emphasize everyone's work but that of GEC!!' was certainly borne out of post-war writings on centrimetric radar. The official DSIR publication on *Science at War* by J G Crowther and R Whiddington (HMSO, London, 1947), lists the Swanage staff engaged on centimetric radar but does not refer to GEC's and EMI's endeavours.

Air Chief Marshal Sir Philip Joubert was unhappy with this state of affairs and in September 1945 wrote to the editor of the *Daily Telegraph* [71].

'Recently there have appeared in the Press long articles on the interesting history of radar.

'Having been closely associated with Radar since 1935 I read them with great interest and *some concern* [author's italics]. In the inevitable abridgment of the material supplied from official sources, the story as printed was incomplete, and the articles presented a rather one-sided picture, with consequent unfairness to certain organisations connected with the development of this unique weapon of war.

'The first omission is in regard to the very great part played by private industry, both in research and in the manufacture of the vital equipment, which saved this country in 1940 during the Battle of the Atlantic. It was in the research laboratories and model shops of private firms, such as GEC, Met.–Vick, BTH, and EMI, that the equipment was produced at a speed and in such numbers as availed to combat the menace of the enemy bomber and submarine.

'At the head of these establishments and among their staff were men of the calibre of Dr Paterson of GEC, and Messrs Blumlein and Browne, of EMI. The death of these members of the EMI team in an aircraft accident while carrying out certain vital experiments was a catastrophe ...'

Marcus Tullius Cicero (106–43 BC) was in no doubt about the pertinent points which must be followed if a reliable history is to be written: 'the first law for the historian is that he shall never dare utter an untruth. The second is that he shall suppress nothing that is true. Moreover, there shall be no suspicion of partiality in his writing, or of malice.'

After the fall of France and the evacuation of the British Expeditionary Force, Churchill said: 'We must be very careful not to assign to this deliverance [from Dunkirk] the attributes of a victory. Wars are not won by evacuation.' Much had to be done to rebuild the UK's defences. New weapons (such as proximity fuse shells[72,73]) would be required in large numbers, together with the scientific means to permit them to be used effectively if success were to be attained. Also during the 'dark days' following the collapse of France, many 'immediate' problems had to be solved. One of these concerned the night landing of bombers following raids on enemy territory. The position was highlighted by Churchill in a memorandum, dated 18th October 1940, to the Chief of the Air Staff (CAS)[74]:

'What arrangements have we got for blind landings for aircraft? How many aircraft are so fitted? It ought to be possible to guide them down quite safely, as commercial craft were done before the war in spite of fog. Let me have full particulars. The accidents last night are very serious.'

Churchill also pursued this issue with the Secretary of State for Air[75]:

'How is it that so few of our bombers are fitted with blind landing appliances? The Minister of Aircraft Production tells me that a number of Lorenz equipments are available.

'The grievous losses which occurred one day last week ought not to be repeated. . . .Pray let me have your observations.'

The Lorenz system was fully developed pre-war for operational commercial airliner service and by the end of 1938 was almost universally employed in the European, South African and Australian regions.

The CAS replied on 25th October[76]: 'There is at present no approved method for the blind *landing* of aircraft, and until one is found the pilot must actually land by seeing the ground itself or a light on the ground.'

He mentioned further that two approved methods of blind *approach* existed. One method, ZZ, used standard radio sets (which were installed in all bombers), and direction finding and radio telephony transmitter sets which existed on the ground. These were operated by trained ground control officers and the sets were available in all bomber stations. However, the success of the method depended very largely on the skill of the control officers who guided the aeroplanes in to land, and the supply of the officers was limited. Only eight aerodromes had such officers, although additional controllers were being trained.

The weakness of ZZ was its inability to operate in visibility of less than 3000 yards because the controller had to see the aircraft at that distance to give the final directions in a time which allowed a pilot to react to them.

The newer method used the Lorenz beam system. A pilot flew along the beam for heading and received automatic warning signals which enabled him to judge the distance to the landing strip. Only eight aerodromes in October 1940 were equipped with the ground installations, including lights, and just under 100 bombers in operational units were fitted with the system.

Fighters could not carry the Lorenz beam apparatus, and, though the pilots might be trained to use ZZ, the Commander-in-Chief did not consider the blind approach technique to be safe for small, fast aircraft. Moreover, fighter aerodromes could not have Lorenz beams for some months, so the newly introduced Beaufighters would have to use bomber aerodromes for blind approaches in the immediate future.

On the point which related to commercial airlines, the CAS noted that a civil airliner pilot had probably accumulated about 20 times the flying experience of the war-trained bomber pilot, and his aircraft operated singly to a schedule. This was not the case with bombers. They could return unexpectedly in large numbers, the CAS observed, and some of them might have damaged equipment, wounded crews, or a shortage of petrol. 'Each must come down in his turn and get clear before the next can be accepted. This takes time. If an early arrival crashes on the landing lane all those waiting in the fog with little fuel [were] in grave danger.'

Actually, the RAF did not have a complete blind landing system in 1940 because such a system had not been given a very high priority in the RAF's call on the limited radio research and development resources available[77]. The position exemplified the need in 1940 for additional radio resources generally and the importance of the Tizard mission.

The Lorenz equipment was for blind approach use only. It was adequate for bringing an aircraft into range of an illuminated landing path provided that this was visible from a height of 50 to 100 ft, and a slant range of 150 yd. For blind landing in conditions of zero or very poor visibility, additional instrumentation, especially an accurate altimeter, was necessary.

Apart from this purpose, a low-level, non-barometric, altimeter was needed for tactical purposes, namely[78]:

1 bombing from a height of 600 to 1000 ft;
2 depth charge attacks from 50 to 100 ft;
3 torpedo releases from 50 to 150 ft;
4 toraplane launching from 500 to 1200 ft;
5 mine laying; and
6 parachute dropping.

Sometime in 1940 EMI began work on a low-level altimeter based on the change of capacitance, between an aircraft and the earth's surface, with the height of the aircraft above the surface. The principle was not new but the method, due to Blumlein, to implement the principle was entirely original.

In the 15th February 1927 issue of *La Nature* a short report was published on an altitude indicator for aircraft. The report stated:

'The US Army Air Corps and the Junkers Company are stated to have evolved independently electrical designs for the determination of the height of an aircraft above the ground. Metal plates connected by a wire are mounted on the bottom wing tips. The two plates constitute one plate of a condenser, the other plate using the earth. The air between constitutes the dielectric. The capacitance (C) of this condenser is very low and increases as the aircraft approaches the ground. If the condenser is connected in an oscillatory circuit these variations in C can be used to produce a signal when the aircraft is within a short distance from the ground, so that the pilot is warned and can take action accordingly.'

H E Wimperis, the Director of Scientific Research, Air Ministry read the report and asked[79] the Royal Aircraft Establishment (RAE) if the W/T Research Section could offer any prospect of apparatus being designed to achieve this purpose. Enquiries[80] showed that the change of capacitance with height principle was first recorded in the Wireless Appendix to the Torpedo Manual for 1916 and that a Flight Lieutenant Chandler and a Squadron Leader Keith had tested apparatus, based on the principle, as an aid to torpedo dropping. Later, in 1920, a Mr Cox-Walker of RAE had carried out a considerable amount of work at Biggin Hill on the method, but the work had not been reported since the 'results were entirely negative'. 'The problem, although not perhaps hopeless appears rather difficult as the changes of C likely to be met with are small and, although quite easily detectable under laboratory conditions, the problem of giving a visual indication in aircraft is likely to be a very difficult matter.'

Further full-scale experiments, in 1928 and 1929 at the Royal Aircraft establishment, using a Bristol fighter 'skeleton' aircraft, were not promising. For heights greater than 15 ft the changes of capacitance were 'so small as to be within the limits of variation of the apparatus used to determine them'. The results obtained were unsatisfactory and it was 'recommended that further work on such an altimeter be discontinued'.

It would seem unlikely that Blumlein knew of the Royal Aircraft Establishment's unpublished work, which was based on the change of frequency of an oscillator with change of tuning capacitance. Blumlein chose to utilise his transformer ratio arm bridge to measure the capacitance between two insulated, widely spaced, conductors in the presence of a third conductor. Figure 14.9a illustrates the capacitances which exist when the two insulated electrodes are mounted at the nose and tail of the fuselage of an aircraft. From the diagram the equivalent circuit of Figure 14.9b follows. This can be transformed to Figure 14.9c in which $\Delta C_0' = C_3 C_4/(C_3 + C_4)$ and similarly for ΔC_1 and ΔC_2. In Blumlein's method[81] $(C_0' + \Delta C_0')$ is measured. This may be rewritten $(C_0 + \Delta C_0)$ where C_0 is a constant capacitance and ΔC_0 is the capacitance which varies as a function of the height of the aircraft above the earth's surface. The change of capacitance is minute compared with the other capacitances of the system and can only be measured by a bridge technique in which these other capacitances can be ignored. Blumlein's double-ratio a.c. bridge using inductively coupled ratio arms possesses this property and it is possible to measure the change of capacitance to the limit where extraneous random capacitance changes and electrical noise prevent further measurements. In practice C_0 is balanced by a zero adjustment when the aircraft is at an altitude (greater than 500 ft) well outside the range of the instrumentation used, and the change ΔC_0 as the aircraft descends is balanced by a second adjustment which can be calibrated directly in feet.

Of the different capacitances, C_1 and C_2 do not enter into the balance condition of the bridge, because they are connected across the source and detector respectively; the bridge is responsive only to changes in C_3, C_4 and C_5 following the zeroing adjustment. The values of these capacitances are very small, being approximately within the range 10^{-14} F to 10^{-18} F, C_1 and C_2 may be one million times larger than these. The basic bridge circuit is shown in Figure 14.10. When C_0 is balanced by C_x, the reading of C_y is an indication of the altitude of the aircraft.

In operation, the out-of-balance alternating potential difference across the bridge is amplified and fed to a phase discriminator. The direct current output of the discriminator is then passed through a milliammeter of special construction. A metal plate is fitted near the end of the scale of the meter so that the pointer and plate form a capacitor, C_y whose capacitance depends upon the position of the pointer. Consequently when the bridge becomes unbalanced, the direct current from the phase discriminator alters the deflection of the meter and hence C_y so as to restore the balance, that is the bridge is self-balancing. The meter is calibrated to read altitude directly.

Measurements of ΔC_0 made on a Lancaster bomber installation are given in Table 14.1.

It is not known whether EMI's low-level altimeter was a private venture of the company or whether a government contract had been awarded to it. Work on the instrument dates from the summer of 1940.

(a)

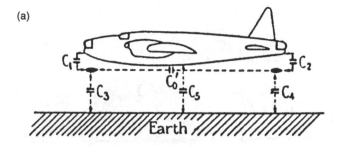

(b)

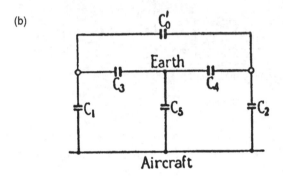

(c)

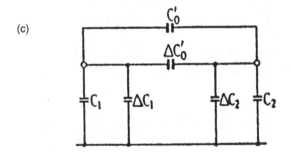

Figure 14.9 (*a*) *Capacitances associated with electrodes on aircraft.* (*b*) *and* (*c*) *Equivalent circuits of electrode capacitances*

Source: *Proc. IEE*, Part III, 1949

Many flight trials were made from early 1941 with the equipment installed in Wellington, Whitley, Halifax, Stirling and Lancaster bombers, and in Sunderland and Catalina flying boats. With the land aircraft some diving tests were carried out, and the altimeter was able to follow a rate of fall of 1000 ft/min without hunting or overswing. 'Landing tests, during which the pilot's vision from the cockpit was obscured, were also undertaken, successful landings being achieved after a glide from 100 ft to 1 ft,

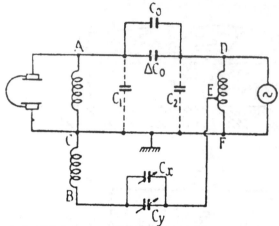

Figure 14.10   A simplified diagram of the altimeter bridge

Source: *Proc. IEE*, Part III, 1949

controlled from readings of the altimeter alone.' In practice, the effective maximum height at which the instrument functioned was about two wing-spans. It worked well in both fine weather and rain.

These favourable results were considered on 21st August 1941 at a conference chaired by Sir Frank Smith, the Controller of Telecommunications Equipment (CTE): it was agreed a tooling-up contract should be placed with EMI 'at once'.[82]

During the late summer of 1941 the necessity for an altimeter for blind landing was 'extremely great'. As no other altimeter, other than the EMI

Table 14.1   Aircraft capacitances

Spacing of electrodes on wings	101 ft
C_0	16 000 μμμF
Conductance	negligible
Zero drift	±3.5 μμμF

Height (ft)	Change of capacitance, ΔC_0 (μμμF)
300	2.2
250	7.5
200	17.5
160	34
100	130
60	390
20	1700
5	4000

type, suitable for this purpose, was at that time in a sufficiently advanced stage of development, there was a view that 'all consideration should be given to accelerating the production of the EMI capacity altimeter'. This urgent requirement led to a tooling-up contract being given to EMI sometime in the early autumn of 1941.

Meanwhile the trials of the Blumlein-designed instrument continued. There seems to have been little to criticise and the conclusions of the trials were consistently favourable[83]:

'the EMI equipment gave very good accuracy very low' (11th August 1941);

'the EMI type had been satisfactorily tested for blind landing' (29th September 1941);

the EMI installation in a Wellington 'was recently tested with satisfactory results at Wyton' (23rd December 1941);

the first model of [the] EMI altimeter 'had just been delivered and had been tested the previous day with good results' (13th January 1942).

These comments are somewhat remarkable when set against the findings of trials on other equipments generally, and other altimeters more particularly, and affirm the soundness of Blumlein's design and the excellence of EMI's engineering.

Contemporaneously with this work, STC was developing a glide path indicator as an aid to aircraft landing in conditions of poor visibility. The company's endeavour led to a comparison[84] being made, at the Bomber Development Unit, Boscombe Down, from the end of 1941, of the two different landing aids. Although STC's equipment was easy to install in aircraft, the total system was more costly than EMI's since the airborne unit could only operate with a ground transmitter which produced a lateral beam. Pilots found the glide path indicator was simple to follow; the behaviour of the aircraft when following the instrument was satisfactory; and sufficient confidence could be entrusted in the indicator to neglect the altimeter during the approach right down to touchdown. The EMI altimeter was bulky, and the installation needed a considerable amount of wiring. Flight tests showed the altimeter was an accurate and sensitive aid to beam approach from the 'inner marker' of the airfield and permitted safe touchdowns to be made: it was accepted for instrument landing purposes.

Of the two aids, the majority of pilots favoured the glide path indicator, but there was a bias in support of fitting both the EMI and STC systems.

With these comments before them, the Radio Aids to Air Navigation Committee, on 30th January 1942, agreed[85] 'there existed operational requirements for both glide path [indicators] and radio altimeters since they were needed for separate purposes', and 'where possible both should be catered for in the layout of future aircraft'. At the meeting, Bomber Command asked that higher priority should be given to the provision of the glide path installations than to the radio altimeters. Furthermore, because 'the probable dates of availability of altimeters [was] so far in the future, Bomber Command stated that they wished all heavies (i.e. Halifax, Lancaster and Stirling bombers] to be provided with the EMI capacity altimeter for operational use next winter (1942–43]....When the TRE Type 4 became available it should replace the EMI in all heavies'.

Events now seem to have dragged on: by 29th June 1942 only 29 altimeters—of which 27 had been allocated—had been completed under EMI's development contract[86]. A further order for 250, given originally on a hand-made basis but later on a semi-tooled basis, was not anticipated to be completed until early January 1943. And the bulk contract for 8000 Type V altimeters (capacity type) was not placed with the Gramophone Company until 17th June 1942[87]. Deliveries were expected to start from the end of October 1942, and from the beginning of November it was hoped they would rise from 25 per week to 250 per week by about April 1943. Previously the CTE had given an assurance that deliveries would commence in the spring of 1942. The delay in placing the contract was felt by the Director of Telecommunications to have 'serious implications' since a decision would have to be made regarding the priority for installation in torpedo bombers (Wellington VIII's and Wellington ICs of Middle East Command), and in aircraft (Halifaxes, Lancasters and Stirlings) of Bomber Command.

It is possible the delay stemmed from the need to assess thoroughly the six types of radio altimeters being progressed in 1941 before a large order was given for one or more of them.

These types of altimeter were:

1 the modified Western Electric Type I altimeter;
2 the Standard Telephones Type II altimeter;
3 the EMI Type V altimeter;
4 the SEI altimeter;
5 the TRE altimeter; and
6 the American altimeter other than Type I.

(Brief details of these are given in Technical note 1, and a review[88] of the available types, at 27th April 1943, is shown in Technical note 2.)

In the USA, the Radio Corporation of America early in the war was developing an altimeter, based on frequency modulation principles, for the US navy. The development came to the attention of the British Air

Commission in Washington, and subsequently the Commission collaborated with the RCA and the USN and observed that the instrument met the RAF's operational requirements. Late in 1942, the altimeter was demonstrated to the RAF by the Commission's engineers and it was found to be superior to the UK's various altimeter developments. The RCA Type AN/APN-1 was adopted by the Allies, and over 30 000 were supplied to the RAF and the Fleet Air Arm[89].

Early in 1943 the Air Staff approached Bomber Command and asked it to reconsider its need for the capacity altimeter since the installation was complicated and the speeds of its aircraft were reduced by 7 m.p.h.[90]. The Air Staff recommendation became official policy in May 1943, except for the fitting of the capacity altimeter into 200 to 350 Wellington VIII and IC aircraft of Middle East Command.

The Middle East demand was 'urgent' in April 1942[91], according to the Assistant Chief of the Air Staff (T). In *The Official Story of Air Operations in the Middle East, from February 1942 to January 1943* (HMSO)[92] the formation of a special squadron of torpedo-carrying aircraft is described.

'By the spring of 1942 the air force over the sea was considerable in numbers, technically well equipped, and efficiently trained.

'It included one Wellington squadron of particular interest, which greatly increased our power of hitting ships at sea. What was wanted was an aircraft that could carry at least one torpedo over many hundreds of miles of sea. Since the only long-range bomber then available in the Middle East was the Wellington, one squadron was given the experiment of fitting torpedoes instead of bombs. The squadron was asked to transform an aircraft that normally bombed from a height of many thousands of feet into an aircraft that could deliver torpedo attacks at night from only a few feet above the sea. . . .

'It was done, largely by men of [squadron no. 38], with a little help from other engineers, on a sandy airfield in the canal zone. The engineering side of it alone meant many changes in the Wellington, for when two torpedoes were hung on it the whole balance of the aircraft was upset. Very soon however the squadron produced a torpedo carrying aircraft that would fly. Then the crews trained themselves to deliver attacks with dummy torpedoes at a small yacht — a difficult technique with which none of them was familiar. . . .Wing Commander J H Chaplin, DSO, DFC, who commanded the squadron, was the man who shaped this new weapon and scored the first success with it.'

Lieutenant M Morgan-Giles, RN, was one of the officers who was engaged in the conversion of the Wellington bombers.

For the above objective an accurate, sensitive and wholly reliable radio altimeter, free from the necessary corrections for temperature and pressure which were associated with non-radio altimeters, was essential, since the torpedo attacks were made from an altitude of just 40 to 50 feet by an aircraft travelling at about 180 m.p.h.

Within three months a long-range torpedo force was added to the RAF's Middle East strength. At first the force appears to have been ineffective. In a memorandum, dated 26th August 1942, to the Chief of the Air Staff, on the use of torpedoes by the RAF, Sir Archibald Sinclair, the Secretary of State for Air noted[93]: '...I must say that I have been depressed myself by our lack of success in recent operations, particularly in the Mediterranean. Our claims are few and not even all of these are substantiated by later information'.

Fortunately this situation did not persist for any appreciable length of time. Towards the end of 1942 the force of Wellingtons and other torpedo-carrying aircraft caused great havoc among the ships which attempted to supply the Axis powers in North Africa.

'Night after night the Wellingtons bombed the ports of Sicily, southern Italy, Tripoli, and Tunisia. The torpedo Wellingtons and the naval Albacores were still roaming the night sky over the sea, searching out the enemy supply ships and sinking them in sheets of flame....In the seven weeks from November 1st [1942], at least 45 ships were sunk or very severely damaged by bombs or air torpedo. Nearly half of that total was hit at sea ...'[94]

The aircraft instrumentation used to effect this destruction is not mentioned in the HMSO booklet. And unfortunately neither Air Commodore J H Chaplin nor Rear Admiral Sir Morgan Morgan-Giles can now (1998) remember the details of the altimeter. However, since

1 it is exceedingly unlikely that a torpedo-carrying aircraft would operate at night over the sea without an accurate low-level altimeter;
2 the EMI/Blumlein capacity altimeter had been tested in a Wellington, at the end of December 1941, and had been found to be accurate and sensitive;
3 in 1942, no other type of altimeter had been flight tested with satisfactory results; and
4 it is known that the EMI altimeter had been dispatched to the Middle East for fitting into Wellingtons,

it may be inferred that it was this altimeter which contributed to the routing of the enemy's supply ships. Without replenishments of fuel and ammunition, Field Marshal Rommel's tanks, vehicles and aircraft could not operate; long-range operations from his bases in Tripoli and elsewhere became impossible and a retreat was inevitable.

A novel use of the capacity altimeter was for the measurement of inter-electrode capacitances in valve factories. During the latter part of the war there was no satisfactory apparatus for the determination of grid–cathode capacitances, of the order of 0.001 μμF, at the speed necessary on a production line. The existing laboratory type of r.f. bridge measurement took several minutes, but when the capacity altimeter, with its self-balancing feature, was adapted for use in valve factories the actual measurement was almost instantaneous. It was only necessary to plug in the valve and observe the reading of an indicating meter. The accuracy was only ±10% because of the logarithmic shape of the meter's scale, but a better accuracy to within ±3% could be determined within a few seconds by balancing the bridge with the linear variable capacitor in the test set.

Technical note 1

Review of radio altimeters, position up to 24th September 1941

(By Radio Department, Royal Aircraft Establishment)

Modified Western Electric Type I

Range 30–1200 ft, holds off at > 3000 ft; total mass 62–70 lb; power consumption 312 W.
Length of feeder to each aerial must not exceed 20 ft.
Approved for Sunderland flying boats only. One altimeter installed at Pembroke dock.
No further development work being undertaken in this country.

Standard Telephones low reading altimeter Type II

Range 20–1200 ft, holds off at > 2400 ft; total mass 60 lb; power consumption 90 W.
Same installation difficulties as Western Electric Type I. Further work on positioning the transmitter required.
Trials satisfactory. One installed on Beaufort at Gosport; one installed at No. 4BAT Flight at Wyton for blind landing trials; one being installed at RAE in a Whitley aircraft. Two on service trials at Pembroke dock for two months.

EMI capacity altimeter Type V

Range 0–100 ft with electrodes fore and aft on fuselage; total mass 90 lb; power consumption 40 W.

Altimeter now passed to the development stage. Expected to take $2-3$ months before developed prototype ready for service trials.

One experimental model available; six more being made, to be followed by 19.

Salford's altimeter

One experimental model constructed. Probable mass 10 lb; power consumption 10 W.

Works on change of radiation resistance of an aerial as it approaches the ground. Aerial mounted from wing to tail to secure necessary length of aerial.

Preliminary tests indicate it might be possible to develop the altimeter to give a reading from $0-c$. 100 ft.

The model so far tested needs experimental work.

TRE 10 cm altimeter

Range c. $10-1200$ ft; total mass $80-90$ lb; power consumption c. 50 W.

It is a development at 10 cm of the frequency change altimeter, on the Western Electric principle.

One experimental model made.

Installation may be extremely difficult. The disadvantage is that the cables must not exceed 6 ft.

American altimeters

No detailed information at present.

Technical note 2

Review of altimeter position at 27th April 1943

American AYD

Range $0-400$ ft, 25 lb, 2.5 A at 27 V; $420-460$ MHz.

No reliable figure for accuracy yet. Almost certainly better than $\pm 10\%$. Latest figure for Wellington installation indicates errors of less than 9 ft.

Difficult to line up and service. Co-axial test gear required. Not voltage regulated. Additional apparatus for voltage regulator appears essential. This will increase mass by c. 20 lb. Appears mechanically and electrically reliable. Installation difficult.

American AYF

Range 0–400 ft, 0–4000 ft in two ranges; slightly heavier than the AYD; probably slightly more power than AYD; 420–460 MHz.

A sample is on its way to RAE for test. No figures of accuracy. American figures suggest accuracy on high rays [*sic*] of not better than at ±50 ft at 1000 ft.

This type has not been tried in this country.

Types II and III

Range 0–1200 ft; 65 lb; 4.5 A at 24 V; 420–460 MHz.

Capable of accuracy of the order of ±30 ft but is unreliable mechanically. It should be noted that results obtained so far on the Type III (improved Type II) indicate an accuracy of the order of 5%, with considerably improved mechanical reliability.

Supply voltage required from 22–30 V. Variable in performance and mechanically unreliable.

Considerable improvement is expected for the next 20 models from RTE and SSC. Test gear and training of personnel is complicated. Installation is difficult. Samples of Type III for test by RAE are expected shortly.

Type V

Indicates from 0–3 times the wing span in feet. Gives accurate reading 0–2 times wing span. Limit depends upon size of aircraft, not much more than 150–200 ft; 80 lb; 2.3 A at 24 V; audio frequency.

Theoretical accuracy is high. In practice probably 5%.

No Service trial figures are available, but this instrument is believed to be reliable. Requires individual calibration for each type of aircraft. Lining up is simple but servicing rather complicated. Regulated between 18–30 V by carbon pile regulator. Limited in application to low range and heavy weight.

Installation difficult.

Type IV

Range 0–150 ft and 0–1500 ft in two ranges; 40 lb; 2.5 A at 24 V; 2400 MHz mean frequency.

Accuracy 0–150 ft is 5%, 0–1500 ft is 12.5%. These figures obtained from TFU Report No. 42.

Supply regulated 22–29 V. Performance given in TFU Report No. 42.
  Appears reliable, and accurate with good constancy of calibration.
  Daily maintenance small. Lining-up requires high degree of skill.
Installation probably easier on account of aerial design.

Radio bomb altimeter

Range 0–500 ft; 4.75 lb; two self-contained 4.5 V dry cells; 2410 kHz.
Designed as an alternative to the flash bomb calibrator, this is not
  strictly an altimeter in the sense that it does not provide a continuous
  indication of height. By recording the time of descent by a stop watch
  the height can be read from charts provided, thus providing spot calibration of the barometric altimeter.

References

1 GLAZIER, E. V. D., and LAMONT, H. R. L.: 'Transmission and
  Propagation' (HMSO, London, 1958), chapter 13, pp. 399–436
2 WATSON-WATT, R A: 'Centimetric waves in RDF', a memorandum,
  March 1942, AVIA 10/54, PRO, Kew, UK
3 YOUNG, P.: 'Power of speed' (George Allen and Unwin, London, 1983),
  p. 72
4 Ref. 3, p. 73
5 'Note on valve development and supply', unsigned, 21st March 1940,
  AVIA 15/648, PRO, Kew, UK
6 BURNS, R. W.: 'Aspects of UK air defence from 1914 to 1935: some
  unpublished Admiralty contributions', *IEE Proc.*, November 1989, **136**, Pt.
  A, pp. 267–278
7 NEALE, B. T.: 'CH—the first operational radar', chapter 8 in BURNS,
  R. W. (Ed.): Radar development to 1945' (Peter Peregrinus Ltd, London,
  1988), pp. 132–150
8 Ref. 5
9 WILLSHAW, W. E.: 'Microwave magnetrons: a brief history of research
  and development' *The GEC Journal of Research*, 1985, **3**, (2), pp. 84–91
10 MEGAW, E. C. S.: 'The high power pulsed magnetron: notes on the
  contribution of GEC Research laboratories to the initial development',
  Report No. 8717, 30th August 1945, 9pp. Research Laboratories of the
  General Electric Company
11 Ref. 5
12 Ref. 5
13 PARKER, P.: 'Electronics' (Edward Arnold, London, 1950), chapter 10,
  pp. 293–334
14 FERRIS, W. R.: 'Input resistance of vacuum tubes as ultra high
  frequency amplifier', *Proc. IRE*, 1936, **24**, (1), pp. 82–104
  North, D. O.: 'Analysis of the effexts of space charge on grid impedance',
  Proc. IRE, 136, **24**, (1), pp. 108–136

15 STRUTT, M. J. O., and ZIEL, A. van der: 'The causes for the increase of the admittances of modern high frequency amplifier tubes on short waves, *Proc. IRE*, 1938, **26**, (8), pp. 1011–1032
16 Ref. 5
17 BOWEN, E. G.: 'Radar days' (Adam Hilger, Bristol, 1987), p. 143
18 ROWE, A. P.: 'One story of radar' (Cambridge University Press, 1948), p. 46
19 Ref. 18, p. 43
20 CLARK, R. W.: 'Tizard' (Methuen, London, 1965), p. 171
21 Ref. 20, p 172
22 Ref. 18, p. 49
23 ESPLEY, D. C., and EDWARDS, G. W.: 'GEC development of centimetric wavelength radar equipment for airborne use', Report No. 8719, 1st September 1945, 5pp, Research Laboratories of the General Electric Company
24 CLAYTON, R., and ALGAR, J. (Eds.): 'A scientist's war: the war diary of Sir Clifford Paterson 1939–45' (Peter Peregrinus Ltd, London, 1991), p. 20
25 Letter: DCD to AMRE, 27th March 1940, AVIA 7/99, PRO, Kew, UK
26 Ref. 23
27 Ref. 18, chapter 7, pp. 54–58
28 Ref. 18, p. 78
29 Ref. 18, p. 79
30 WATSON-WATT, R. A.: 'Three steps to victory' (Odhams Press, London, 1957), chapter 47, pp. 272–275
31 'Cavity magnetron. Messrs Randall, Boot and Sayers', T166/10, PRO, Kew, UK
32 Ref. 10, p. 4
33 Ref. 24, p. 53
34 Ref. 18, p. 78
35 Ref. 24, p. 71
36 Ref. 24, p. 57
37 Dr. PATERSON, D. C.: letter to A. P. Rowe, 18th July 1940, AVIA 7/597, PRO, Kew, UK
38 Ref. 24, p. 72
39 Ref. 24, p. 88
40 Ref. 24, p. 122
41 Ref. 24, p. 168
42 Ref 23, p. 3
43 Ref. 23, p. 4
44 JAY, K. E. B., and SCOTT, J. D.: 'History of the development and production of radio and radar', Part II, pp. 197–198, CAB 102/641, PRO, Kew, UK
45 Ref. 24, p. 61
46 Ref. 24, p. 63
47 Minutes of the 1st meeting of the 'Committee on 10 centimetre wave AI to provide liaison between DCD, MAPRE, RAE, EMI, and GEC', AVIA 7/251, PRO, Kew, UK
48 Note by A. P. Rowe, 18th July 1940, AVIA 7/137, PRO, KEW, UK

49 WALMSLEY, Dr T.: letter to A. P. ROWE, 11th January 1941, AVIA 7/2011, PRO, Kew, UK

50 See AVIA 7/1186, PRO, Kew, UK

51 See AVIA 15/635, PRO, Kew, UK

52 CORK, E. C.: 'Ultra short wave cable measurements', 23rd November 1940, AVIA 7/137, PRO, Kew, UK

53 WALMSLEY, Dr T.: letter to A. P. Rowe, 18th January 1941, AVIA 7/137, PRO, Kew, UK

54 WALMSLEY, Dr T.: letter to A. P. Rowe, 29th January 1941, AVIA 7/2011, PRO, Kew, UK

55 WALMSLEY, Dr T.: letter to A. P. Rowe, 11th January 1941, AVIA 7/2011, PRO, Kew, UK

56 LEWIS, Dr W. B.: letter to Secretary MAP, 24th January 1941, AVIA 7/137, PRO, Kew, UK

57 WALMSLEY, Dr T.: letter to A. P. Rowe, 22nd March, 1941, AVIA 7/2011, PRO, Kew, UK

58 Archives of the Central Research Laboratories, EMI, Hayes, UK

59 Memorandum, ACAS (R) to VCAS, 7th June 1941, AIR 20/1536, PRO, Kew, UK

60 DEE, P. I.: 'The present state of centimetric AI equipment', 13th June 1941, AVIA 7/1084, PRO, Kew, UK

61 Ref. 24, p. 191

62 Ref. 24, p. 237

63 Ref. 24, p. 246

64 BLUMLEIN, A. D.: 'Improvement to klystron (resonator lies partly inside and partly outside vacuum)', British patent no. 574 708, September 1940

65 CALLICK, E. B: 'Metres to microwaves' (Peter Peregrinus Ltd, London, 1990), p. 84

66 Ref. 58

67 BLUMLEIN, A. D.: 'Tests on model voltage doubler modulator', 13th October 1941, EMI Archives

68 Ref. 24, p. 191

69 BLUMLEIN, A. D.: 'The delay-line circuit for supplying power pulses to magnetrons', British patent no. 589 127, October 1941

70 BATT, R.: 'The radar army' (Robert Hale, London, 1991), pp. 53, 55, 76

71 JOUBERT, Air Chief Marshal Sir Philip: 'Radar pioneers', *Daily Telegraph*, 5th September 1945

72 BURNS, R. W.: 'Early history of the proximity fuse (1937–1940)', *IEE Proc.*, May 1993, Part A, 140, (3), pp. 224–236

73 BURNS, R. W.: 'Factors affecting the development of the radio proximity fuse 1940–1944', *IEE Proc. — Sci. Meas. Technol*, January 1996, **143**, (1) pp. 1–9

74 Memorandum, W. S. Churchill to CAS, 18th October 1940, quoted in Ref. 2, p. 533

75 Memorandum, W. S. Churchill to Secretary of State for Air, 20th October 1940, PREM 3/20/4, PRO, Kew, UK

76 Memorandum, CAS to W. S. Churchill, 25th October 1940, PREM 3/20/4, PRO, Kew, UK
77 Report on 'Blind landing system', MAP and AM, CAB 81/22, PRO, Kew, UK
78 Minutes of meeting 'To discuss the performance of various types of radio altimeter and the possibility of introducing it to the Services', 26th September 1941, AIR 14/630, PRO, Kew, UK
79 WIMPERIS, H. E.: letter to Chief Scientist, RAE, 7th February 1928, AVIA 13/291, PRO, Kew, UK
80 Letter, Chief Scientist, RAE, to Secretary Air Ministry, 11th February 1928, AVIA 13/291, PRO, Kew, UK
81 WATTON, W. L., and PEMBERTON, M. E.: 'A direct-capacitance aircraft altimeter', *Jour IEE*, 1949, **96**, Part III, pp. 203–210, 212–213
82 Minutes of CTE's conference, 21st August 1941, AVIA 7/1421, PRO, Kew, UK
83 See AIR 20/2354, AIR 20/3471, AIR 14/630 and AIR 20/3472, PRO, Kew, UK
84 Comparative report on STC glide path indicator and EMI capacity altimeter, January 1942, AIR 14/630, PRO, Kew, UK
85 Minutes of 3rd meeting of the Radio Aids to Air Navigation Committee, 30th January 1942, AIR 14/630, PRO, Kew, UK
86 Minutes of 81st meeting of the Development and Production of Scientific Equipment Committee, 29th June 1942, AIR 20/ 3472, PRO, Kew, UK
87 Minutes of 1st meeting of the Committee on Radio Altimeters, 24th June 1942, AVIA 7/1522, PRO, Kew, UK
88 See AVIA 7/1522 and AVIA 7/1421, PRO, Kew, UK
89 'History of radio engineering under CTS British Air Commission 1940–1945', AVIA 38/1247, PRO, Kew, UK
90 Notes, 1st January 1943, AIR 20/2355, PRO, Kew, UK
91 Minutes of 73rd meeting of the Development and Production of Scientific Equipment Committee, 20th April 1942, PRO, Kew, UK
92 ANON.: 'RAF Middle East' (HMSO, London, undated), p. 50
93 Memorandum, Sir Archibald Sinclair to CAS, 26th August 1942, AIR 8/635, PRO, Kew, UK
94 Ref. 92, p. 142

Chapter 15

The Battle of the Atlantic

In volume 5 of *History of the Second World War*, Churchill wrote:[1] 'The Battle of the Atlantic was the dominating factor all through the war. Never for one moment could we forget that everything happening elsewhere, on land, at sea, or in the air depended ultimately on its outcome, and amid all other cares, we viewed its changing fortunes day by day with hope or apprehension.' The battle caused Churchill more anxiety than any other battle. 'The only thing that ever really frightened me during the war was the U-boat peril', he said.

Churchill's concern was well founded. In the First World War, U-boats sank 4837 ships totalling 11 135 000 tons[2] and early in the Second World War, from June to October 1940, Axis submarines sank 274 ships (of 1 395 000 tons)[3]. It seemed then that history would repeat itself and that grave consequences for the well-being of the United Kingdom and its Allies would arise if the U-boat could not be defeated.

The situation regarding the crucial need for success in the Battle of the Atlantic could be stated simply: imports were needed to feed the population and maintain the war production. If these fell below the levels necessary to sustain the country, then defeat was a possibility. Huge amounts of imports were required. In 1941, for example, estimates showed that more than 36 million tons of dry cargo, and not less than 720 tanker-cargoes of oil had to be imported to maintain the war effort. But with the prevalent shipping losses only 28.5 million tons of imports and 660 tanker-cargoes of oil would be available. By the end of the year there would be a deficit of 7 million tons of supply imports and 2 million tons of food; and the oil stocks would be down to a dangerously low figure of 318 000 tons. Furthermore, the combined efforts of all the UK and the Commonwealth shipyards could not fabricate more than 1 million tons of shipping per year[4].

Victory did not come quickly or easily. As late in the war as March 1943 —43 months after the start of hostilities—Grand Admiral Doenitz's fleet sank 627 000 tons of shipping in one month. In the middle of that month, 40 U-boats attacked two convoys and sank 21 ships for the loss of only one submarine[5]. The British Admiralty subsequently believed that the U-boats came nearest to victory at this point in the war. Yet, just two months later, in May, Doenitz was obliged to withdraw temporarily his U-boats from the North Atlantic because their sinkings were greater than could be tolerated. In a single month, 41 German submarines were destroyed. This figure may be compared with the numbers, 23 and 35, of U-boats sunk during the whole of 1940 and the whole of 1941 respectively[6].

In a report to Hitler, Doenitz wrote: 'We are at present facing the greatest crisis in submarine warfare since the enemy, by means of location devices, makes fighting impossible and is causing us heavy losses.' Later, Hitler ascribed the subsequent stagnation in the U-boat war to a single invention of the Allies. He probably had in mind the centimetric radar set known as Air Surface Vessel (ASV) radar which was a version of H_2S radar. Analysis of the Battle of the Atlantic shows that operational use from about March 1943 of this radar in Coastal Command's aircraft, had the most profound effect in achieving success. The radar led to victory over the U-boats.

In his book *Three Steps to Victory*, Sir Robert Watson-Watt described[7] Blumlein as the 'key man in H_2S'. It was the EMI-engineered radar which was installed in Coastal Command's planes, and as, according to Sir Robert, the size of the H_2S team was 'always small' the role played by Blumlein in the development of this radar set was of central importance.

A full appreciation of the impact of ASV Mark III radar on the war against the submarine can only be grasped from a knowledge of the great difficulties encountered: consequently a brief survey of the measures taken to defeat this formidable weapon will now be narrated.

In 1914 the German fleet included 21 U-boats, but by the end of the war, a total of 373 had been put into service[8]. They caused great havoc among Allied ships. In one week, in September 1916, three U-boats operating between Beachy Head and the Eddystone lighthouse sank 30 ships, even though there were in that region 49 destroyers, 48 torpedo boats, 70 Q ships, 468 armed auxiliaries and a large number of aircraft[9].

From 1st February 1917, U-boats were permitted by the German government to sink, on sight, any merchant ship in the war zone. As a consequence, shipping losses increased alarmingly: in February they were more than 450 000 tons, in March more than 500 000 tons, and in April approximately 850 000 tons. Twenty-five percent of all merchantmen which left Britain were sunk before their intended return[10].

Strangely, the Admiralty was reluctant to introduce the convoy system, but under pressure from Mr Lloyd George (the Prime Minister) and Sir Maurice Hankey (Secretary to the War Cabinet), the Admiralty changed its mind and convoys were gradually organised[11]. The first convoy for inward bound shipping sailed in May 1917, and the first convoy for outward bound shipping was formed in August 1917. The effect was dramatic: losses in convoys fell to 1% of those among independent and during the final 17 months of the war, only 154 ships were sunk in convoy, although 16 657 escorted voyages were made[12]. Furthermore, of these ships, less than six were sunk when the convoy had both air and sea protection[13]. This was one of the most important lessons to be learnt from experience of the First World War, it was to be a lesson which had to be painfully relearnt during the Second World War.

An additional feature of the 1914–18 conflict, which was repeated in the 1939–45 war, was that vast forces, both air and sea, had to be mobilised to protect the convoys. From 1st May 1918 to 12th November 1918 these forces comprised[14]:

Sea	Air
> 300 destroyers and escort vessels	190 land planes
35 submarines	216 sea planes
4000 auxiliary vessels	75 airships

After the cessation of the First World War, Germany, by the Treaty of Versailles, was denied the right to construct or acquire submarines. The spirit of this restriction was partly circumvented in 1922 when the German Submarine Construction Office was founded, in The Hague, with the approval of Admiral Behnke, Commander-in-Chief of the German navy, under cover of a Dutch firm named Ingenieurkantoor voor Scheepsbouw Ltd (I.v.S). This office was directed by a former Chief Constructor of the Germaniaweft, Kiel, and included a German naval representative. Its objective was to provide an efficient U-boat construction staff able to keep abreast of all technical developments by means of practical work for foreign navies. A secret Berlin company, Mentor Bilanz Ltd (MB), provided the link between I.v.S. and the German Admiralty. Later, in 1928, MB was replaced by the new company, Ingenieur Buro für Wirtschaft und Technik, to make preparations for an eventual speedy and effective rebuilding of the German U-boat fleet[15].

Practical experience of U-boat construction was obtained when, in 1927, an agreement was concluded with the King of Spain for the technical section of Mentor Bilanz to build a 750-ton U-boat in Cadiz. A small 250-ton U-boat was built in Finland in 1930 with the permission of the Finnish government. It was the prototype for the first 25 German coastal U-boats to be assembled.

On 16th March 1935, Germany repudiated the Treaty of Versailles. Three months later, on 18th June, a Naval Agreement between Germany and Great Britain was effected, according to which Germany was given the right to possess a tonnage of submarines equal to that of Britain. Preliminary work on U-boat construction was so far advanced as a consequence of these operations that 11 days after the signing of the agreement, that is on 29th June, Germany commissioned in Kiel its first U-boat since 1918. By the end of 1935, 14 U-boats had been commissioned and when the Second World War commenced in September 1939, Germany had 49 operational U-boats and a further 8 were undergoing training and trials. During the ensuing hostilities, an additional 1101 U-boats were fabricated. Doenitz's strategy for German victory was simple. In 1940 he said:[16] 'I will show that the U-boat alone can win this war ... nothing is impossible to us.' He maintained this view throughout the war.

Following the outbreak of war on 3rd September 1939, the Admiralty quickly appreciated the need for convoys. Unfortunately the means available for their defence were totally inadequate.

1 Coastal Command had a few squadrons of Anson aircraft and two squadrons of Sunderland flying boats. The Ansons had a range of only 510 miles and so were used principally to patrol the North Sea; the Sunderlands had a range of 850 miles.
2 No very long-range aircraft, able to give continuous convoy coverage across the Atlantic, were in service.
3 The anti-submarine bombs available to Coastal Command aircraft had a very short delayed-action fuse and successful bombing attacks on surfaced U-boats depended on direct hits being obtained.
4 Airborne depth charges had not been perfected and were not available in sufficient numbers until 1941.
5 The Air Surface Vessel (ASV Mark I) radar which had been developed by September 1939 had detection ranges of just three miles and six miles when flown at 3000 ft and 6000 ft respectively. It was of very low performance and almost useless for anti-submarine work, but was 'better than nothing'.
6 Sea escorts were so scarce that some convoys had just one destroyer or sloop to give defensive cover and, until October 1940, outward bound convoys in the Atlantic could only be escorted to 17° W (c. 300 miles west of Ireland)[17].
7 None of the destroyers or sloops on escort duties had a sufficient endurance to remain with a convoy to Newfoundland. Refuelling at

sea had been neglected by the Royal Navy in peacetime and equipment did not become available for regular use until about June 1942.

8 Radio telephony apparatus for use by ships at sea was non-existent in 1939 and signalling at sea between escort ships and between the escorts and the ships of the convoy had to be either by visual methods or by wireless telegraphy, which was slow in operation.

9 Only two ships (HMS *Rodney* and HMS *Sheffield*) had been fitted with radar by September 1939. (The aircraft carrier HMS *Ark Royal* had no radar when it was sunk towards the end of 1941.)

10 During the First World War, and by 1917, a low-frequency direction-finding network had been established which covered the Mediterranean Sea, the Eastern Atlantic and the North Sea. There were 11 British stations, 11 French stations and 21 Italian stations. By 1929 this network had been reduced to one station, at the Lizard. However, by February 1939, the Admiralty's Operational Intelligence Centre was again in full working order with stations at Flowerdown, Scarborough, Bermuda, Malta, Trincomalee, Aden and Hong Kong

Figure 15.1   Adcock antenna direction-finder and receiver hut

Source: Imperial War Museum

(Figure 15.1). No ships had been fitted with high-frequency direction-finding equipment by the outbreak of war.

11 Much effort been expended from 1917 on the development of an under-water detection system (ASDIC) and considerable reliance in its ability to locate submarines had been engendered from the results of sea trials. Indeed, in the report of the 'Defence of Trade Committee 1936', the following statement occurs: '. . . it is considered that war experience will show that with adequate defences, the operation of sub-marines against merchant vessels in convoy can be made unprofitable'. However, ASDIC had several limitations: its undersea range was only 1.3 km; it could not at first give depth information; it could not detect surfaced U-boats; and it could not provide continuous detection from the time a U-boat was first located to the time of the launch of depth charges. Nevertheless, such was the confidence of the naval staff that in 1937 they reported to the Shipping Defence Advisory Committee that 'the submarine should never again be able to present us with the problems we were faced with in 1917'[17].

12 At the outbreak of war, the Admiralty's encoded and enciphered signals were far from being really secure. By September 1939 the War Office and Air Ministry were using a Type X machine for the trans-mission of secret communications. This was a cipher machine of an improved Enigma type with a Type X attachment: it was a fully safe and efficient machine. The Admiralty did not use this machine, but relied on cipher books for its naval cipher and code books for its administrative code. The former were utilised by officers and the latter by ratings. Regrettably, the German decoding and deciphering service, B-Dienst, was able to read the administrative code traffic extensively in September 1939. (A code operates on complete words or phrases; a cipher operates on single letters.)

When the Second World War commenced, the German Navy had 57 operational U-boats, of which 30 were of the short-range type, suitable only for patrolling duties in the North Sea. Of the 27 ocean-going types, 17 had sailed for the Atlantic in August 1939[18].

At this time, the amount of shipping registered in Great Britain was 21 million tons and comprised about 3000 deep-sea cargo ships and 1000 coasters. The average number of ships at sea on any given day was 2500[19]. For the protection of these ships there were just 150 escort destroyers and a few squadrons of Coastal Command aircraft.

Initially, the U-boats were reluctant to attack ships in convoy, and by the end of 1939 only four escorted ships had been sunk. During the same period 102 independents were sunk. These were ships which had a speed greater than 15 knots or less than 9 knots. The latter ships were too slow for the escorted convoys; the former were less susceptible to attack by submerged submarines.

The fall of France in June 1940 had a profound effect on the Battle of the Atlantic. Admiral Doenitz now had access to naval bases at Brest, St Nazaire, Lorient, La Pallice and Bordeaux. As a consequence, the routes from the U-boat ports in the Baltic to the patrol areas in the North Atlantic were reduced by 450 miles, the U-boats could remain on station for longer periods of time than hitherto, and the Baltic naval ports could concentrate on U-boat construction rather than on refitting, maintenance and construction. In addition, the fall of France allowed the Germans to establish a Focke-Wulf 200 (Kondor) airbase near Bordeaux.

These acquisitions led to heavy shipping losses. In the first two months of their being established in France, the Kondors sank 30 ships totalling 110 000 tons. The U-boats, as mentioned previously, sank 274 ships (1 395 000 tons) in the period June to October 1940.

Doenitz's U-boat operations were aided by the successes of B-Dienst. By April 1940 it was able to read between 30% and 50% of all British naval cipher traffic and it could read the Admiralty's administrative code messages extensively[20]. Moreover, the Merchant Navy Codes and Merchant Ship Code, which were used from January 1940, were a prolific source of intelligence to B-Dienst.

The Admiralty's problems were compounded when Italy entered the war on 11th June 1940. At that time only one ship in the Mediterranean had been fitted with radar. Furthermore, the Admiralty's naval strength had suffered losses at Dunkirk and during the Norwegian campaign, and there was a need to retain a substantial fleet in Home waters because of the threat of invasion.

During the first year of the war, the most urgent British needs were to increase the numbers of escort vessels and Coastal Command aircraft available for convoy protection and reconnaissance duties, to develop and produce reliable radars for airborne and ship-borne use, to break the German naval Enigma messages, to evolve high-frequency direction-finding equipment and to design and manufacture ship-borne radio telephony apparatus.

To meet the first of these requirements, the government in July and August 1939 had ordered 56 Flower Class corvettes. These were escort vessels, based on a whaler, designed by Smith's Dock, Middlesborough, and cost £90 000 each. Originally they were intended for coastal duties, but they were soon employed in the North Atlantic. From about May 1940 the corvettes began to be commissioned in fairly substantial numbers. The limitation of the corvette was its low maximum speed of 16 knots: a surfaced U-boat could outrun a corvette. Nonetheless, for close convoy support, the corvette was a useful vessel: it was not suitable for chasing contacts astern of a convoy. Additionally, 50 old destroyers were obtained from the USA in 1940.

Further strengthening of the Home Fleet's offensive capability came when it received in the last quarter of 1940 substantial reinforcements of radar equipment (including Type 281, a development of Type 280). Type 281 operated on a wavelength of 3.5 m and had a power output of 350 kW[21]. Aircraft could be detected at 100 miles when flying at 20 000 ft and the set could be used for surface and AA gunnery.

In September 1940, Air Service Vessel radar (ASV Mark II) (Figure 15.2) was fitted in two squadrons of Coastal Command's Hudsons (which reconnoitred the North Sea), in Whitleys (for the North Western approaches), and in Sunderlands and Wellingtons (for the Bay of Biscay and the nearer reaches of the Atlantic[22]. The first attack on a U-boat by a Coastal Command aircraft was made on 19th November 1940 when a Sunderland, in the Bay of Biscay, dropped a depth bomb; it failed to

Figure 15.2   The photograph shows the transmitter antenna (on top of the fuselage) and the receiver antennas (mounted on the starboard and port wings and fuselage) of the ASV Mark II radar

Source: Imperial War Museum

explode. The first authentic case of damage to a U-boat occurred on 10th February 1941.

Towards the end of 1941 the number of aircraft engaged on anti-submarine duties was about 400, of which 300 had ASV Mark II radar. Since the number of U-boats at sea seldom exceeded 50 in any one month, the number of aircraft searching for U-boats was about 8–10 per submarine. But even with these forces, success was elusive. The first sinking did not take place until 30th November 1941 (which was 26 months after the outbreak of war) when a Whitley (with ASV Mark II radar) sank U 206 in the Bay of Biscay. This was followed by the sinking of U 451, in the Straits of Gibraltar, on 21st December 1947.

A version of ASV Mark II radar was adapted for ship use and coded Type 286 radar. By midsummer 1940, 97 escorts out of the total of 247 in Home waters had been fitted with this set.

The first ship-borne centimetric radar (Type 271) went to sea in HMS *Orchid*, a corvette, in March 1941. By July 1941, 26 ships had been fitted and by September 1941, it had been accepted for general service. The antennas of these radars were, depending on the type of vessel fitted with 271, 32 ft to 48 ft above sea level. As a result the range of 271 was quite small. A report[23] dated 10th December 1943 showed the ranges, in yards, that could be anticipated (Table 15.1). The conclusion reached was: 'In rough weather (seas 3 or higher) contact can be obtained most easily in the 2000 to 3000 yards range, but even at these ranges, contact cannot be guaranteed'. Considering the vastness of the Atlantic Ocean, the range of 271 was suitable for close convoy support only. The use of 271 by hunting groups of ships necessitated additional information so that the searched areas could be much restricted in extent. This information was to come from two sources; namely, from the breaking of the German naval Enigma ciphers, and from land-based, and later ship-based, h.f. direction-finders.

On 4th March 1941, the Lofoten islands were raided by British commandos and the German armed trawler *Krebs* captured. The material obtained from it enabled the Government Code and Cipher School at Bletchley Park to read the whole of the German signal traffic for February 1941 at various dates from 10th March 1941; for April 1941 at various dates

Table 15.1   Type 271 radar ranges in yards

Ship	All seas	Sea state	
		Seas 0, 1, 2	Seas 3, 4, 5
Corvette	4700	3900	3400
Frigate and destroyer	4600	4500	4100

between 27th April and 10th May; and much of the traffic for May with a delay of three to seven days. Further progress was made when two German weather ships, the *München* and the *Lauenberg*, were captured on 7th May 1941 and 28th June 1941 respectively, and when U 110 was captured on 9th May 1941. These successes allowed Bletchley Park to read the June traffic almost currently and the July traffic currently. To speed up the process of decipherment, theory led to the construction of machines called bombes. By the end of June, the number of anti-Enigma bombes in use at the Government Code and Cipher School was six, of which one was employed for naval work. After the first week of August 1941 to the end of the war, all but two days of Home Waters traffic was read—mostly within 36 hours[24].

Further intelligence about the disposition of Doenitz's U-boats came from the establishment of a network of land-based h.f. direction-finding stations, and from the installation of h.f. direction-finders on board naval ships. When Brest was occupied by the Germans, following the fall of France in 1940, the French Admiralty's radio messages were found, almost intact, including a collection of short-wave bearings taken on the U-boats by British and French direction-finding stations[25]. Also, when the British submarine HMS *Seal* was captured in May 1940, the Germans were able to examine the radio messages which the *Seal* had received from the British Admiralty. These showed that the Admiralty had on a number of occasions supplied warships with the bearings of U-boats. According to the war diary of Doenitz for 23rd January 1940: 'As far as can be ascertained, the enemy's errors in fixing by direction finding vary with the range from the coast. At 300 miles the average error is 60 to 80 miles. Hitherto, the best fix, immediately off the coast of France, was 30 miles out. The worst error amounted to 320 miles at a range of 600 miles'.

With practice and the expansion of the British direction-finding service, the accuracy of position-fixing improved and by October 1940, the U 47 and other 'meteorological' U-boats were reporting that even up to longitude 25° W they were nearly always pursued by British forces within one or two hours of making short signal weather reports. Counter-measures were introduced by the Germans—such as greater radio discipline, changing of radio frequencies, and extensive adoption of a 'short signals' procedure—but these 'seemed unavailing against the efficient British direction finding technique [The Germans] had to assume that all [their] radio messages and short signals would be fixed and probably exploited'.

At the end of 1941, the shore-based network comprised[26]:

16 direction-finding stations (covering the North Atlantic), in the United
Kingdom, Ireland and Newfoundland
5 direction-finding stations in the Mediterranean region
3 direction-finding stations in the West Indies, and
3 direction-finding stations covering the South Atlantic.

During 1941, following the introduction by Doenitz in June 1941 of his
'wolf packs', the use of wireless telegraphy by U-boats before an attack was
as follows[27]:

1 a U-boat sighting a convoy made a high-frequency signal to its base;
2 the U-boat base then ordered, by high-frequency signals, other U-
  boats to join in an attack;
3 the U-boat which made the original report continued to shadow and
  report changes in course and speed of the convoy by h.f. signals;
4 when the wolf pack had gathered, the shadowing U-boat was ordered
  to transmit signals on medium frequencies so that other U-boats could
  be guided to the convoy, even if their navigation had not been
  accurate.

Thus it followed that if the positions of these U-boats could be determined, even approximately, by shore-based direction-finders, convoys
could be diverted from encountering hostile attacks. According to the
Admiralty, evasive routing was, until the end of 1941, 'our most powerful
counter measure'[28]. It was generally effective, but when it failed, 'the
results were apt to be disastrous'. Furthermore, it was obvious that if the
shadowing U-boat could be kept submerged for a few hours, evasive action
by the convoy would not be reported and an attack in force would not
develop. The only ways by which this could be achieved were the use of
either reconnaissance air patrols or pairs of escort vessels, fitted with h.f.
direction-finding equipment, operating as striking forces in the danger
areas.

In July 1941, the first h.f. direction-finding set, Type FH3, was fitted in
a ship, but the results, based on aural reception, were disappointing. A
later version, Type FH4, which had a visual indicator, was commissioned
in October 1941 in HMS *Culver* and was an improvement on Type FH3[29].
Introduction of this equipment (Figure 15.3) into naval service was slow
and it was not until 1943 that there were sufficient sets for two or three
escort ships per convoy. FH4 was to prove of considerable importance in
locating U-boats. In his book *Three Steps to Victory*, Watson-Watt referred to
h.f. direction-finding equipment as being one of the steps[30].

During 1941 low-power high-frequency voice sets were introduced for
convoy escorts. They were used without restriction and permitted the

Figure 15.3   The h.f. direction-finder receiver

Source: Imperial War Museum

commander of the escort group to dispose his naval forces to meet an expected attack, and also to instruct the commodore of the merchant vessels to order a change of course when a threat appeared ahead of the convoy.

Further technical developments in 1941 were the fitting of catapults to some ships and the utilisation of net defences. The catapults enabled a fighter, such as a Hurricane, to be catapulted, by means of 12 rockets, from the deck of a cargo ship, to intercept F-W 200 aircraft. After an engagement the fighter had to be flown to the nearest land or be ditched in the sea. Net defences were first used by the Admiralty in the 1870s to protect capital ships, at anchor in open roadsteads, from torpedoes released by torpedo boats[31]. In the Second World War nets were employed from

August 1941, and by the end of the war, more than 700 ships had been fitted with them. The nets covered between 60% and 75% of a ship's side and reduced its speed by about 17%[32]. Statistics compiled after the war showed that 15 ships were saved by their nets out of 21 ships which had been contacted by torpedoes[33].

Other significant developments in 1941 were the introduction of the Lend–Lease Bill (of March 1941), the development of 'snowflakes', and the denial of bases in the Canary Islands to the Germans. The Lend–Lease Bill permitted the transference to the Royal Navy of ten US Coastguard cutters; allowed British warships to be repaired and refitted in United States yards; enabled British troops to be relieved by US troops in Iceland; and made provision for US naval forces to escort shipping of any nationality to and from Iceland. 'Snowflake', an illuminant, was issued to merchant ships from May 1941, when it was found that star shells gave insufficient illumination at night[34]. Snowflakes were fired by rockets, unlike star shells which were fired from a gun. The flash from the firing of the latter could cause temporary blindness to ships' 'lookouts' and the officers on a bridge. From September 1939, two German supply ships had been based at Las Palmas (in the Canary Islands) and had been used at night to refuel U-boats. Strong protests to the Spanish government led to a cessation of this facility. In May 1941 a German supply ship was sunk, and in June 1941 five others were destroyed.

Throughout 1941 additional naval ships became available from the shipyards. These included 2 aircraft carriers, 10 six-inch gun cruisers, 19 destroyers, 38 Hunt Class destroyers, 69 corvettes and 27 submarines[35].

Despite all these measures, shipping losses continued to be very high and caused much anxiety. For the period 3rd September 1939 to 6th December 1941 the losses were as shown in Table 15.2[36]. (See also Figure 15.4.)

These sinkings were made possible partly by the breaking of the Admiralty's codes and ciphers and partly by the listening gear aboard the U-boats. In the autumn of 1939 it was known that German submarines had sonic listening apparatus only and were without ASDIC type

Table 15.2   Shipping losses (in tons)

Period	British	Others	Totals
3rd September 1939–9th April 1940	339 000	349 000	688 000
10th April 1940–17th March 1941	1 677 000	637 000	2 314 000
18th March 1941–6th December 1941	1 134 000	430 000	1 564 000
Grand totals	3 150 000	1 416 000	4 566 000

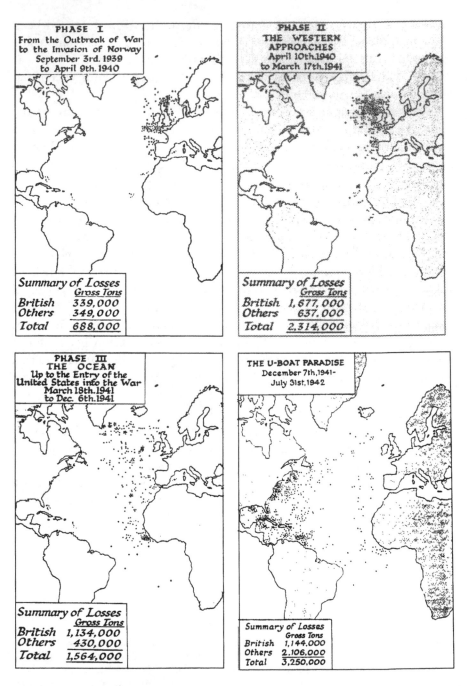

Figure 15.4   Maps showing the locations of Allied shipping losses for various periods of World War II

Source: 'The Second World War', Vols. 3, 4 and 5, by W. S. Churchill (The Reprint Society, London, 1950)

equipment and receivers for detecting supersonic radiations[37]. By the summer of 1940, considerable attention had been given to the subject of ASDICs by the German Headquarters and efforts were being made to find ways to reduce the echoes from their boats. Echo-ranging gear for U-boats, 'S-Gerät', was first mentioned in June 1940. Approximately one year later, in the autumn of 1941, U 570 was captured and renamed HMS *Graph*. It was fitted with a short-range supersonic echo-ranging device, AEG type S-Gerät, intended primarily for the location of mines at distances of up to 400 m. The equipment used a magnetostriction receiver, sharply tuned to a single frequency, but this was not very efficient for the interception of ASDIC (ultrasonic) 'pinging', unless the frequency of the anti-submarine vessel's ASDIC was the same as that of the AEG receiver The *Graph* also possessed sonic listening apparatus known as Gruppenhorchgerät (GHG) — a multi-unit hydrophone system. It comprised 48 Rochelle salt receivers, arranged in a circle, attached to the boat's bottom well forward near the foreplanes. Each receiver included a one-valve head amplifier and was connected to an electrical compensator which enabled the bearing of a continuous sound to be obtained with an accuracy of about 1° at ranges of up to 20 miles under favourable conditions. GHG was fitted to all U-boats and was very vulnerable to depth charge attacks. For the detection of pinging, GHG was considered to be worthless.

Although Coastal Command was much strengthened during the first 33 months of war, its successes were few. Only about 9 U-boats had been sunk by aircraft by the end of May 1942. Doenitz was quite scornful of the efficacy of Coastal Command to sink his submarines and remarked[38]: 'The aircraft can no more eliminate the U-boat than a crow can fight a mole.'

One of the problems which faced aircrews was the difficulty of making a bombing run at night on a surfaced U-boat. The minimum range of ASV Mark II radar enabled an aircraft to approach to within less than one mile from a submarine, but the final run-in had to be made visually in conditions of poor visibility. A solution to this problem was provided when the Leigh light (a high-powered searchlight) was fitted, from May 1942, to some Wellingtons of Coastal Command[39]. By June 1942 four aircraft had been provided with the Leigh light. Some successes were achieved, but the principal effect of the introduction of the Leigh light was to persuade U-boat crews to surface during daylight hours, to recharge their submarines' batteries, rather than risk sudden exposure at night.

During 1942 Rommel captured a serviceable Wellington, fitted with ASV Mark II radar, in Tunisia. He used it with great effect for reconnaissance against Allied shipping. Later it was recalled to Berlin, by the German High Command, for tests[40]. The outcome was the development of a countermeasure to ASV Mark II radar, namely a listening receiver, which was produced by the Paris firm of Metox. This Metox receiver covered the frequency range of 120 to 600 MHz and had a range of about 10 miles. Its bearings on Allied aircraft equipped with ASV Mark II radar were accurate to $\pm°$[41]. The antenna of this German search receiver (FuMB) was originally a demountable vertical wooden frame antenna, mounted in clips at the side of the bridge or on the periscope shaft. Various frame shapes were used, but usually a diamond shape with a horizontal crossbar was used. All frames had separate horizontal and vertical antennas to receive horizontally and vertically polarised transmissions; they had to be unshipped for diving. Later, a permanent type of antenna, resembling a drum on a vertical axis, was fitted on the port side of the conning tower; it could not give any indication of bearing[41].

Following Metox's introduction, Coastal Command's crews began reporting in the latter half of 1942 the phenomenon of 'disappearing radar contacts'. Indeed, at a War Cabinet Chiefs of Staff Committee meeting held on 22nd December 1942, Admiral Sir Dudley Pound, the First Sea Lord, reported[42]: 'Practically no contacts with U-boats were now being obtained by aircraft fitted with 1.5 m ASV ... '.

Until 1943, no very long range aircraft were available to give continuous air reconnaissance cover to the convoys crossing the Atlantic Ocean. There was a gap in the mid-Atlantic, approximately 800 miles from land, which could not be reached by Coastal Command's aircraft, and so shipping losses in this region were particularly heavy. Essentially what was required to circumvent this problem was an unsinkable aircraft carrier. Churchill in December 1942 suggested that a portion of an Arctic ice-field should be detached and towed to the mid-Atlantic, where it could be used for this purpose[43]. However, a few months earlier, in September, an English eccentric, Geoffrey Pyke, had proposed that a material comprising 86% water and 14% wood pulp in a frozen state—which he called Pykrete—should be used to construct a 2 000 000 ton aircraft carrier which would have dimensions of 2000 ft (length) × 300 ft (width) × 200 ft (depth). Pyke named his project 'Habbakuk'[44] after the book in the Old Testament: 'Behold ye among the heathen, and regard and wonder marvellously, for I will work a work in your days, which ye will not believe, though it be told you' (Habbakuk 1:5).

Both Lord Louis Mountbatten and Mr Winston Churchill were ardent supporters of Pyke's notions. Churchill said: 'I attach the greatest importance to the examination of these ideas. The advantages of a floating island, or islands, if only used as refuelling depots for aircraft, are so dazzling that they do not at the moment need to be discussed'[45].

According to Pyke's proposals, the sides and deck of the aircraft carrier would be 40 ft thick. It was considered that this would give the whole interior of the vessel immunity from 'any then known air or underwater attack'. Habbakuk would be able to move at 7 knots, and would carry 200 Spitfires or 100 Mosquito bombers. Its complement would be 3590 officers and men.

Lord Cherwell, Churchill's Scientific Adviser, did not share the Prime Minister's enthusiasm and suggested that a cellular concrete structure should be considered[46]. But Churchill would not be swayed by Cherwell's arguments and in a footnote to Cherwell's memorandum on the subject wrote: 'I have long thirsted for the floating island. . . . Don't get in its way'. Cherwell resigned from Mountbatten's Habbakuk Committee in January 1943.

Although much testing of material was undertaken in this country and abroad, and the ideas of Pyke were subjected to very considerable scrutiny and discussion, nothing came of the Habbakuk project. On 6th July 1943 R Freeman, Chairman of the Main (Technical) Committee noted[47]: 'Successful construction of a ship in Pykrete cannot at present be ensured. To achieve this required further design, experiment and trial'. A month earlier, on 10th June 1943, a committee, with Churchill in the chair, had concluded that ' . . . a ship could not be constructed until 1944/45 and . . . the cost per ship would be about £6 million, instead of less than £1 million as originally estimated'[48].

Actually, by June 1943, the U-boat had been effectively defeated and there was no need for Habbakuk. Moreover, the number of aircraft carriers available for convoy escort duties had increased considerably in 1943. In October 1942, the Admiralty had ordered that six grain ships, then under construction, should be fitted out as merchant aircraft carriers (MACs). Six tankers were also provided with aircraft carrier type decks (about 400 ft by 62 ft), and by the summer of 1943, these ships were in operation. Eventually a total of 15 merchant aircraft carriers were constructed. Their flight decks severely limited the type of aircraft which could be accommodated, but it was found that the older type of Swordfish could operate from them[49]. These planes had caused much damage against the Italian fleet in Taranto and so were far from being useless. Also in 1942–43, some light escort aircraft carriers had become operational.

Table 15.3   Escort vessels required

	British	American
Required	725	590
Available	383	122
Shortage	342	468

Table 15.4   U-boat construction

Date	500 t	740 t	1000–1200 t	1600 t	Others	Total
June 1941	155	34	4	—	34	227
Jan. 1942	145	60	6	—	9	220
June 1942	191	52	17	5	6	271
June 1943	205	40	17	9	—	271
Feb. 1944	137	19	8	4	—	168

In March 1942 a report indicated the numbers of British and American escort vessels that were required[50] (Table 15.3). Additional escorts were needed to counter the increasing number (Table 15.4) of U-boats deployed by Doenitz[51].

Of the submarines in service, 37 were based at Lorient, 25 at Brest, 26 at St Nazaire, 16 at La Pallice and 19 at Bordeaux. When in harbour, the boats were protected by reinforced concrete shelters, the outside walls and roof of which were 8 ft 2 in and 11 ft 6 in thick respectively[51]. They were almost impregnable to Allied bombing attacks.

Among the measures taken in 1942 to defeat the U-boat, mention must be made of operational research — one of the three steps to victory of Watson-Watt. In January 1942, Professor P M S Blackett (who became a Nobel Prize winner in physics after the war) transferred from Coastal Command to the Admiralty and applied the methods of operational research to a variety of topics associated with the Battle of the Atlantic. His section showed that operational research could suggest important changes of tactics. Some of these were:

1 increasing the size of convoys from 32 to 54 ships could lead to a reduction of their losses by 56%[52];
2 increasing the number of escorts from six to nine per convoy could reduce losses by 25%[52];
3 one very long-range Liberator (based in Iceland) in 30 sorties — an average service lifetime could save more than six ships, whereas the same plane used for bombing Berlin would drop less than 100 tons of bombs in its lifetime, which would kill not more than 24 people and destroy a few houses[52];
4 the policy of fitting new ships, of 11.5 to 15 knots, and over 8000 tons, with nets appeared to be profitable in that the value of the cargo and tonnage saved exceeded the gross cost[53];
5 certain positions in a convoy were 6.4 times as hazardous as the remaining positions, and so required greater protection[53];

6 the chance of a 'kill' by an aircraft attacking a U-boat was very much greater when the bomb (Mark II) was released within 30 seconds of the U-boat diving[54];

7 the effectiveness of ASDIC searches could be summed up in the empirical formula $k(1/T_1 - 1/T_2)$ which gives the average number of contacts to be expected in the interval from time T_1 to time T_2 after the U-boat has dived[55].

Nonetheless, notwithstanding the huge air and sea forces employed, the development of airborne and ship-borne radars, the breaking of the German naval Enigma ciphers, the utilisation of land-based and ship-based h.f. direction-finders, the use of h.f. radio telephony, 'snowflakes', nets, Leigh lights, catapult-armed merchant ships, merchant ship aircraft carriers, the evolution of ASDICs able to give depth information, the introduction of operational research, the improvement of depth bombs and anti-submarine weapons, the consideration of new tactics for air and sea attacks on submarines, the development of refuelling at sea, and other matters, the threat to the security of the United Kingdom in March 1943 was, according to the Admiralty, greater than at any other time of the war. The situation could be stated simply: the rate of shipping losses, of 627 000 tons per month in March 1943, was perilously close to the maximum rate of shipbuilding of all the British, Commonwealth and American shipyards combined. If the losses exceeded new construction for a limited period of time, then the Battle of the Atlantic would be lost.

A new detection and location system was urgently required. The ship-borne anti-submarine radar and ASDIC sets in use were short-range systems; the h.f. direction-finders installed on ships, when used together by two or more vessels, could give the approximate location of a U-boat at greater ranges, but during the time taken for an escort vessel to travel to this position, the U-boat would generally be in a different place; and the airborne radar of Coastal Command was now (1943) almost worthless for offensive attacks on U-boats because of the Metox receiver.

The development of the new radar system (ASV Mark III) came about in a rather indirect way.

On 2nd September 1941, Lord Cherwell sent a memorandum and report[56] to the Prime Minister. The report was an analysis of the effectiveness of the Allies' bombing offensive over Germany and showed that '... two-thirds of the [air] crews believed they had found their targets, whereas actually one-fifth had really dropped their bombs within five miles thereof. In full moon, about one-third were within five miles of the target, and in new moon, only about one-twentieth'. Thus, bombers were being shot down and bomber crews were being killed and injured for very little result As Cherwell noted in his memorandum: 'The report makes depressing reading'.

Clearly, a method of navigation was needed which would aid bomber crews in finding their targets. (The GEE and OBOE navigation system did not become operational until the summer of 1942 and the end of 1942 respectively.)

The problem of identifying enemy towns and cities was one which had first engaged the attention of government scientists in June 1940, that is soon after the fall of France. There was a view then, expressed in a paper[57], that the detection at a distance of the magnetic fields generated by power lines, distribution networks and electrical machinery should be possible. Consequently if electronic equipment suitable for use in aircraft, could be developed for this purpose, there was a prospect that towns and cities could be more easily located at night. Tizard, the Chairman of the Committee for the Scientific Survey of Air Defence, was not too sanguine about the idea: 'Whether one will get any more accurate information than one can get by the use of ASV apparatus, I rather doubt, but I suggest that you [Sir George Lee, the Director of Communication Development, Ministry of Aircraft Production] might ask Swanage [the site of the Ministry of Aircraft Production's research establishment] for their views'[58].

Dr W B Lewis, the Deputy Superintendent at Swanage, in a letter dated 12th July 1940, agreed with the conclusions of the paper and opined that it 'should be possible to detect towns at up to 20 km', although any observations obtained would not permit direction-finding to be achieved. He noted that 'some time ago' Dr E G Bowen had made some preliminary observations of electromagnetic radiation in the 5 m to 7 m region, using a cathode ray oscilloscope 'locked' to the 50 Hz mains frequency, and had viewed components of noise which appeared stationary on the screen. These were associated with power stations and towns. 'It [was] not thought that such radiation would be appreciable beyond about two miles from its source, but these observations [differed] from those of the magnetic field in that a rough d/f was possible.' Bomber Command liked the suggestion of employing radio equipment to detect towns and so, in early September 1940, experimental work began at Swanage.

While these discussions were taking place, the Post Office Research Station, Dollis Hill, published, in July 1940, Research Report No. 11240 on 'A method of locating from an aircraft the position of high voltage overhead power lines'. Briefly, its conclusions were, first, an aircraft flying at 5000 ft could detect a power line at a horizontal (ground) distance of one mile; and secondly, an aircraft at 12 000 ft could detect a power line vertically below its position.

These results seemed to offer the hope that enemy power stations could be located by simply following the routes of the high-voltage lines. 'If enemy super power stations could be put out of action a valuable contribution towards the crippling of his activities would result'[59].

Rowe, the Superintendent of the MAP Research Establishment in a letter dated 24th September 1940, was not optimistic about the outcome of their investigation of the method for several reasons. First, the title of the Post Office report was much more ambitious than the subject matter covered, which was concerned with merely the possibility of detecting at a distance the magnetic field from power lines; secondly, the determination of the direction of the field was not possible; thirdly, the regions of the strongest fields were not necessarily those associated with high-voltage lines because current magnitude was an important factor; fourthly, experiments on the use of leader cables as navigation aids, made 'some time ago' at the Royal Aircraft Establishment, showed the method to be impractical; and fifthly, other work on navigational aids was in hand. He concluded: 'It is not felt that this [method] provides any practical means of locating and flying along power lines'.

Of the other methods of detecting towns at night from the air, it seemed in September 1940 that the use of Air Surface Vessel (ASV) radar had some potential. Hanbury Brown, in 1938, while working on radar development at the Bawdsey Research Station, had investigated the reflection of 4 m and 10 m electromagnetic waves from the ground. Using a 1 kW transmitter which fed 2 to 4 μs pulses into a horizontal dipole mounted on an Anson aircraft, Hanbury Brown had observed the output of a tuned receiver as the Anson flew over various types of country. At 4 m, and with the aircraft flying at 2000 ft above the sea, reflections were obtained from Yarmouth at a distance of eight miles; and a medium-sized tanker gave receivable reflections at a distance of five miles. At 10 m, and flying at 1000 ft over open country in Suffolk, a small town could be 'seen' at five miles; and at a distance of seven miles from Norwich large buildings gave three or five times the response of normal returns[60].

Cherwell in his memorandum of 2nd September 1941 referred to this method: 'The use of apparatus similar to ASV to tell when the bomber is above a built-up area is to be examined . . .'[61].

Following receipt of Cherwell's report, Churchill, the next day, wrote[62] to the Chief of Air Staff: 'This is a very serious paper, and seems to require your most urgent attention. I await your proposals for action'. The Chief of Air Staff replied[63] on 11th September 1941 and said the 'use of apparatus similar to ASV for identifying built-up areas' was being investigated. Churchill, who never lacked ideas, wrote again, on 22nd September 1941, to the Chief of Air Staff and suggested the use of radio marker beacons which would be dropped, before dark, in open country near targets. After dark the beacons would 'speak' and reveal their positions to a bomber force[64].

On 26th October 1941 a 'Sunday Soviet' was held at the Telecommunications Research Establishment (TRE) to discuss the question of the detection of towns. But as A P Rowe mentions[65] in his book *One Story of Radar*: 'The day ended sadly, for I recall that we went to our homes tired and without an idea'. Fortunately, this was not the end of the matter, otherwise the Battle of the Atlantic and the Allies' bombing offensive over Germany would have had, in 1943, an entirely different outcome. The next day, the Director of Radar (Air Ministry) wrote[66] to the Director of Communications Development and observed: 'It is understood you are arranging for immediate trials and a report on the possibility of using for this equipment a) our present ASV equipment, if necessary slightly modified, and b) telecommunication receiving equipment which might indicate built-up areas by the reception of fields set-up by machinery in factories, transmission lines, generating stations, etc.'

A week later a TRE report[67], dated 5th November 1941, on the trials which had been completed concluded: 'It is therefore our opinion that the Director of Radar's requirements cannot be achieved in a short time by any of the methods so far envisaged, and furthermore, that it is very doubtful if even a long term research would yield results of operational significance. In conclusion it is felt that enough is now known on the subject to make further such requirement redundant'.

On the same day the Chief of Air Staff in a memorandum[68] to Churchill mentioned that: 'Preliminary trials have already been carried out. I will report further on this when the results are available, but the prospect is not encouraging'.

Churchill sought Cherwell's comments. In a very brief note[69], dated 11th November 1941, Cherwell reported; 'A later report shows that ASV methods are much less discouraging than they seemed when CAS reported'. With this news the Prime Minister maintained his pressure on the Chief of Air Staff[70]: 'Is it true that ASV methods are more favourable by later reports?' he enquired on 14th November 1941. The CAS replied[71] on 17th November 1941: 'The latest reports are still unfavourable. ASV does not yet provide a reliable method of identifying built-up areas, but research and experiments are continuing'.

Cherwell's reaction to this statement was to inform[72] Churchill, on 20th November 1941, of the basis of his (Cherwell's) note of the 11th. 'I understand that CAS, when he says that the latest reports on ASV are still unfavourable, had in mind experiments with the standard equipment. I was referring to experiments, still in an early stage, with a short-wave equipment of the ASV type, which are in fact very promising.' On the same day the CAS[73] informed Churchill: 'I have since heard that a technique based on the same principles as ASV but using a very much shorter (centimetre) wavelength gives most promising results in preliminary tests. ... [A] decision to use centimetre ASV in bombers could only be taken after the most careful consideration of the effects of losing this equipment to the enemy'. When Cherwell referred to centimetric ASV radar he had in mind GEC's Air Interception (AI Mark VII) radar system which operated at the wavelength of 9.1 cm (as described in Chapter 14).

With hindsight it seems surprising that the 'Sunday Soviet' of 26th October 1941 'ended sadly', 'without an idea' on the way forward, given the well known pre-war observations of ground and sea targets with metric ASV radar; GEC's development of AI centimetric radar; the centimetric radar experiments at Worth Matravers which had led to radar returns being received from the town of Swanage as well as from surface vessels in the English Channel; and the 'lead' given by Cherwell, in September–October 1941, regarding the use of Air Surface Vessel radar.

The breakthrough came early in the same week, when Dee and his team, working on applications of centimetric radar, and Skinner and his team, working on basic problems associated with microwave technology, discussed the problem further[74]. Their thoughts led to an investigation of 'town finding' using the GEC centimetric AI radar set at Christchurch airfield.

Within a week of the 'Sunday Soviet' a Blenheim, V 6000, fitted with an early centimetric AI system, was prepared for flight trials. The aircraft had a perspex nose enclosing a parabolic antenna 28 inches in diameter, which rotated continuously about an inclined axis: the beam of the 9.1 cm radiation was approximately 15° wide with its axis inclined at about 10° downwards from the horizontal[75]. On Saturday, 1st November 1941, the first flight (with Dr B J O'Kane of GEC, and G S Hensby of TRE as observers), over Southampton, showed that with this arrangement the reflected radiation from towns gave detectable signals at ranges up to 40 miles. The coverage of the rotating beam in azimuth was limited to ±6° by the nose structure. A range versus bearing display of the ground being overthrown was presented on the screen of the cathode ray tube.

Further flights, during which the c.r.t. displays were photographed, clearly demonstrated that:

1 large towns such as Southampton, Bournemouth and Wolverhampton were easily detectable at distances up to 35 miles from a height of 8000 ft;

2 aerodromes with hangars gave echoes as great as those from large towns;

3 hills gave no observable effects; and estuaries gave the best results at low altitudes.

The photographic evidence caused much excitement. Rowe has recalled: 'When the still wet prints were laid on my table I am remembered to have said: "This is the turning point of the war".'

However, there was a possible difficulty in using centimetric radar. Since aircraft hangars, for example, gave responses as large as those from towns, would an aircrew observer be able—to use Watson-Watt's colourful phraseology[76]—'to "read as he ran" the riddle of the many little green blobs on an otherwise nearly dark radar screen?' O'Kane has recorded that the AI equipment gave 'a very crude picture'. Furthermore, the centimetric equipment was 'somewhat unreliable': 'One quite often spent several days trying to make the equipment work' (O'Kane). Clearly for operational use by the RAF, reliable equipment which gave a meaningful image when operated by non-university Service personnel was required.

Events now moved rapidly. At an important meeting[77] held on 23rd December 1941, chaired by the Secretary of State for Air and attended by some very senior officers of the Ministry of Aircraft Production, the War Cabinet Office, Bomber Command, and the Telecommunications Research Establishment, 'to discuss H_2S' (the code name given to the proposed blind bombing apparatus), the possibility of using the AI type of equipment for blind bombing of towns was first reported. The Secretary of State ordered six flights to 'determine whether the signals obtained in separate flights could be definitely associated with specific ground objects'.

During the discussions the question was raised as to which transmitter valve should be used. It was an issue which was not to be resolved until August 1942. Two high-power valves capable of generating centimetric waves existed, namely, the klystron and the magnetron. Of these, the magnetron was ideal for the intended application but its possible use posed a dilemma. Tests showed the valve to be indestructible to the explosive charges used in aircraft when equipment had to be destroyed. Consequently because the highly secret valve was not known to the enemy, there was a high probability that the Germans would soon recover a magnetron from a crashed RAF bomber, which would then enable them to design and manufacture centimetric Air Interception radar equipment. On the other hand, the principle of the klystron had been published in 1939 but the klystron's power output was smaller than that of the magnetron, and, furthermore, the valve was not yet available because the necessary production techniques had not been finalised. Lord Cherwell opined that the use of the klystron would give a range of 15 to 20 miles, though Dr Dee of TRE doubted whether any evidence existed for this opinion.

On this contentious issue the Secretary of State ruled[77] that any 'klystron production should be treated as additional and nothing should be allowed to hold up production of the magnetrons'. The urgent need to strengthen the effectiveness of Bomber Command's offensive over Germany by the use of H_2S radar with magnetrons led Lord Cherwell to stating that: 'We might have to rob GL [radar sets] for [their] magnetrons'. The alternative, he said, was to develop H_2S in a simplified form 'not as a navigational aid so much as a device to ensure hitting a built-up area as such'. This suggestion, it was felt, could be implemented and put into operation several months earlier than a comprehensive H_2S system and had considerable merit[77].

On 29th December 1941 representatives of RDQ, RDL and TRE visited[78] Boscombe Down (the Aircraft and Armament Experimental Establishment) and inspected the new Lancaster, Stirling and Halifax four-engine heavy bombers which were now being introduced to replace the aging Wellington and Whitely two-engine bombers for the air offensive against Germany. Of these aircraft, the Halifax offered the easiest solution to the problem of fitting a rotating parabolic mirror which was likely to be 30 inches in diameter and 18 inches deep.

Several positions on the aircraft were considered: the front gun turret; the underside of the fuselage forward of the bomb doors; the underside of the fuselage aft of the rear under turret; the wings; and the under turret position. For technical reasons the latter position was the unanimous choice of those present at a meeting[79] held on 5th January 1942 with design engineers of Handley Page.

The next day Rowe wrote to the Director of Communications Development and requested that contracts should be placed with Nash and Thompson, and Metropolitan Vickers for the development and construction of two hydraulically operated, and two electrically operated, scanning assemblies respectively: and with Handley Page for the development and construction of two perspex cupolas.

A few days earlier, on 1st January 1942, Dr (later Sir Bernard) A C B Lovell was ordered by Rowe to take charge of the H_2S programme[80]. In his book *Echoes of War*, written about 50 years later, Lovell states (p. 107): 'Shortly after the Secretary of State's meeting on 23 December 1941 contracts were placed with EMI for the *manufacture* of 50 complete H_2S units' (Lovell's italics). The statement seems to enhance TRE's and

Lovell's roles in the H_2S programme, while denying EMI any part in the *development* of H_2S. The following narrative, based on primary source documents held in the Public Record Office, shows that the earliest operational RAF H_2S system, and the Coastal Command version, ASV Mark III, were *developed and engineered* (with the exception of the scanner and antenna) by an EMI group led by Blumlein. The systems were manufactured by EMI's Gramophone Company. (If a contract had been given to EMI for *manufacture* only, there would have been no need for any involvement of EMI's Central Research Laboratories.)

EMI's involvement in this programme stemmed from a meeting[81], held on 15th January 1942, between Blumlein and TRE representatives. It was 'emphasised that as no apparatus at present [existed] which [had] been constructed specifically for this project [H_2S], it was impossible at this stage to define the electrical units sufficiently well to enable the construction of a production prototype'. Nevertheless, Blumlein felt 'that if a few hundred equipments were required at an early date (as distinct from, and in addition to, full-scale production), the first design might be taken as a basis for [the] collection of components and ordering, providing that the possibilities of some wastage would be tolerable'. On the question of whether a klystron or a magnetron should be used, the TRE representatives said a comparison of the results of some laboratory experiments ('made within the past week') with the results of flight experiments, in which the power radiated had been reduced, indicated there was a 'good prospect' that 'sufficient range' would be obtainable with a 9PK2 klystron. It was recommended by A P Rowe of TRE that Messrs EMI Ltd 'be given a development contract to construct three sets of electrical units suitable for airborne trials of the H_2S project'. The specification would be provided by TRE.

The vexed question of which transmitter valve should be used was raised again at a meeting held on 26th January 1942, and attended by representatives from the Ministry of Aircraft Production, EMI, and TRE, under the chairmanship of Sir Frank Smith, the Controller of Telecommunications Equipment, to discuss the development and production of H_2S. Among the constraints which might affect the progress of the H_2S programme, the following were noted[82]:

1 BTH and MOV had each been requested to plan for an annual production of 25 000 magnetrons;
2 BTH expected to start production in July 1942 and MOV in December 1942;
3 klystron production would 'most likely prove to be a bottleneck' because the manufacturing technique had not yet been developed; and

4 sample Sutton type klystrons which were superior in some respects to the BTH type were available and would probably be satisfactory for experimental work.

With these limitations in mind it was noted/agreed;

1 'a gamble should be taken on the klystron proving satisfactory';
2 the development of klystrons by EMI should start at a once;
3 EMI would undertake the development of experimental H_2S units using klystrons;
4 EMI should be asked to develop 200 sets, by semi-tooled methods, 'at once';
5 the 200 sets should be considered as prototypes of the bulk production of, say, 1500 equipments by EMI;
6 TRE would develop the magnetron units;
7 TRE would undertake trials of the experimental H_2S units using klystrons and magnetrons.

The prospect of an effective navigation and bombing aid being effected in a comparatively short time provided much enthusiasm. The development and subsequent manufacture of a successful H_2S system would have essentially the same effect on the UK's bombing offensive as an increased bomber production. Moreover the cost of an H_2S system even in its complicated form was likely to be equivalent to a small fraction of the construction cost of a bomber. Accordingly the progression of H_2S was to be treated as demanding the highest priority. Indeed, of the many projects being considered by TRE only two others—AI Mark VIII radar and Mandrel—had the same priority rating.

The pressing need for some form of navigation/bombing aid had, as noted earlier, led Lord Cherwell to suggest the development of a simple type of H_2S for bombing use only. If it could be implemented the production and installation problems would be greatly eased, but its operational application would demand a much greater accuracy of a pilot's navigation than that required with the complicated version of H_2S. That is, it was only if that navigation were accurate and the pilot arrived within a few miles of his destination that his bombing objectives could be immediately identified. On the other hand, the complicated H_2S set had the advantage that it would enable the aircrews to select their target(s) with a guaranteed accuracy.

From the Air Staff's viewpoint, the choice between the two systems would be decided from two considerations[83]. First 'the system [would have to be] accurate enough to guarantee that bombs would fall within the industrial or other areas selected as a target'; secondly, 'if possible bombs should fall on an agreed spot in that target, if and when required'. If either of these requirements were met it would be an advance on the current system of operating without a radar aid.

For development purposes, EMI was given an indication—based on the number of valves (of all types) which would be employed—of the complexity of the H_2S system during a discussion[84] held at the end of January 1942, between Blumlein and AD/RPV. The ingenious Blumlein wanted to use 45 valves per equipment to provide all the facilities which he envisaged would be needed; but the TRE view that the number should not exceed 30 prevailed. Agreement was also reached on the intermediate frequency (13.5 MHz) and valves (V R 65) for the i.f. amplifier of the H_2S set. EMI/Blumlein had in mind two Ediswan V.1120s for the modulator but were persuaded that one E.1271 would meet requirements.

Meanwhile, during the period of the above talks on contractual and other matters, RAF aircrews were assessing the performance of the blind navigation apparatus which had given such promising results, in November and December 1941, when operated by O'Kane and Hensby. Since they were both university graduates with a good knowledge of radar and electrical instrumentation, it was to be expected that their observations and interpretations of 'the riddle of the many little green blobs' that appeared on the cathode ray tube screen would be superior to those of untrained inexperienced aircrew members. Consequently tests had to be carried out by aircrews to assess the likely merit of blind navigation equipment when employed operationally.

The flight tests commenced on 5th January 1942 and lasted until 11th February 1942[85]. Two aircraft were used: a Blenheim IV V.6000, fitted with a horizontal scanning system in the nose of the aircraft, and a range—azimuth display; and a Wellington IC T.2968, which contained apparatus giving a true p.p.i. presentation which had been designed for ASV use. The aircraft were flown by five pilots.

Some targets were readily identified from the blips on the screen of the cathode ray tube, but on other occasions known towns remained unidentified. Aircrew comments included the following:

1 'It is obvious that over a known course the operator had no difficulty in homing on one of the blips selected.'
2 'Two unsuccessful attempts were made [to home on to Yeovilton]. It seems that this failure was due to the fact that the run was unfamiliar...'
3 'Visibility was exceptionally good and the track was pinpointed the whole way.'

4 'Operator homed aircraft successfully on Boscombe Down from the South, but was unable to recognise it from East or West.'
5 '... visibility made it impossible to find [objective]'.
6 'The object was to see if the gear would pick up towns. In this it was disappointing.'
7 'It appears that the operators only knew their position over a pre-arranged track.'

From these and other findings the conclusion was drawn that blind navigation apparatus, when used over a track with which the operator was familiar from previous flights, was an aid to navigation. But: 'over unknown country the observers [were] unable to recognise their positions. They [were] also unable to recognise known objects when approaching them from an unusual direction'. Clearly, the apparatus employed was unsatisfactory, since many night raids over Germany would be against unseen or unknown towns and cities, and it was possible that these targets would, on some occasions, be approached from different directions. Much technical development work would have to be undertaken to overcome the limitations of the present apparatus.

Fortunately the genius of Blumlein thrived on challenges. His reputation in 1942 rested not simply on the 100-plus patents which he had produced, but on the solutions which he had provided from 1924 to many complex engineering problems. His work on telephony, electrical measurements, monophonic and stereophonic recording and reproduction, high-definition television (which used techniques and sub-systems congruent to those employed in radar), electronics, transmission lines and antennas, gave Blumlein an experience which then (1942) could not be rivalled by any other person in the UK. Certainly, he was senior, in engineering acumen and experience, to any member of TRE, and in terms of engineering innovation and achievement no member of the establishment could match his proven record.

Blumlein and his small EMI team set about their task with vigour and inventiveness. On 4th February 1942 O'Kane visited TRE for a meeting with EMI and TRE representatives. He later recorded in his diary[86]: 'They [EMI], *as usual*, [authors italics] were full of ingenious ideas using about 100 valves to give distorted waveforms etc.'

TRE's confidence in EMI and Blumlein's team was exemplified during discussions about the types of electrical and scanning apparatus needed for two different applications of centimetric airborne radar; namely, radar (H_2S) for the bombing offensive, and radar (ASV) for the detection of U-boats. Both applications could use the same cupola, scanner and scanner drive, but as regards the electrical equipment either the H_2S units could be used unchanged for ASV purposes, or a separate set of ASV units could be developed.

The choice between the two possibilities would determine the maximum range at which surface vessels could be detected. It was envisaged early in 1942 that the H_2S radar transmitter valve would be a klystron, giving about 1 kW, rather than a magnetron. Such a power output would permit large ships to be observable at a maximum range of not greater than 20 miles. The range on submarines would probably be not more than five miles. However, if a magnetron transmitter valve were used, the corresponding ranges would be about 60 miles, and perhaps 15 to 20 miles. Of the two equipments, the former would comprise four units and the latter six units.

From a technical point of view the ideal would be to develop a dual-purpose set of the six-unit type which could be used unchanged for either H_2S or ASV. This would permit suitably equipped aircraft to be available for operational use by either Bomber or Coastal Commands. The objection to such a proposal lay in the ruling that magnetrons must not be flown over enemy territory. If this ruling were negated, then the EMI development effort could be redirected along the lines needed for the engineering of a six-unit combined ASV/H_2S set instead of the four-unit H_2S equipment.

An alternative suggestion would be to employ the EMI apparatus for H_2S and to develop an ASV set based on the AI Mark VIII design. Five hundred of the AI Mark VIII sets were expected to be manufactured between August and December 1942 by GEC. The main difficulty with this suggestion was to arrive at a satisfactory aircraft installation layout of mounting trays and cabling, that is an arrangement which would allow either radar set to be employed. But if the ASV and H_2S systems were engineered by the same industrial firm, the attractive possibility existed that certain units could be common to both systems and the aircraft installation scheme would be eased.

A third possibility was to have separate independently developed H_2S and ASV sets. Ferranti had a development contract for the design of an ASV radar, but the design was not so closely based upon the AI Mark VIII design that the company could make immediate use of the GEC AI Mark VIII drawings. Moreover, Ferranti had been rather slow in other development work and it was believed that ASV equipment would not be available from them at as early a date as would be possible under the previously mentioned arrangements.

With these factors in mind, Rowe opined[87]: '(a) The case for the [dual-purpose set] programme is overwhelming except for the undesirability of use of the magnetron over Germany. If this decision cannot be revoked, the following recommendations would appear the next best means of expediting the use of 9 cm ASV. (b) A contract should be placed with Messrs EMI Ltd for development and production of 1500 ASV electrical equipments based on the AI Mark VIII design but using as many units as possible in common with the H_2S production. (c) If (b) proves unacceptable due to the time factor, it may be noted that as an interim measure some ASV equipments could be provided by modification of early Mark VIII AI production. It would be necessary to place a contract with Messrs EMI for the same number of display units for common use of H_2S and ASV.'

A meeting to consider the commonality of units for H_2S and ASV was held on 17th March 1942 at TRE. It was attended by Blumlein and Browne of EMI and Dee, Curran and Lovell of TRE. It was agreed the ASV and H_2S systems[88] would comprise seven and five units respectively, and of these, four units would be common to both radars. Both systems would use the same scanner, antenna, cupola and magslip assembly. This scheme was discussed at a meeting held on 24th March 1942, attended by representatives of the Air Ministry, the Ministry of Aircraft Production, TRE (Rowe, Dee and Robinson) and EMI (Shoenberg and Blumlein), and chaired by Sir Frank Smith (Controller Telecommunications Equipment). The wisdom of having a common ASV/H_2S set was agreed. It was noted that the Chief of Air Staff viewed the development of the equipment as a matter of great urgency; that GEC 'could make no serious contribution to ASV without grave interference with the AI Mark VIII programme'; and that 'Messrs EMI would be the most suitable contractor for this work'.

Ten days later Dee, in a progress report, noted that design of the prototype set of electrical units was well in hand at EMI with the exception of the klystron T^2R unit. The research necessary for the implementation of this unit had proved unexpectedly difficult and was only just reaching completion [89].

The great need to determine the procedures necessary to obtain the earliest possible deliveries of the special ASV equipments for bomber type aircraft led to discussions with representatives of MAP and TRE at the Gramophone Company on 14th April 1942. Contracts for 200 H_2S equipments and 1300 H_2S equipments had been placed with EMI and the Gramophone Company respectively, but since the 200 prototype H_2S sets were to be constructed by hand, the arrangement did not conduce to an early delivery date. Indeed Shoenberg had thought EMI would take until Christmas to produce the 200 sets. However, production on a limited tooling basis would result in quicker deliveries than could be obtained by fully hand-made methods, even were sufficient skilled labour available.

With this method of production, there was a risk that modifications to the designs would be needed from time to time as testing progressed. If the hand-made policy were continued, design changes could be accommodated easily prior to tooling-up. Additional bureaucracy would be a feature of such a limited tooling production procedure. The Director of Radio Production would have to obtain from the Controller of Telecommunications Equipment his agreement that the Director of Contracts be notified that the Ministry of Aircraft Production would accept responsibility for the cost of such components as they were ordered even if later they proved to be of an incorrect type when the design was finalised. Furthermore, a named officer from TRE would have to certify the transfer of design information, including changes, from EMI to the Gramophone Company.

Given the urgency of the ASV/H$_2$S programme it was agreed that the DRP and the DC would cancel the existing contracts with EMI and the Gramophone Company and issue a new contract to the Gramophone Company for 'the manufacture and a supply of 1500 complete ASV Systems, i.e. 1500 of each unit required for ASV/H$_2$S equipments'.

All of this referred to the fully developed and engineered ASV/H$_2$S set. However, Lord Cherwell's suggestion of a simpler apparatus, if operationally successful for bombing purposes only, could be manufactured and introduced into service much more quickly than the apparatus which EMI was developing.

Trials using two types of 'split aerial system' were conducted by O'Kane and Hensby early in 1942. The two antennas used comprise[90]:

1 two 4λ paraboloids (c. 15 inches diameter) mounted side by side at a small angle, and
2 one 4λ paraboloid carrying a dipole which could be rotated eccentrically about the axis of the mirror.

The performances of both antennas were 'roughly the same'; and the limitation of the scheme was such that only ground objects within azimuth limits of $\pm 20°$ with respect to the axis of the aircraft could be displayed. Within these limits the maximum range (c. 10 miles) of the apparatus was on average about one-third of that of the more complex scanning system. An AI Mark VII radar with the addition of a linear time base and antenna switch provided the electrical units.

Approximately 20 flights at heights up to 10 000 ft over well known targets in the Salisbury area were made. The main conclusions reached were:

1 the H_2S scanning system offered the likelihood of successful target selection and accurate location with some possibility of selective bombing within the target area;
2 the H_2S split antenna system was unlikely to provide successful bombing of a specified target unless a very high navigational accuracy were available by other methods, and successful selective bombing within the target area was most improbable;
3 the H_2S scanning system could be used for an ASV radar application, but the alternative system had insufficient range;
4 the H_2S split antenna system might lead to the bombing of false targets placed by the enemy in open country, whereas the scanning system, which provided a map of an extended region, avoided this defect;
5 the range of the H_2S split antenna system would be seriously reduced if a klystron were employed, since the range obtainable even with the magnetron was already near the operational minimum.

These cogent points could lead to only one conclusion[91]: the balance of advantage in favour of the scanning system as against the split antenna system was 'clear and big', and all 'efforts should be concentrated forthwith at the highest priority on the scanning system of H_2S'.

Work on the installation of H_2S type equipment into the two Halifax four-engine bombers commenced shortly after 27th March 1942 when the first of these aircraft, V 9977, landed at Hurn aerodrome. Handley Page had fitted a perspex cupola in the under turret position, and all was now ready for the assembly and interconnection of the various electrical units and scanning antenna into a coherent system. The aircraft had to be cabled and fitted with racks to take the units provided by TRE. Of these, the centimetric unit was the magnetron transmit–receive system currently being manufactured by GEC for the AI Mark VII radar. Since the model shop facilities at the GEC Wembley research laboratories were far superior to anything that existed in TRE[92], a large proportion of the engineered units (with modifications by TRE) of this radar formed the basis of TRE's experimental H_2S equipment. The essential new development by TRE was the provision of circuits to enable the p.p.i. time base to be synchronised to the rotating Nash and Thompson scanner[93].

TRE's task in the early months of 1942 was constrained by the available AI Mark VII units and their objective was simply to reproduce the Blenheim V 6000 results using the Halifax with its all-round-looking scanner and p.p.i. type display.

Approximately two weeks later, on 12th April 1942, Halifax R 9490 landed at Hurn. The first flight test, on 17th April, using the V 9977

Halifax was not a success. 'The [TRE] equipment worked but very poorly. At 8000 ft altitude the range on towns was only four to five miles and the gaps and fades in the p.p.i. pictured implied that there must be something horribly wrong with the polar diagram of the scanner[94]. Strenuous efforts were made to improve the magnetron H_2S in this aircraft but 'little progress' was effected. Lovell, on 22nd May, flew in the Halifax. 'Very depressing; picture is extremely bad at the moment', he noted in his diary[95]. This bleak position did not augur well for the EMI equipment, which used the lower power output klystron valve rather than the magnetron valve as the transmitter.

Churchill, on 23rd April 1942, was informed that two experimental sets had been fitted in Halifax aircraft, and airborne trials were already being carried out. 'They [were] proceeding satisfactorily' he was told. Actually only one set had been fitted.

With typical enthusiasm, Churchill, a few days later, wrote[96] to the Secretary of State for Air: 'I hope that a really large order has been placed and that nothing will be allowed to stand in the way of getting this apparatus punctually. If it fulfils expectations it should make a big difference in the coming winter'.

Following the depressing flight trial of 22nd May it seemed that some restraint on the optimism everyone felt concerning H_2S should be expressed. The Secretary of State for Air was given a realistic assessment of the situation by Llewellyn, of the Ministry of Aircraft Production:

'I, however, think that it is as well not to be too optimistic about the early arrival of H_2S. The instrument is still only in its early development.

'We shall do well if we produce a small number of hand-made sets which you can use in bombers this year. . . .

'I had a long talk with Dr [*sic*] Shoenberg of EMI Ltd this morning [28th May 1942] and am satisfied that both that Company and TRE are pressing forward with this apparatus to the greatest extent possible with something so new in which accurate working is essential.'[97]

The requirement for H_2S/ASV was imperative. Even in its unsatisfactory state, as exemplified by the V 9977 flight demonstrations, it was thought to have some worth. From the Air Staff's viewpoint the imperfect equipment meant that their original expectations had to be tempered by reality. Now, 19th May 1942, the requirement was for a system which fulfilled the following conditions[98]:

1 'That the system should be accurate enough to guarantee that bombs would fall within an industrial or other area selected as a target.'

2 'That the Air Staff would be satisfied in the first instance if the range of the device enabled the aircraft to home on a built-up area from 15 miles at 15 000 ft.'

Thus the prime need for H_2S was as a bombing aid rather than as a navigation aid. 'It was agreed that details in design to enable it to be used as a navigational aid to determine a specific area or target could be incorporated during the later stages of development and operational trial.'[98]

The restricted Air Ministry directive was demonstrated successfully in May 1942. On 14th May 1942 EMI's prototype apparatus, after bench testing at the Hayes works, was taken to Hurn aerodrome by Trott and Cutts, of Blumlein's team, for installation in Halifax R 9490. The aircraft had been fitted with an electrically operated scanner designed and manufactured by the Metropolitan Vickers Company. In addition R 9490 had been partly cabled and fitted with mounting trays for the assembly of the H_2S system. The completion of the assembly required a further week after which Trott and Cutts returned to Hayes.

At the end of May 1942 TRE evacuated its various buildings at Worth Matravers and Hurn airport and re-established itself in Malvern and at Defford aerodrome[99].

On 26th May Trott, Cutts and White, of EMI arrived at Defford to engage in flight trials. Bad weather—frequent heavy showers and squalls—made immediate flight testing impossible and opportunity was taken to complete one or too small matters in the installation. The inclement weather led to certain parts of the apparatus (particularly the receiver) and some of the cabling being drenched with rainwater, which leaked in through the centre gun turret. Consequently a few more days were spent in rearranging these parts of the equipment[100].

The first flight trial, with Halifax R 9490 operating at 10 000 feet, took place on 2nd June 1942, the observers being the three EMI engineers and O'Kane, who was now EMI's liaison with TRE. 'In general, the [EMI] apparatus functioned correctly and, apart from alterations which [EMI] already had in mind for the prototype models, there seemed little to criticize'. The range, however, was poor, being about six miles maximum, and the polar diagram seemed unsuitable as there was a marked gap at about two miles. This had been expected, as the TRE equipment in the other Halifax, V 9977, had suffered from this defect, and some work had already been done by TRE towards designing a suitable modification of the reflector. (The design/specification of the cupola, scanner and antenna was the responsibility of TRE.)

The need to remove the scanner from the aircraft and incorporate a modification provided an opportunity for the EMI engineers to recheck carefully some high-frequency adjustments and to replace the crystal detector, and White was able by 6th June to improve the signal-to-noise ratio by 9 dB. The following day, since the scanner was not in working order, an opportunity was taken to increase the power output of the klystron transmitter. EMI's test gear now gave a reading of 750 instead of the 225 which had been measured previously.

In an enthusiastic report to Churchill, dated 5th June 1942, Lord Cherwell wrote[101]:

> 'Preliminary trials with H$_2$S (admittedly on a Blenheim instead of a heavy bomber) have been extremely satisfactory. Flying above 10/10 cloud without any sight of the ground for one and a half hours, the bomber made 10 attacks on a pre-determined site at Southampton. His track was chartered by 3 GL [gun-laying radar] sets on the ground and the point marked at which he signalled that he proposed to drop the bomb. Knowing his height, it could easily be calculated just where the bomb would have fallen. All 10 would have dropped within five furlongs (from Charing Cross to Westminster Abbey) of the target.
>
> 'When we remember that, according to Justice Singleton, only a fraction of our bombs at present drop within five miles of the target, the tremendous improvement is self-evident. To use this apparatus would be equivalent to multiplying a bomber force by a large factor. The scientific men have done an excellent job of work, but I have the impression that time could be gained if some extra pressure could be put on in other directions, e.g. if they could telephone directly to somebody such as Renwick, who could cut red tape, and get comparatively small things done without delay.'

(The reference to Sir Robert Renwick, who had responsibility, in the Ministry of Aircraft Production, for the RDF Chain and Gee, stems from the request made to him, at the end of May, that he should add to his workload 'the task of accelerating the production of and rendering monthly reports on: the Air Position Indicator, H$_2$S, Beam Approach Equipment, Long Range Track Guide Beams, Radio Air Altimeter, and Monica'. It was felt that in the field of production Sir Robert's 'qualities of drive and tact and his engineering skill and resources [were] of especial value'[102].)

Churchill was gratified to receive the good news from Cherwell. In a communication dated 7th June 1940, to the Secretary of State for Air, he stated[103]:

> 'I have learnt with pleasure that the preliminary trials of H$_2$S have been extremely satisfactory. But I am deeply disturbed at the very slow rate of progress promised for its production. Three sets in August and 12 in November is not even beginning to touch the problem. We must insist on getting at any rate a sufficient number to light up the target, by the autumn, even if we cannot get

them into all the bombers, and nothing should be allowed to stand in the way of this.

'I propose to hold a meeting to discuss this next week and to see what can be done. The relatively disappointing results of our second big raid makes it doubly urgent.'

Sadly, the welcome news about H_2S was gravely marred by the most awful tragedy which occurred on 7th June. On that day Halifax bomber V 9977 crashed in the Wye Valley with the loss of life of all 11 occupants on board. Blumlein was one of those killed.

References

1 CHURCHILL, W. S.: 'Closing the ring' (Cassell London, 1954), p. 20
2 ANON.: 'The Battle of the Atlantic' (HMSO, London, 1946), p. 8
3 HEZLET, Vice-Admiral Sir A.: 'The electron and sea power' (Peter Davies, London, 1975), p. 198
4 HINSLEY, F. H.: 'British Intelligence in the Second World War' (HMSO, London, 1955), Vol. 2
5 Ref. 3, p. 231
6 BURNS, R. W.: 'Radar development to 1945' (Peter Peregrinus Ltd, London, 1988), p. 260
7 WATSON-WATT, R. A.: 'Three steps to victory' (Odhams Press, London, 1957), p. 408
8 PRICE, A.: 'Aircraft versus submarine' (Janes, London, 1973), p. 29
9 MACINTYRE, D.: 'The Battle of the Atlantic' (Methuen, London), p. 19
10 Ref. 2, p. 19
11 LLOYD GEORGE, D.: 'War memoirs' (Odhams Press, London, 1938), p. 677
12 Ref. 11, p. 703
13 Ref. 2, p. 31, and Ref. 9, pp. 21-22
14 Ref. 9, p. 24
15 HESSLER, Fregatten Kapitan G.: 'German U-boat building policy, 1922-1945', ADM 186/802, PRO, Kew, UK
16 Ref. 2, p. 5
17 ROSKILL, Captain S. W.: 'The war at sea 1939-45' (HMSO, London, 1954), Vol. 1
18 Ref. 4, Vol. 1, p. 333
19 Ref. 9
20 Ref. 4, Vol. 2, p. 635
21 COALES, J. F., and RAWLINSON, J. D. S.: 'The development of UK naval radar', chapter 5 in Ref. 6
22 BOWEN, E. G.: 'Radar days' (Adam Hilger, Bristol, 1987), p. 112
23 ANON.: 'Operational use of radar 271 in anti-submarine warfare', ADM 219/29, PRO, Kew, UK
24 Ref. 4, pp. 337-338

25 ANON.: 'The U-boat war in the Atlantic', ADM 186/802, PRO, Kew, UK

26 Ref. 3, p. 202; and Ref. 25

27 ANON.: 'Use of ships fitted with H/F D/F in the Western Approaches', ADM 220/69, PRO, Kew, UK

28 ANON.: 'The Battle of the Atlantic', ADM 223/220, PRO, Kew, UK

29 ANON.: 'H/F D/F in convoy protection', ADM 220/69, PRO, Kew, UK

30 Ref. 7

31 BURNS, R. W.: 'The background to the development of early radar, some naval questions', chapter 1 in Ref 6

32 ANON.: 'Admiralty net defence', a report, 14th January 1943, ADM 219/68, PRO, Kew, UK

33 Ref. 2, p. 34

34 Ref. 2, p. 32

35 ANON.: 'U-boat warfare, 1940–45', ADM 199/2496, PRO, Kew, UK

36 CHURCHILL, W. S.: 'The Grand Alliance' (Cassell, London, 1952), pp. 129–131

37 ANON.: 'Listening gear aboard U-boats', ADM 219/562, PRO, Kew, UK

38 Ref. 2, p. 47

39 JOHNSON, B.: 'The secret war' (BBC, London, 1978), pp. 214–218

40 Ref. 21, p. 113

41 Ref. 35

42 Minutes of a conference held on 22nd December 1942, War Cabinet, Chiefs of Staff Committee, PREM 3/10, PRO, Kew, UK

43 CHURCHILL, W. S.: memorandum to General Ismay for Chiefs of Staff Committee, 7th December 1942, PREM 3/216/6 PRO, Kew, UK

44 Press release: 'Habbakuk', 28th February 1946, pp. 1–4, Admiralty, K. 17052, Imperial War Museum, London

45 Ref. 43

46 CHERWELL, Lord: memorandum to the Prime Minister, 10th December 1942, PREM 3/216/6, PRO, Kew, UK

47 Report of the Main (Technical) Committee, 6th July 1943, pp. 1–7, PREM, 3/216/2, PRO, Kew, UK

48 Minutes of a meeting held on 10th June 1943, pp. 1–2, PREM 3/216/2, PRO, Kew, UK

49 Ref. 2, p. 51

50 Ref. 17, Vol. 2, p. 9

51 Ref. 35

52 Ref. 9

53 Ref. 32

54 ANON.: 'Analysis of attacks with Mark 24 mine on U-boats', 19th November 1943, ADM 219/66, PRO, Kew, UK

55 ANON.: 'Asdic search for U-boats' 5th December 1943, ADM 219/57, PRO, Kew, UK

56 CHERWELL, Lord: memorandum to the Prime Minister, 2nd September 1941, PREM 3/10, PRO, Kew, UK

57 ANON.: 'Detection of magnetic fields generated by power lines, distribution networks and electrical machinery', c. June 1940, AVIA 7/195, PRO, Kew, UK

58 TIZARD, Sir H.; a letter to Sir George Lee, 3rd July 1940, AVIA 7/195, PRO, Kew, UK

59 WALMSLEY, T.: letter to A P Rowe, 12th September 1940, AVIA 7/195, PRO, Kew, UK

60 BROWN, R. H.: 'Notes on the reflection of 4 m and 10 m waves, observed by aircraft during flight' 26th September 1940, AVIA 7/195, PRO, Kew, UK

61 Ref. 56

62 CHURCHILL, W. S.: memorandum to the Chief of Air Staff, 3rd September 1941, PREM 3/10, PRO, Kew, UK

63 Chief of Air Staff: memorandum to the Prime Minister, 11th September 1941, PREM 3/10, PRO, Kew, UK

64 CHURCHILL, W. S.: memorandum to the Chief of Air Staff, 2nd September 1941, PREM 3/10 PRO, Kew, UK

65 ROWE, A. P.: 'One story of radar' (Cambridge University Press, 1948), pp. 116–117

66 TAIT, Air Vice-Marshal V. H.: memorandum to the Director of Communications Development, 27th October 1941, AVIA 7/195, PRO, Kew, UK

67 PRIEST, D. H.: memorandum to the DCD, 5th November 1941, AVIA 7/195, PRO, Kew, UK

68 Chief of Air Staff: memorandum to the Prime Minister, 5th November 1941, PREM 3/10, PRO, Kew, UK

69 CHERWELL, Lord: memorandum to the Prime Minister, 11th November 1941, PREM 3/10, PRO, Kew, UK

70 CHURCHILL, W. S.: memorandum to the Chief of Air Staff; 14th November 1941, PREM 3/10, PRO, Kew, UK

71 Chief of Air Staff: memorandum to the Prime Minister, 17th November 1941, 3/10 PRO, Kew, UK

72 CHERWELL, Lord: memorandum to the Prime Minister, 20th November 1941, PREM 3/10, PRO, Kew, UK

73 Chief of Air Staff: memorandum to the Prime Minister, 20th November 1941, PREM 3/10, PRO, Kew, UK

74 Ref. 65

75 TRE memo No. 12/106, 'H$_2$S', 23rd April 1942, DERA, Malvern, UK

76 Ref. 7, p. 404

77 Minutes of a meeting held on 23rd December 1941 to discuss H$_2$S, AIR 19/303, PRO, Kew, UK

78 DEE, P. I: letter to DCD, 9th January 1942, AVIA 7/3574, PRO, Kew, UK

79 Meeting with Design Engineers of Handley Page, 5th January 1942, AVIA 7/3574, PRO, Kew, UK

80 LOVELL, Sir B.: 'Echoes of war. The story of H$_2$S radar' (Adam Hilger, Bristol, 1991)

81 Memorandum on 'H$_2$S', 16th January 1942, AVIA 7/3574, PRO, Kew, UK

82 Notes of meeting held on 26th January 1942, AVIA 7/3574, PRO, Kew, UK

83 AIR 19/303, PRO, Kew, UK

84 WALMSLEY, T.: letter to P I Dee, 3rd February 1942, AVIA 7/3574, PRO, Kew, UK

85 'Summary of comments by aircrews on the present performance of "B N'"', 26th February 1942, AVIA 7/3574, PRO, Kew, UK

86 O'KANE, B. J: wartime diary, archives of the Imperial War Museum, London, UK

87 ROWE, A. P.: memorandum to DCD. 16th March 1942, AVIA 7/3574, PRO, Kew, UK

88 Notes of a meeting at TRE, 17th March 1942, AVIA 7/3574, PRO, Kew, UK

89 DEE, P. I: 'Progress report', 26th March 1942, AVIA 7/3574, PRO, Kew, UK

90 Ref. 75, p. 5

91 SKINNER, A. H. M.: memorandum, 23rd May 1942, AIR 19/303, PRO, Kew, UK

92 Ref. 80, p. 57

93 Ref. 80, p. 101

94 Ref. 80, p. 103

95 Ref. 80, p. 108

96 CHURCHILL, W. S.: letter to the Secretary of State for Air, May 1942, AIR 19/303, PRO, Kew, UK

97 LLEWELLIN, J. J.: memorandum to the Secretary of State for Air, 28th May 1942, AIR 19/303, PRO, Kew, UK

98 Ref. 80

99 Ref. 65, chapter 14, pp. 128–134

100 'Flight trials', 1942–43, AVIA 7/1678, PRO, Kew, UK

101 CHERWELL, Lord: memorandum to W S Churchill, 5th June 1942, PREM, 3/10, PRO, Kew, UK

102 Air Ministry to the Prime Minister, 31st May 1942, PREM 3/10, PRO, Kew, UK

103 CHURCHILL, W. S.: memorandum to the Secretary of State for Air, 17th June 1942, PREM 3/10, PRO, Kew, UK

Chapter 16

The crash and its aftermath

On Saturday, 6th June 1942, Blumlein and his senior staff met Dee, Lovell, Curran, O'Kane and Squadron Leader E C Sansom (a specialist navigation officer who had been at Bomber Command Headquarters but who, now, from 28th May 1942, was stationed at TRE, Defford) for talks on points relating to the prototype H_2S equipment. The next day Blumlein and his colleagues, F Blythen and C O Browne, travelled from the hotel in Tewksbury, where they had been staying, to the airfield at Defford. According to Curran the events which then unfolded are as follows[1]:

'I found myself senior scientist on the testing team with Geoff Hensby there as on several similar previous tests. From EMI came Blumlein, Brown [*sic*] and Blythen. Geoff Hensby's in-built experience made him essential to the team and Brown and Blythen wanted to go, partly to see for themselves conditions of operation and performance of H_2S. The units were a mix, some built by EMI and some by TRE. These three and me, together with seven RAF crew members, meant the allowed total of 11 was reached. Then out of the blue Blumlein said to me he would love to go along. I felt I had to give him my place and, further, offer my flying suit as time had virtually run out, adding that I had seen an H_2S system not unlike the one to be tried. So off went all 11 of them ...'

O'Kane's recollections of Sunday, 7th June, differ slightly from Curran's account[2].

'There was, I would confirm, no question of anyone "giving up a place" as the flight was laid on for the "benefit" of Blumlein and his colleagues. Unlike the other EMI team members, White, Trott and Houchin, I do not think they had flown with any version of the system. It was also Squadron Leader Sansom's first flight after his arrival from Bomber Command. I have always regarded myself as lucky to be alive as I had intended to help Hensby to explain what was going on but decided not to do so because I considered there were enough people on the aircraft already. I do not recall anyone trying to limit numbers. Sam [Curran] may well have intended to fly but certainly, in my view, did not stay on the ground in order to let Blumlein do so.'

Lovell[3] recalled that on the Saturday evening he flew, with Hensby, from Defford and observed responses from Gloucester, Cheltenham and several other towns which were clearly displayed at greater ranges than he had seen before. Consequently, 'it was natural that, on the Sunday, the EMI team decided to see for themselves how the magnetron equipment was performing'.

On the fateful day, shortly before 3.00 p.m. Halifax V 9977 taxied to the end of the runway to await take-off instructions. From the control tower, a Mr D Moseley received an urgent phone call stating that the aircraft had a power failure to the radar set. With Mr R Hayman, an inspector, he went to the aircraft and, with engines still running, found a faulty cable socket connection on the front of the voltage control panel: the socket had not been screwed up fully home. When this was connected the aircraft took off.

What happened next is recorded in O'Kane's diary[4].

'Blumlein, Brown, Blythen and Sansom went up with Hensby in V 9977. Airborne at 14.57 hours, As they were not down —

1700   Lovell rang control. Not landed.
1730   Rang control. No news and no contact. No interest.
1745   Rang control. No news. Rang 'A' Flight Officers mess, control: nobody about.
1800   Rang G/C King [the CO of Defford]
1810   W/C Horner [an officer on King's staff] rang up and wanted to know what was up. Told him.
1900   Nothing done.
1930   F/Lt Reynolds (another officer on King's staff] had just left control having informed the group.
1945   First news of crash.'

The bomber had crashed at about 16.15 hours in the Wye Valley near Welsh Bicknor, Herefordshire and all 11 personnel on board had been killed. These included, in addition to Blumlein, Browne and Blythen of EMI, Sansom of TRE, Hensby and Pilot Officer C E Vincent of Lovell's team, and the five RAF members, namely, the pilot, Pilot Officer D J D Berrington, the navigator, Flight Sergeant G Millar, the wireless operator/air gunner, AC2 B C F Bicknell, the flight engineer, assumed to be LAC B D G Dear, and the second Pilot, Flying Officer A M Phillips[5].

Later that evening Rowe sent a telegram to EMI which read:

PRIORITY MESSRS E.M.I. SOUTHALL MIDDLESEX
DEEPLY REGRET TO INFORM YOU THAT MR A. D. BLUMLEIN MR C. O.
BROWN AND MR F. BLYTHEN LOST THEIR LIVES ON 7TH JUNE 1942 AS THE
RESULT OF AN AIRCRAFT ACCIDENT LETTER FOLLOWS PLEASE ACCEPT
MY PROFOUND SYMPATHY
SUPERINTEND [*sic*] TRE GREAT MALVERN

White telephoned the dreadful news to G E Condliffe of EMI, Hayes, and Group Captain King drove Lovell and O'Kane to the wreckage of the Halifax to recover what they could of the most secret H_2S equipment. Only the indestructible magnetron transmitter valve was recognisable[6].

Next day Shoenberg and Condliffe were driven to Defford and with White, Trott and some RAF Officers were taken to the scene of the disaster. Condliffe officially identified the bodies of his former EMI colleagues[7].

Mrs Blumlein learned of the crash on 8th June 1940[8].

'I was alone in London with the two children. Actually I was putting clean sheets on the bed [because] Alan was coming home that day.... It was 10 [o'clock] in the morning and I saw two cars [drive up]. One [had] Shoenberg and Mrs Shoenberg, the other had Condliffe and his wife. I knew what they had come for when I saw them; and Shoenberg's face would have told anybody anything. They didn't know a lot then, only that there had been this crash. [Shoenberg] said to me: "There is just a chance that one of them may have bailed out." So I said: "It wouldn't have been Alan would it?", and he [confirmed] that it wouldn't have been Alan. Apparently Alan was in the tail of the [aircraft]. I think someone recognised him but the others were [unrecognisable]. We had three coffins but they [contained just] three masses of bones.'

On her husband's final visit to Cornwall to see his wife and children, Mrs Blumlein recalled:

'The last time he left us he was rather late and he hurried off to the bus — only to return to give me the keys of the car — I said he should not have bothered and the boys and I went out to see him off. I can see him now — hurrying up the road with his case. The boys kept calling "Goodbye Daddy" and I silently said "Oh Alan, turn around in case you never see them again" but of course being late he went on — and on and on — right out of our lives. He told me once that if anything ever happened to him I was to accept the facts and face up to them and carry on and marry again within six months. "I shall be gone like a candle blown out", he said "but you will have to go on living." He believed his afterlife lay in whatever good or bad he had done in the world — and in his children — but I know he is still somewhere.

'When Mr Shoenberg told me of the crash I said, "Oh, what was he thinking when he knew they were crashing." Our dear Isaac Shoenberg said, "Of you and the children of course".'

During his last flight, Blumlein had worn the gold, engraved watch which EMI had presented to him at the end of 1934. He was very proud of the watch, but it was not recovered from the wreckage and handed back to Mrs Blumlein. Only the few belongings which he had left in the hotel were returned. Happily, there was a sequel.

More than twenty years after the tragedy, Mrs Blumlein received a telephone call from EMI and was invited to have lunch with a few senior staff members. Astonishingly, she was told that one of EMI's engineers had Blumlein's watch. 'I didn't know this engineer and apparently he did not [wish to face] me. He kept this watch in his drawer for years.... I could never understand why the man did [that] ... of course, it was smashed to pieces, but you could see the engraving on the back, and I was pleased I had got it. I don't know how this man [acquired the watch].'

On the accident itself the records of the Air Historical Branch 5 of the Ministry of Defence state simply[9]:

'HalifaxV 9977 Flying accident on 7 June 1940.
Caught fire in air 2/3000 ft lost height into ground.
Failure of starboard outer engine in flight which subsequently caught fire.
The aircraft crashed into the ground from a height of approx 500 feet.
The engine failure and resulting fire, due to fracture in fatigue of an inlet valve stem, C of S.
Caught fire in an attempt to restart engine. Fire extinguishers failed to operate and evidence suggests bottles not filled.
O C: Possibly attempted to restart engine supplying power for special equipment to enable experiment to be continued.
Fire extinguisher bottles probably left makers empty; steps to ensure inspection on aircraft and periodically [?].' (The last word is uncertain.)

The reason for the fatigue fracture and the failure of the crew to control the fire and escape have been the subject of an exhaustive report by W H Sleigh, a former Chief Engineer, Royal Radar Establishment Airstation, Pershore. When he retired in 1984 he decided to investigate the 'worst single accident from 1926 to 1977 in the history of the radar establishment's flying units' since 'no proper records appeared to exist that gave a logical explanation for the basic cause of the aircraft's destruction'.

Halifax Mark 2 V 9977 was one of 2145 aircraft constructed at English Electric, Preston in mid-1941 and completed its flight acceptance tests from the local Samlesbury airfield on 25th August 1941. The aircraft was powered by four Rolls-Royce Merlin Mark 20 engines. Each engine had

12 cylinders and each cylinder had two inlet and two exhaust valves. Thus the total number of tappets which had to be adjusted during servicing was 192.

From the evidence in the TFU Installation Workshop notebooks of an aircraft engineering inspector at Hurn and Defford, and the records of Rolls-Royce, it is known that the tappets on the starboard outer engine (No. 198341) had been checked ten flying hours prior to the accident.

It is presumed that during the inspection, which probably occurred sometime in the period of 1st to 16th May when the aircraft was at Hurn, the engine fitter failed to ensure the positive locking of one of the many tappet adjustment stud lock nuts. This allowed the tappet to unscrew, thereby causing excessive loading on the valve, since, with the Merlin inlet valves, the cam loads increased markedly with wide tappet clearances. The extra loading led to fracture in fatigue of the inlet valve in the area of the collet location.

As a consequence, the ignited mixture in the cylinder would have had unrestricted permanent access to the flame trap elements protecting the mass of charge mixture in the induction manifold. The eventual breakdown of this element would have led to ignition of the mixture in the manifold; the flame would have travelled back through the supercharger to the carburettor and initiated a major fire. In addition, the rapidly expanding gases in the induction manifold would have caused fractures of the aluminium castings of the induction manifold and supercharger, thereby exposing the wing fuel tanks to the fury of the fire. At the time of the fire the Halifax had about 1600 gallons of fuel on board.

Sleigh has posited that the 'scattered' cockpit layout of the Halifax, with its remote main fuel system management cocks in the centre of the fuselage, well aft of the crew station, was a 'fundamental inhibition to flight safety, especially with pilots "new-to-type" '. By contrast, the cockpit layout of the Lancaster, 'with its neatly grouped and easily accessible system controls facilitated the handling of major emergencies'.

The pilot, Berrington, of V 9977 on 7th June 1942, was well experienced, with 3300 flying hours, many of them on Lancasters, as an aircraft captain. His total experience on Halifax aircraft amounted to just 13 hours. Sleigh has concluded that this lack of experience; the shortcomings in basic aircraft design to incorporate, even by 1939 aircraft engineering standards, certain elementary safety features in both cockpit and engine bay design; and the questionable qualification of the crewman occupying the flight engineer's station, militated against a swift execution of fire drill. The fire could not be extinguished and burnt through the starboard mainplane. The aircraft rolled over at 350 ft above the River Wye, crashing inverted, in a ball of fire, some 130 yards from the river.

Of the other factors which had a bearing on the accident, Sleigh has identified the captain's failure (as events suggest) to ensure each crewman carried a parachute; the captain's acceptance of a crew complement above that considered safe in the event of an emergency; the failure of the Air Ministry's Design Directorate to identify the shortcomings referred to above; and, possibly, the absence in RAF Station Orders of clear directives relating to some of these points.

And so, human error, the cause of many accidents, resulted in 11 tragic deaths. Just one small nut, out of 192, left untightened, led to the death of probably the greatest British electronics engineer of the twentieth century. Blumlein was a mere 38 years of age. His colleagues Blythen and Browne, too, were excellent engineers and were only 30 and 37 years of age respectively.

Blumlein's death, 'on active service', was noted in the obituary column of the *Daily Telegraph* of 10th June 1942; and the deaths of Blythen and Brown 'as the result of an accident' were recorded the following day. The phrase 'on active service' rather than 'as the result of an accident' may have been suggested by EMI's Directors. One of the London newspapers had published a paragraph about the crash which had referred to the 'Television expert engaged in hush-hush work'. This 'careless talk' had angered the Directors, who feared the EMI works might be bombed. They had the safety and well-being of their workforce to consider and had taken much care to suppress any publicity relating to the company's war-time activities; these had to be closely guarded. Earlier in the war, several small bombs had fallen on the Hayes site, so it was essential that news of 'hush-hush work' should not be divulged.

(Two years later, on the afternoon of 7th July 1944, a V1 flying bomb exploded on an air raid shelter at the Hayes site, killing 37 occupants and injuring many more, The explosion was very close to the Central Research Laboratories, where all EMI's most secret R&D work was undertaken, and caused much grief to dozens of local families. For a short period wartime production in the vicinity of the blast was brought to a standstill.)

The burial service of Blumlein, Browne and Blythen was arranged by EMI and was held at Golder's Green Crematorium at 11.30 a.m. on 13th June 1942. It was organised at short notice—Blumlein's mother learnt of the service late on the afternoon of the 11th—and no notice could be published in newspapers on sale on the 12th. EMI had been unable to confirm the details of the service because of the need for an inquest and because the victims of the crash were from three separate establishments.

'Nusie', as Blumlein's mother was known, has left an account of the burial service[10].

'We drove direct to the Golder's Green Crematorium, and found Doreen [Blumlein], Philip [Wake, Mina's husband], and Vera waiting for us there, and a great crowd [had] gathered who followed us into the chapel. The arrangements were very good and the whole service beautiful, [with] very fine organ music which Doreen and Mrs C O Browne [had] chose[n].

'The three coffins were already there covered with wreaths. Alan's had only Doreen's lovely red roses. The service was taken by an RAF chaplain — a friend of Mr Browne's. ... The 121st Psalm, "I will lift up mine eyes" [was read], and the hymn "Fight the good fight" was sung at the end. ... The coffins were covered with purple and edged with pale gold and the flowers showed up so beautifully on them.

'Mina [ADB's sister] introduced, at their request, the Directors of [EMI] to me. I was glad to have an opportunity to meet Mr Shoenberg and to thank him for his kindness and encouragement and appreciation of Alan. He said, very feelingly: "I loved him like a son. There is no one who can take his place in the work he was doing."

'Mr Clark, the Chairman of [EMI] took me to a seat and had much to tell me of the place Alan held in the esteem of his fellow workers and in the world of science and engineers. He said the firm had received letters from members of the Government which they would have liked to pass on to his family, but they were so strictly private. They, the Directors, had written to ask leave to photograph these letters and keep them for us to see after the war. One thing he quoted from Sir Archibald Sinclair's was "His death is more than a loss, it is a national loss."

'There were hundreds of floral wreaths — very many from RAF flying grounds where Alan had often been — more simple than in pre-war times but all beautiful.'

(The letters are not in EMI's excellent archives, and Simon Blumlein, ADB's eldest son, has no recollection of having seen them. It is possible the letters are with Clarke's papers held by the University of Wyoming, USA.)

The sum of money raised by EMI's staff for flowers was so great that the Directors imposed a limit on the amount which was spent on the floral tributes. With the approval of the contributors the balance was sent to the RAF's Benevolent Fund.

Twelve days after the tragedy, on Friday 19th June, a memorial service, arranged by EMI, for the three engineers was held. EMI had decided to have the service because the company had been unable to make public the details of the cremation service and it felt that many persons who had been associated with Blumlein, Blythen and Browne would wish to pay their last respects to them. The old parish church of Hayes, Middlesex, was filled with scientists, representatives of government departments, colleagues and friends. Among the family mourners were

Blumlein's widow, mother and sister; his father-in-law and mother-in-law, Mr and Mrs W H Lane; his brother-in-law Captain Thurston T Lane; and his sister-in-law and husband, Lieutenant and Mrs E W Tyler.

The deaths of the three members of the Central Research Laboratories was the subject of discussion of EMI's Finance Committee on 12th June 1942. It was reported (*vide* Minute 429) that the following amounts would be available in the Company's hands:

Special insurance	A D Blumlein	£5000
	C O Browne	£5000
	F Blythen	£2500
Workmen's compensation	A D Blumlein	£600
	C O Browne	£530
	F Blythen	£100

and that the following amounts were payable to the legal personal representatives by the pension schemes:

A D Blumlein	£2364
C O Browne	£1461
F Blythen	£622

The Committee decided to recommend to the Board of Electric and Musical Industries Ltd that life annuities should be purchased for the widowed mother of Blythen and the widows of Blumlein and Browne, and for their children, at a total cost of £25 355. These generous arrangements meant that financial impoverishment would not be an immediate problem.

The apparent triviality of the cause of the accident led Lord Cherwell[11] to write to Churchill and enquire: 'If (as stated in the Secretary of State's Minute of the 11th August) a new Halifax flying over England with every precaution is apt to crash owing to the slackening-off of one locking nut in the engine, perhaps this defect may be partly to blame for the high casualty rate which has been suffered by this type of aircraft?' In operations over Germany, Halifax bombers suffered nearly double the attrition rate of Lancaster four-engine bombers. These rates were the subject of a report[12] on 'Investigations into Halifax losses' for the period 1942 to 1944, but, of course, the slackening-off of a nut on a single Rolls-Royce engine did not, *per se*, reflect adversely on the design of the Handley Page aircraft.

The satisfactory functioning of the EMI equipment and the great promise which seemed to be offered by the H_2S principle led to several very important questions. Policy decisions had to be taken at a high level to determine whether the TRE type or the EMI type of H_2S radar should be chosen; the klystron/magnetron issue had to be resolved; an assessment had to be made regarding the numbers of sets to be constructed by model shop methods and by full production facilities; and the War Cabinet had to decide on the allocation of the completed equipments to Bomber Command and Coastal Command.

The first two meetings convened after the accident to consider these matters were held on 4th and 5th July. At the second meeting it was agreed that 200 sets could be produced by the end of 1942. Since EMI had undertaken to make by model shop methods two sets per week from the beginning of September and a total of 50 by 31st December 1942, it was necessary for the Research Prototype Unit (RPU) of TRE to assemble 150 by this date. J Sieger, who was in charge of the RPU, 'elected to produce the set as designed by EMI and not copies of the existing TRE lash-up'[13]. Sir Robert Renwick, who chaired the meeting, informed Churchill of this on the 6th and said that 'the set to be made should be the EMI set and not the TRE one, as the former is well engineered and some 75% of the drawings are in existence'.[14]

The meeting further agreed that aircraft would be required during the months of September, October, November and December at the rates of 1, 3, 4, and 4 per week respectively. This would give a total of 48 fitted aircraft by the end of the year. It would be the responsibility of TRE to produce a fitting, testing, maintenance and training programme to accord with the schedule.

The contentious klystron/magnetron issue was the main topic of discussion at a meeting held on 15th July and chaired by the Secretary of State for Air[15]. There were essentially two factors to be considered. First, could use of the klystron valve lead to results which would be acceptable to Bomber Command; and secondly, if it did not, should the enemy be allowed to learn the secret of the magnetron valve for the sake of the operational advantages which would accrue from its use in only two squadrons? In arriving at a consensus, the meeting had to bear in mind the conclusions of the H_2S trials using klystrons and magnetrons, carried out in the periods 5th to 6th, and 12th to 15th July respectively, which showed ranges of 8 to 12 miles and 38 to 40 miles, for the klystron and magnetron equipments; and the fact that during evasive actions, targets were likely to be lost when the lower-power valve was utilised. If the magnetron were chosen, it would be inevitable that an aircraft with H_2S apparatus would be shot down and that the almost indestructible magnetron would be located in the wreckage. Since it was, in the opinion of Watson-Watt, highly unlikely that the Germans had yet developed a magnetron valve, such a find would almost certainly enable them to discover the principles of its construction.

On the other hand, the radar pioneer felt it would take them 12 to 18 months to develop the valve for operational use.

These weighty considerations led to the meeting agreeing that the 200 sets to be made by model shop methods should contain the magnetron and not the klystron. Churchill was informed, by Renwick, of this decision on 16th July and was told that a complete schedule of components and 60% of the drawings of the mechanical details of the units had been obtained from EMI[16]. A month later Renwick reported[17] to the Prime Minister: 'Considerable progress can be reported. . . . All departments are working well together, and the EMI are being most co-operative. . . . One satisfactory result of the urge upon the crash-programme is the probable bringing forward of the bulk production by the Gramophone Company by some two or three months.'

Certainly the need for a self-contained aircraft navigation system was great. The 'GEE' navigation and bombing aid, which was first used operationally by the RAF over Germany in March 1942, was from August 1942 being jammed by the enemy[18]. Moreover, GEE had a relatively short range, about 350 miles, and so could not be used on deep penetration raids over enemy-held territory: 'OBOE', the accurate navigation and bombing aid which TRE was developing, did not become operational until December 1942. As with GEE, the range was limited because the beacon carried by the aircraft had to be interrogated by ground stations. Just one aircraft could be controlled every ten minutes. H_2S was the only system which would enable large numbers of aircraft to fly deep into Germany without the need for UK ground-based radio/radar stations.

Good progress continued to be made by EMI and on 19th September 1940 the first flight test of EMI's prototype H_2S equipment on Halifax W 7808 took place. On the aircraft were Squadron Leaders Alexander and Gilfillan, Pilot Officer Ramsey, Dr O'Kane of GEC, and Mr I J P James of EMI. The Halifax took off from Defford and flew for approximately three hours in 3/10 cloud at 10 000 ft to 30 000 ft over Cheltenham, Gloucester and Coventry. According to James[19], 'We had outstanding views on the p.p.i. [the Plan Position Indicator] of the river Severn'.

Ten days later W 7808 flew with EMI's production version of H_2S for the first time. Flight Lieutenant J H Sawyer was the observer and his report[20] noted:

'Obtained excellent results on the usual echoes of Gloucester and Cheltenham. Ranges 20 – 25 miles; no gaps evident. . . . With H_2S alone we selected Gloucester (within 25 miles) and directed aircraft towards the town. Wing Commander Homer then took violent evasive action but the echo was held oriented and we finally homed dead over the centre.'

Much flight testing of H_2S by the Bomber Development Unit was undertaken during the autumn of 1942. The average flight navigator had to be trained in the use of H_2S; an assessment had to be made of the likely serviceability of the equipment in operations; maintenance procedures had to be devised and personnel trained to undertake the repair of equipment; and an estimate of the effectiveness of bombing with the aid of H_2S had to be determined.

An analysis of the first 21 bombing operations, in the autumn of 1942, led by the Path Finder Force (PFF) against German targets, showed that, regardless of weather and the success of the PFF, the average percentage bombing success within two miles of the target was 15%. However, trials by the Bomber Development Unit (BDU) elicited the fact that when blind bombing with H_2S, almost 100% of bombs fell within one mile of the aiming point. In actual operations over enemy territory it was anticipated this figure would fall to about 70% since the average navigator working under operational conditions would be less accurate than one of the BDU's skilled navigators. In addition the then current (end of 1942) serviceability figures indicated that about 30% of H_2S sets were likely to fail before reaching the target.

Taking all the relevant factors into consideration, Bomber Command[21] was optimistic that bombing with the aid of H_2S in the whole bombing force would be about three times the effectiveness (23%) obtained if it were installed only in the Pathfinder Force, and over four times the effectiveness (15%) of bombing in late 1942.

Furthermore, the use of H_2S in the whole force would enable bombing to be carried out under bad weather conditions when target marking by the PFF would be impracticable. In late 1942 operations could not be carried out on about 30% of nights because of inclement weather conditions over the targets: consequently H_2S would permit a more intensive bombing offensive. Such bad weather operations would have the extra advantage of protecting aircrews from German night fighter and anti-aircraft gun activity since these defences would be seriously handicapped.

Also H_2S would assist British bombers in avoiding heavily defended areas en route to their targets, thereby minimising losses.

All of this, of course, held the promise of a truly efficacious and deadly aerial onslaught against German towns and cities. Unfortunately there was a grave overriding policy constraint on the employment of apparatus containing magnetron valves. This policy had been considered by the British Joint Communications Board in London and its decision had been sent by telegram to the Commonwealth Joint Communications Committee in Washington. Subject to the acceptance of the policy by the Combined Chiefs of Staff, the position was as follows:

'Until 1st March 1943, or until such prior date as unrestricted use [was] announced by the Combined Chiefs of Staff, this [centimetric] equipment [using magnetron valves] should not be used over enemy territory or under circumstances involving risk of enemy capture.'

Clearly, this whole issue demanded the attention of the War Cabinet Chiefs of Staff Committee and a conference was held on 22nd December 1942. It was chaired by the Prime Minister and was attended by Sir Stafford Cripps; Air Chief Marshal Sir Charles Portal, Chief of Air Staff; Sir Robert Watson-Watt, Ministry of Aircraft Production; Admiral of the Fleet Sir Dudley Pound, First Sea Lord and Chief of Naval Staff; Lieutenant General Sir Hastings Ismay, Office of the Minister of Defence; and Lord Cherwell.

The prime matter for resolution was this[22]: because only a few H_2S/ASV sets existed (in December 1942), which was of greater importance, the value to be gained by the use of H_2S/ASV by the Pathfinder squadrons during January and February, or the possible adverse effect on the war at sea of the early capture by the enemy of this equipment? Sir Charles Portal, on behalf of the Air staff, put forward the views given earlier and Sir Dudley Pound explained the Admiralty's anxiety to avoid an early compromise of H_2S/ASV radar.

In the Battle of the Atlantic (at December 1942) practically no contacts with U-boats were being obtained by Coastal Command aircraft because the enemy was using receivers (Metox) to listen to the 1.5 m ASV Mark II radar signals sent out by these aircraft. If 10 cm ASV radar were used, there was a good prospect the U-boats would be caught unawares and heavy submarine losses would accrue. But if the new radar sets were used in bombing missions over Germany, there was a very real possibility a magnetron valve would be quickly captured or reconstructed from a crashed bomber and any advantage gained against the U-boats consequently annulled.

Pound argued that by withholding the use of H_2S until March 1943, Bomber Command would not be greatly penalised, whereas several months might be provided in which the country had the full advantage of the use of the centimetric radar in the anti U-boat campaign. Even a small number of additional submarine 'kills' made a great difference to the war at sea, since each U-boat was estimated to sink some 30 000 tons of shipping. And it was known that many U-boats were being built or fitted out in the submarine construction yards of Hamburg, Kiel, Danzig, Bremen and Vegesack. On 1st September 1942 the actual numbers were 84, 34, 31, 30 and 19 respectively, a total of 198.

The First Sea Lord received almost no support for his view[23]. Watson-Watt thought it was quite inconceivable 'the enemy could have neglected all the opportunities he had been given for learning that we were using 10 cm waves': therefore 'it should be assumed that the enemy knew we were using 10 cm waves'. Furthermore, he pointed out that the fitting of radar sets in U-boats, 'of which there was some evidence', would obviate the need for a listening device to detect ASV radar radiations.

Lord Cherwell, too, was unsupportive. He felt there was a tendency to expect too much from 10 cm ASV. 'A reasonable expectation was the doubling of our U-boat kills in the Bay of Biscay.... While 10 cm ASV would not make a big difference to the war at sea, H_2S would make a big contribution to our bomber offensive', he opined.

The majority view of the conference was that, on balance, H_2S should be released in January 1943 for use by the Pathfinder squadrons of Bomber Command. The First Sea Lord was invited to report on the conclusion and the arguments upon which it was based to A V Alexander, the First Lord of the Admiralty. Thereafter, the policy about the use of H_2S would have to be referred to the Defence Committee or War Cabinet, if the First Lord so desired.

Naturally Alexander regretted the conference's conclusion and the lack of any support for the Admiralty's position. In a memorandum to Churchill he said: 'It would be some set-off if [the Admiralty] could use the apparatus very intensively for a time in attacking U-boat bases', and sought Churchill's aid. But Churchill would not overrule the wish of the conference: 'The best targets must be chosen', was his response.

Sir Robert Renwick, as requested by the Prime Minister, reported regularly on the manufacturing and installation position of the H_2S/ASV equipment. In his first fortnightly report, dated 8th December 1942, he noted that EMI had delivered 50 sets of units and installation had been completed on 11 Halifaxes. Two weeks later Renwick was able to say that EMI had delivered a further 11 sets, making a total 61, but none had been produced by TRE's Research Prototype Unit. The first set from RPU was delivered on 14th January 1943. Two days earlier the Air Ministry had given authority to Bomber Command to begin operations, with H_2S over enemy territory, without restriction.

The first raid during which H_2S was used took place on the night of 30th/31st January 1943[24]. Further raids occurred on 2nd/3rd, 3rd/4th, and 4th/5th February, as detailed in Table 16.2, and from these an assessment was made of the efficacy of H_2S in aiding aircrews. Only a few H_2S aircraft were involved, since by the 15th January 1943 just 12 Halifaxes and 15 Stirlings had had H_2S apparatus installed.

Table 16.1   Target programme

	End Sept.	End Oct.	End Nov.	End Dec.
EMI	8	15	30	50
RPU	—	25	80	150
Totals	8	40	110	200

Table 16.2   Raids using H_2S

Night	Target	No. of H_2S aircraft	Weather en route	Weather at target	Visual identification of target	Losses
30/31 Jan.	Hamburg	13	poor	poor	impossible	0
2/3 Feb.	Cologne	10	poor	poor	impossible	1
3/4 Feb.	Hamburg	11	poor	poor	impossible	0
4/5 Feb.	Turin	8	cloud	no cloud	moderate to good	0

On the four operations, navigators were able to identify a large number of landmarks and to obtain ranges and bearings from which to fix their positions without difficulty. These landmarks included towns, islands, all coastlines, rivers and lakes. By means of H_2S, navigators were able to maintain accurate track and timing, locate their targets without difficulty and avoid defended areas en route. (See Figure 16.1.)

As a bombing aid, H_2S was found to be invaluable as a means of securing tactical freedom and ensuring accurate timing. Targets were identified without difficulty and once seen on the screen, could be kept in view and attacked from any desired direction.

Headquarters, Bomber Command[25], was delighted with its new equipment:

'H_2S in its present form fully meets Air Staff requirements and has exceeded expectations in that towns have proved easy to identify both by shape and relative positions. In addition to the exceptional value of H_2S for identification and bombing of the target, its great navigational value has been proven beyond all doubt.... In fact the problem of accurate navigation under almost any weather conditions is solved by H_2S when operated by a trained navigator.

'The ease with which targets have been identified and attacked proves that if this device were introduced into as many heavy bombers as possible it would greatly increase the destructive power of the bomber force and considerably reduce the restrictions imposed on operations by adverse weather conditions.'

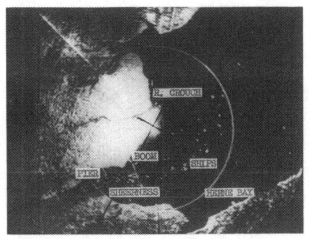

Figure 16.1 An H₂S c.r.t. image of the Wash

Source: EMI Archives

With such great enthusiasm for H₂S/ASV it was to be expected that both Bomber Command and Coastal Command would want this radar fitted to their aircraft (Figure 16.2). Slowly the numbers increased. The position, shown in Table 16.3 at 23rd February 1943, and a forecast, was given in a report by the Secretary State for Air[26].

Figure 16.2   The EMI type 184A H₂S display unit fitted into an aircraft

Source: EMI Archives

Table 16.3   H₂S/ASV fitting

	Total a/c fitted	a/c being fitted	Fitting forecast—week commencing					
			23/2	2/3	9/3	16/3	23/3	30/3
Sets available for fitting to aircraft, i.e. 50% of production			7	8	10	10	10	10
Spare sets to squadrons (allocated weekly)	(24)	(14)	7	8	10	10	10	10
BOMBER COMMAND								
Halifax, 35 Sq.	12	3	3	3	2	—		
Lancaster, 156 Sq.	—	—	—	2	4	6		
Stirling, 7 Sq.	15	1	3	3	—	—		
Halifax, BDU	2	—	—	—	—	—		
COASTAL COMMAND								
Halifax	—	—	—	—	—	—	10	2
Wellington XII	8*	11	1	—	4	4	—	—
Halifax, very long range	—	—	—	—	—	—	—	8

* includes 1 Wellington VIII

Table 16.4 H_2S *use*

	Night of				
	30/31 Jan.	2/3 Feb.	3/4 Feb.	4/5 Feb.	Totals
Successful, %	46	60	55	88	60
Technical failure, %	39	30	18	—	21
Failures not attributable to H_2S, %	15	—	27	12	17
Aircraft missing, %	—	10	—	—	2

A most heartening early feature of the H_2S bombing raids was the substantial reduction in bomber casualties, since blind bombing with H_2S put the enemy's searchlights, guns and night fighters at a considerable disadvantage. In 1000 bombing sorties on Wilhelmshaven, Cologne, Milan and Lorient, carried out during the three nights 12th/13th, 13th/14th, and 14th/15th February, 25 bombers were lost[27]. This loss rate of 2.5% may be compared to previous loss rates of up to 7% for Halifax aircraft.

Another encouraging feature was the steady increase in the successful use of H_2S, as shown in Table 16.4[28]. Correspondingly the percentage of failures showed an equally steady decline.

The first three months of 1943 was a period of grave concern for the Admiralty. Notwithstanding all its endeavours to contain the U-boat menace, Allied shipping losses had continued to increase. In 1940 and 1941 the gross tonnages of shipping sunk by U-boats were 2 103 000 and 2 133 000 respectively. But in 1942, following the attack, on 17th December 1941, on Pearl Harbour by the Japanese, the total tonnage of Allied merchant ships sunk by German submarines was 6 252 000. As noted previously, even in March 1943, 43 months after the commencement of hostilities, the losses for that particular month were a staggering 627 000 tons.

Slowly, as more and more H_2S/ASV sets were manufactured, the number available for anti-submarine reconnaissance and attack by aircraft of Coastal Command increased (Figure 16.3). The figures for the first quarter of 1943 are shown in Table 16.5.

These, and later aircraft, transformed the Battle of the Atlantic. In May 1943, 41 U-boats were sunk, 14 by surface ships, 22 by aircraft and one by aircraft and ships working together[29]. This total was more than the total number of German submarines destroyed in either 1940 or 1941, the figures for which were 23 and 35 respectively. In July 1943, 31 U-boats were sunk by aircraft and 5 were sunk by surface ships. The May losses were greater than could be tolerated and therefore Doenitz withdrew,

Figure 16.3   A Wellington of Coastal Command fitted with a nose-mounted ASV Mark III scanner

Source: DERA, Malvern

temporarily, his U-boats from the North Atlantic. Shipping losses (Tables 16.6 and 16.7 and Figure 16.4) fell dramatically, never to increase for the remainder of the war to the levels of the early years of the conflict.

Captain S W Roskill, who wrote the official history of *The War at Sea*, said[31]: 'We had a narrow escape from defeat in the Atlantic'. The narrowness of this escape from possible subjugation to victory in the battle may be gauged from three German quotations. In April 1943, Goebbels (Hitler's Propaganda Minister) told the German people: 'In the U-boat war we

Table 16.5   ASV Mark III available for anti-submarine aircraft

Week commencing	Number of ASV Mark III sets fitted	Aircraft type
23 January 1943	1	Wellington
9 March 1943	4	Wellington
16 March 1943	4	Wellington
23 March 1943	10	Halifax
30 March 1943	2	Halifax
30 March 1943	8	Halifax (very long range)

Table 16.6   Annual U-boat losses

	1939	1940	1941	1942	1943	1944	1945	Total
Losses	9	23	35	85	238	250	158	798

Table 16.7   Shipping losses[30]

Period	British	Others	Total
7 December 1941 to 31 July 1942	1 144 000	2 106 000	3 250 000
1 August 1942 to 21 May 1943	1 974 000	1 786 000	3 760 000
22 May 1943 to 18 September 1943	45 960	161 627	207 587
19 September 1943 to 15 May 1944	119 854	194 936	314 790

have England by the throat'[32]. The following month Doenitz reported to Hitler[33]. 'We are at present facing the greatest crisis in submarine warfare, since the enemy, by means of location devices, makes fighting impossible and is causing us heavy losses'. And in January 1944, Hitler remarked, in a New Year Message: 'The apparent stagnation in the U-boat war is to be ascribed to a single invention of the enemy'.

The contribution of ASV Mark III radar-equipped aircraft is apparent from an analysis of U-boat losses (Table 16.8).

The First and Second World War offensives to counter the deadly attacks of enemy submarines were of epic proportions. Apart from the very extended timescales of these battles, particularly that of the 1939–45 War, vast sea and air resources were required, and losses on both sides were enormous. During the 51 months of the Great War, Germany built 365 U-boats, lost 178 and caused the Allies to lose 4837 merchant vessels totalling 11 135 000 tons gross. In the Second World War, covering 68 months, 2775 Allied merchant ships aggregating 14 573 000 tons gross were sunk for the loss of 781 U-boats[34].

A May 1945 Joint Statement, by the President of the United States of America and the British Prime Minister, in which the U-boat losses are given as 700, shows the proportion of these assigned to the two Allies[35] (Table 16.9).

To effect these successes, hundreds of warships were needed and many thousands of aircraft sorties were flown. On 1st October 1943 the numbers of the Royal Navy's destroyers and escort destroyers, escort vessels and minesweepers were 288, 325 and 220 respectively. Coastal Command provided long-range anti-submarine escorts to more than 1250 ocean convoys, involving 123 022 sorties; and flew 125 854 sorties on convoy

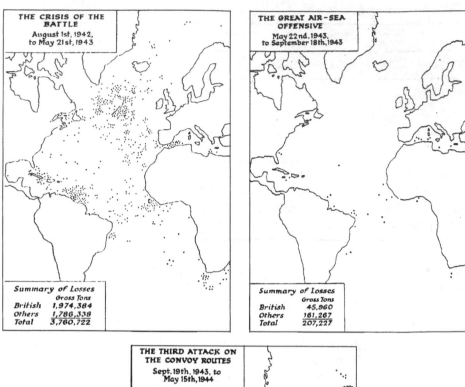

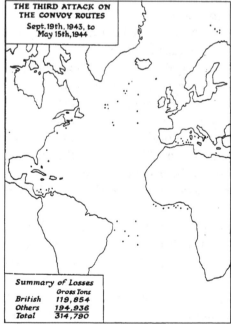

Figure 16.4   *Maps showing the locations of Allied shipping losses from 1st August 1942 to 15th*
May 1994

Source: 'The Second World War', Vols 4 and 5, W. S. Churchill (The Reprint Society,
London, 1950)

Table 16.8 U-boat losses

Losses by year	Ships	Aircraft	Ships and aircraft	Submarines	Mines	Collision	Scuttled	Marine casualty	Other
1939	6	—	—	1	2	—	—	—	—
1940	11	1	3	2	3	1	—	—	2
1941	25	3	2	2	—	3	—	—	—
1942	33	38	6	2	3	2	—	—	1
1943	57	145	14	6	2	9	1	1	3
1944	69	96	25	7	12	10	11	6	14

Table 16.9 Breakdown of losses

Losses by	Numbers	Percentages
British and British-controlled forces other than US	464	66.3
UK and US forces under British control	145	20.7
Other causes	91	13

protection duties; and Bomber Command flew 18 725 sorties and laid 47 307 mines in its mining campaign in the Baltic, the Skagerrak and the Kattegat. In addition, Bomber Command dropped 46 700 tons of bombs on naval land targets, including harbour installations, and U-boat and other ship construction facilities[36].

Victory in the Battle of the Atlantic was not obtained cheaply. Many ships and aircraft were destroyed. Bomber and Coastal Commands lost 1769 aircraft as a result of purely maritime operations and the Royal Navy, from 3rd September 1939 to 1st October 1943, lost 120 destroyers and escort destroyers, 40 escort vessels and 19 minesweepers. Not all of these losses, however, were a consequence of the Battle of the Atlantic[37].

Against this background and the crucial importance of the ASV Mark III radar, the question arises: What was the contribution of Blumlein? It is indisputably known that the version of ASV Mark III radar adopted for anti-submarine patrol aircraft was developed and engineered by EMI and that the 'brilliant' Blumlein was a 'key man' in an 'always small H$_2$S team'[38].

E A Newman, who worked under Blumlein on H$_2$S, has recorded[39]:

'The circuitry of H$_2$S was almost entirely due to Blumlein, with Frank Blythen as his right-hand man. Frank Blythen, incidentally, was himself a brilliant creative engineer. Browne was effectively an EMI admin. type.... I might say here that the transmission part consisting of the power output (first a klystron and then a magnetron) was rather a trivial part of the whole.... [The] real cleverness in H_2S was the system and the circuitry which was, I think, all Blumlein.... The EMI H_2Ss worked on ASV from the beginning and they worked well. We tested one of the early models over the Wash and it was very good.'

In an interview, Newman, who 'did a lot of wiring of H_2S and therefore saw much of Blumlein', stated that he had 'no doubt' that the waveform generator, p.p.i., receiver techniques, the idea of using magslips, and the 'great bulk' of the H_2S project were due to Blumlein with assistance from Blythen and White. Nind, too, has confirmed that the H_2S circuits were due to Blumlein.

After the crash, Newman said: 'Since I was responsible ... for all development in three of the H_2S units, and since I was closely associated with people dealing with the remaining units, and since I was in charge of making up the training system ... I can definitely assert that the changes made to Blumlein's basic conceptions were nil, and to his detailed circuitry very few.'

In considering the vital role of the H_2S/ASV radar, it is worth recording Lord Cherwell's view[40]: 'It is very satisfactory to find this universal enthusiasm on the part of the users of H_2S. It is most fortunate that you [Churchill] had your meeting last July [1942] to stimulate it, otherwise we should not have had it for another four or six months, nor for that matter any short wave ASV.'

H_2S/ASV radar was one of the very great wartime technological developments. Both the bombing offensive over enemy territory and the battle against the U-boats were enormously aided by its use. Thousands of Allied servicemen and merchant seamen were spared death as a consequence of the radar's effectiveness, and vast quantities of food and war supplies were saved for the Allied cause.

Many persons were associated with the H_2S/ASV programme. Credit for initiating the action which led to H_2S/ASV radar must be accorded to Lord Cherwell. Without his exposure of the ineffectiveness of the RAF's bombing offensive, and Churchill's powerful and vital support, the development programme would never have commenced in November 1941, Dee, of TRE, suggested using centimetric air interception radar for 'town finding', and a GEC AI Mark VII radar was adapted by TRE for this purpose; preliminary flight trials were conducted by O'Kane of GEC and Hensby of TRE; Lovell was appointed TRE's project leader; EMI was given a development contract; and various organisations—the Air Ministry, the Ministry of Aircraft Production, the TFU, the BDU, and TRE—played important roles in the realisation of the H_2S/ASV radar.

But, above all, it was equipment developed and engineered by EMI which was selected for the manufacturing programme. And it was EMI-constructed apparatus which showed the early operational worth of the H_2S/ASV system.

Blumlein's role in the programme was supreme. It was his genius and the talents of his group which enabled the 'very crude images', produced by the 'unreliable', adapted, GEC AI Mark VII radar, to be replaced by excellent, meaningful, images capable of being interpreted by the average RAF/Coastal Command navigator/bomb aimer. As Watson-Watt mentioned in his book *Three Steps to Victory*, the 'brilliant' Blumlein was a 'key man' in an 'always small H_2S team'.

References

1 CURRAN, Sir S.: 'Fateful substitution', *Daily Telegraph*, 2nd June 1977
2 O'KANE, B. J.: letter to Sir Bernard Lovell, 19th August 1990
3 LOVELL, Sir B.: 'Echoes of war' (Adam Hilger, Bristol, 1991)
4 MOSELEY, D.: letter to W H Sleigh, 6th August 1984, DERA, Malvern, UK
5 SLEIGH, W. H.: 'Halifax Mk 2 V.9977 (Prototype H_2S radar). Fatal accident 7 June 1942', 1985, DERA, Malvern, UK
6 Ref. 3, p. 27
7 TROTT, F. R.: memorandum to Dr White, 33rd April 1967, EMI Central Research Laboratories archives
8 BLUMLEIN, Mrs D.: taped interview, National Sound Archives, London, UK
9 Ref. 5, p. 7
10 BLUMLEIN, Mrs J. E.: letter, 16th June 1942, private collection
11 CHERWELL, Lord: memorandum to the Prime Minister, F 227/23, archives of Nuffield College, Oxford, UK
12 'Investigation into Halifax loss', AIR 14/1794, PRO, Kew, UK
13 AVIA 7/3574, PRO, Kew, UK
14 RENWICK, Sir R.: memorandum to the Prime Minister, 6th July 1942, PREM 3/10, PRO, Kew, UK
15 Notes of a meeting on the development and operational use of magnetron and klystron valves, 15th July 1942, AIR 14/1293, PRO, Kew, UK
16 RENWICK, Sir R.: letter to the Prime Minister, 16th July 1942, PREM 3/10, PRO, Kew, UK
17 RENWICK, Sir R.: letter to the Prime Minister, 19th August 1942, PREM 3/10, PRO, Kew, UK
18 CHERWELL, Lord: memorandum on 'H_2S' to the Prime Minister, 26th August 1942, F 227/23, archives of Nuffield College, Oxford, UK
19 JAMES, I. J. P.: letter to the author, 9th March 1981, personal collection

20 'Diary for Halifax 7711 and Halifax 7808 and experimental H₂S gear, 13th September–29th September, 11th October 1942, AVIA 7/1678, PRO, Kew, UK

21 'Notes on the value of H₂S to Bomber Command', 3rd January 1943, AIR 14/1294, PRO, Kew, UK

22 Minutes of a conference, War Cabinet, Chiefs of Staff Committee, 22nd December 1942, PREM 3/10, PRO, Kew, UK

23 'Anti-U-boat warfare', a memorandum, no. 30, 6th December 1942, CAB 86/3, PRO, Kew, UK

24 ACAS (Ops): letter to Air Officer Commander-in-Chief, 12th January 1943, AIR 14/1294, PRO, Kew, UK

25 Memorandum on 'Operational use of H₂S', 9th February 1943, H/Q Bomber Command, PREM 3/10, PRO, Kew, UK

26 Secretary of State for Air: Report on 'H₂S and ASV fitting programme and operational use', 23rd February 1943, CAB 86/3, PRO, Kew, UK

27 Air Ministry: memorandum to the Prime Minister, 15th February 1943, PREM 3/10, PRO, Kew, UK

28 H/Q Bomber Command: letter to Sir Robert Renwick, 7th February 1943, PRO, Kew, UK

29 'German U-boat losses in World War II', Naval History Division, Office of the Chief of Naval Operations, Washington, DC, 1963

30 CHURCHILL, W. S.: 'The Second War' (Cassell, London, 1951, 1952) Vols. 4 and 5

31 ROSKILL, Captain S. W.: 'The war at sea 1939–45' (HMSO, London, 1954), Vol. 111, Part 1, Chapter 2

32 ANON.: 'The Battle of the Atlantic' (HMSO, London, 1946), p. 64

33 Ref. 31, p. 15

34 HESSLER, Fregatten Kapitän G.: 'The U-boat war in the Atlantic', 27th July 1950, ADM 186/202, PRO, Kew, UK

35 T. L. R.: memorandum to the Prime Minister, 9th May 1945, PREM 3/413/7, PRO, Kew, UK

36 'War at sea in relation to OVERLORD, part played by RAF', AIR 20/934, PRO, Kew, UK

37 AIR 20/934, PRO, Kew, UK

38 WATTSON-WATT R. A.: 'Three steps to victory' (Odhams Press, London, 1957), p. 408

39 NEWMAN, E. A.: taped interview with the author, personal collection

40 CHERWELL, Lord: memorandum to the Prime Minister, 19th February 1943, PREM 3/10, PRO, Kew, UK

Chapter 17
Genius

In his book of 1753, *On the Interpretation of Nature*, Diderot wrote[1]: 'We have three principal means: observation of nature, reflection, and experiment. Observation gathers the facts, reflection combines them, experiment verifies the result of the combination. It is essential that the observation of nature be assiduous, that reflection be profound, and that experimentation be exact. Rarely does one see these abilities in combination. And so, creative engineers are not common.'

In general, the word 'genius' is used in two separate, although closely related, senses to describe intellectual ability. In one sense, it has a simple clinical definition as a high measure of performance in a standard intelligence test; in the other, it describes the exceptional creative ability that leads to great and lasting achievement.

For the latter interpretation—which is the one that enriches humankind, and can be examined through history—genius involves originality, creativity and the ability to think and work in new fields, to give the world something it would not otherwise have had. Genius in this sense differs from talent both quantitatively and qualitatively. Talent refers to an inherent aptitude for some particular type of activity and implies a comparatively quick and easy acquisition of a given skill.

The nineteenth-century British scientist Sir Francis Galton initiated the systematic study of genius[2]. He formulated the theory that genius is an extreme degree of three combined traits—intellect, zeal and power of working—that are shared by all persons in various grades. He advanced his views in his book *Hereditary Genius*, of 1869, and presented clear statistical evidence that genius, as measured by outstanding achievement, tends to run in families. Since then, the contribution of environmental factors has been considered and the consensus appears to be that genius is a function of both hereditary and environmental influences. The initial potentiality for exceptional accomplishment comes from hereditary, but whether or not this potentiality is brought to fruition depends, to some extent, upon education and opportunity.

Blumlein was a remarkably creative and fecund engineer, whose type of genius has had few equals this century. Although his lifespan of 38 years was tragically short, he stands very favourable comparison with Brunel (probably the greatest British engineer of the nineteenth century) Edison (who has been called the greatest American inventor ever) and Michael Faraday (the greatest nineteenth-century experimentalist, anywhere). It can be demonstrated that Blumlein's contributions to human knowledge and well-being exceed, in quality and quantity, those which had been achieved by Faraday at the age of 38 years.

The comparison must not be taken too far. Faraday was an experimental scientist who lacked a formal university education. For him, empiricism necessarily played an important role in his investigations. In one respect, though, the influence of opportunity, their careers had a common factor. Both Faraday and Blumlein were fortunate in commencing their work when the subjects of their studies were in a very nascent state. Faraday's practical investigations on electromagnetic induction, the conduction of electricity in liquids and in gases, and the relationship between light and magnetism, owed little to other experimentalists. And when Blumlein began his work on electrical monophonic recording and reproduction, and on electrical stereophonic recording and reproduction, only the Western Electric Company and Bell Telephone Laboratories respectively had pursued or were pursuing similar enquiries. Again, Blumlein's contributions to high-definition television, wide-band electronics, slot antennas and radar were all made when the subjects were in their early formative periods.

Nevertheless as L Wolpert has stated[3], it is necessary to be suspicious of good fortune, or luck, or serendipity in science and engineering; 'One cannot but be struck by the fact that it is almost always the best scientists [and presumably the best engineers] who are the luckiest. Recall Louis Pasteur's irritable response to someone who thought he might have been lucky: "Fortune favours the prepared mind".'

All of us are surrounded by countless problems, but it is the gift of the great researcher to choose those which are important and are likely to be amenable to solution. Selecting the right problem to investigate is an intellectual and creative act, which may or may not be influenced by what has been determined previously. Similarly selecting the right components to synthesise into an engineering system having, possibly, a practical utility is an intellectual and creative act. Both Marconi and Baird had a genius for creating systems by adapting the artefacts of others. P P Eckersley, the

BBC's Chief Engineer in the 1920s, who was familiar with the work of these inventors, has said: 'Neither Baird nor Marconi were pre-eminently inventors or physicists; they had, however, that flair for picking about on the scrap-heap of unconnected discoveries and assembling the bits and pieces to make something work and so revealing possibilities if not finality'. C S Franklin, one of Marconi's engineers, once opined that 'Marconi may not have been a great scientist but he was a great man . . .'. His work led in 1909 to the award of the Nobel Prize in Physics.

Sometimes it is important, for originality, to be in ignorance of the endeavours of others; then, the constraints which may shackle them and progress are absent. As Einstein said: 'Imagination is more important than knowledge'.

Both Faraday and Blumlein were aided in their creativity by their environments. Faraday, as Director of the Royal Institution, London, was ideally placed to carry out his experiments and he enjoyed a freedom of enquiry and an access to laboratory resources which, although not unique, were certainly rare in nineteenth-century Britain. Blumlein's work was bounded by industrial and commercial considerations. Nonetheless, he enjoyed excellent facilities: his genius was recognised at an early stage by Shoenberg, and he was given great freedom of movement.

Genius can be manifested in several ways. There is no simple formula for this sort of ability. Thomas Edison is reputed to have said: 'Genius is one per cent inspiration and 99 per cent perspiration.' For him, this was a fair description of the way he worked. Trevelyan Miller, in his biography of Edison, relates an anecdote about Edison's method[4].

'Edison was once asked how many laboratory experiments it was necessary for him to make to perfect the storage battery. "Goodness only knows", he replied. "We used to number our experiments consecutively from one to 10 000. When the got up to 10 000 we turned back to one and ran up to 10 000 again, and so on. I don't know how many experiments we made up — I lost track of them — but it was not far from 50 000. It took 10 years and cost $3 000 000".'

This anecdote tells much about Edison's approach to a problem. He did not question whether something could be done, only how it could be done. His inventions and discoveries came from practical experiments, 'invariably by chance, thus reversing the orthodox concept of pure research leading to applied research'[5].

Nikola Tesla, who worked for Edison in 1882, described his method of working as follows[6]: 'If Edison had a needle to find in a haystack he would proceed at once with the diligence of a bee to examine straw after straw until he found the object of his search'. This attitude did not suit Tesla, and he left the company in 1883 following a dispute with Edison.

Edison's attitude was probably a consequence of his education. Like the great experimentalist Faraday, Edison left school at 12 years of age, to become a newsboy on the railway. He once reminisced[7]: 'At the time I experimented on the incandescent lamp, I did not understand Ohm's law'. He observed: 'I do not depend on figures at all. I try an experiment and reason out the result, somehow, by methods which I could not explain. . . . I speak without exaggeration when I say that I constructed 3000 different theories in the development of the electric light, each one of them reasonable and apparently likely to be true. Yet in two cases only did my experiments prove the truth of my theory'. Edison thus exemplifies the definition of genius as an infinite capacity for taking pains.

Tesla was another type of genius altogether, well grounded in mathematics and science; to him, the amount of time Edison spent on his procedures was deplorable. 'I was a sorry witness of such doings, knowing that a little theory and calculation would have saved him 90 per cent of his labour.'[8]

Edison's lack of a formal education certainly handicapped him when it came to arriving at viable designs, systems and structures. He filed no fewer than 389 patents for electric light and power, 195 for phonographs, 150 for telegraphy, 141 for storage batteries and 34 for the telephone.

Blumlein's approach had more in common with other well educated engineering geniuses, including such figures as Tesla, Isambard Kingdom Brunel and Charles Proteus Steinmetz. All were prolific engineers, but all used careful design and mathematical analysis to plan their work whenever possible. One of Brunel's earliest designs, produced in 1829 when he was just 26 years old, was for a suspension bridge across the River Avon. Brunel's second design for this crossing was selected and pronounced to be the most mathematically exact of all those tendered—even though the competition included plans by the great bridge builder Thomas Telford.[9]

Similarly, Blumlein based his circuit designs on a rigorous application of the theory of electronics. Gains, frequency responses, impedances, pulse widths and shapes were all determined theoretically and as E A Newman, one of his assistants, has recalled[10]: if the experimental values did not correspond with the calculated ones 'there was trouble'.

The majority of Blumlein's patents are disparate in their subject areas: not for Blumlein scores of patents on a particular invention. His original and very important work on stereophony and on slot antennas led to just two patents. Blumlein's work in these fields epitomises Abraham Lincoln's opinion that: 'Towering genius disdains a beaten path. It seeks regions hitherto unexplored.'

Blumlein, Brunel and Edison were all prolific and versatile in their engineering work. Edison's 1093 patents deal with improvements in telegraphy, electric lighting, power systems, phonographs, batteries, magnetic ore separators and moving pictures. Brunel designed and supervised the building of docks, railways, ships, tunnels, cuttings, buildings and bridges of many types. Blumlein's work embraced the fields of telephony, electrical measurements, sound recording and reproduction, television, electronics, antennas, cables, camera tubes and radar. Tesla and Steinmetz were also wide-ranging in their interests.

In one respect Brunel's work was more difficult than that which faced Blumlein and Edison. Brunel's designs were for large-scale, costly structures and so by their nature extensive prototype exercises and testing were precluded. Nonetheless, such was Brunel's genius and facility with engineering principles that many of his structures were of an individual and original nature.

Like Blumlein, Brunel had an enormous mental vitality and advanced his work with bold concepts. His timber ship, the SS *Great Western* of 1836, was the first large steamship to regularly ply the transatlantic route (even though Dr Dionysius Lardner of the University of London had thought such a feat would be impossible); Brunel's SS *Great Britain* of 1839 was the first large ship of its size to be driven by a propeller; and his SS *Great Eastern* of 1853 was the first large vessel to be constructed with a double hull, something previously unknown. Its size was not exceeded for 40 years.

Neither Marconi nor Baird shared Blumlein's or Brunel's boldness of concept. Both Marconi and Baird had a 'blind spot'. Marconi's dominant urge was to extend his system of wireless communications, but he failed at first to appreciate the importance of continuous-wave operation of transatlantic working, and thereby failed to visualise the benefits of radio telephony. Several of Marconi's contemporary inventors took a different view, but Marconi did not share their optimism. He considered the Morse code to be quite suitable for ship communications and for transoceanic signalling, and saw no real need for a wireless telephone. His approach was pragmatic and he was not interested in pursuing scientific investigations in fields which had a doubtful commercial viability. This blind spot was unfortunate initially for the Marconi companies and the furtherance of communications. Luckily for Marconi, some of the important early work on radio telephony was undertaken by two of his rivals, de Forest and Fessenden, and neither inventor had access to financial or engineering resources or skills comparable to those of the Marconi companies.

In Baird's case, he ignored for too long the inevitable move towards high-definition television and the use of cathode ray tubes in receivers. He remarked in October 1931 — six years after he had demonstrated a crude system of television—that he saw 'no hope for television by means of cathode-ray bulbs', and that 'the neon tube [would] remain as the lamp of the home receiver'. He was sceptical about the success of the use of short waves in television 'because they covered a very limited area'. Unfortunately for Baird and the Baird companies both EMI in the UK and RCA in the USA, among others, were vigorously investigating the utilisation of the high frequency and very high frequency bands for high-definition television transmission. Contemporaneously they were developing not only cathode ray tube television receivers but also electronic camera tubes. Their view was that high-definition television had to be based on an all-electronic system, rather than one which used some form of mechanical scanning, since only such a system would lead to images having an entertainment value.

Although Blumlein's and Brunel's originality of thought notably advanced complex electrical engineering and civil engineering systems and structures respectively, they each had the facility to look afresh at existing designs and configurations. When, for example, in 1934 Blumlein patented the applications of a circuit which comprised in its most basic form just one valve and one resistor (the cathode follower), valve amplifiers had been used for more than twenty years. The circuit, and its transistor equivalent, the emitter follower, subsequently became one of the most important circuits in the whole corpus of electronics. It provides an elegant and simple solution to the problem of matching two dissimilar loads.

Again when, in 1838, Brunel designed and supervised the construction of the Maidenhead bridge across the Thames it had the flattest arch of any brick structure in the world, despite the fact that bricks had been used for building purposes for hundreds of years. The bridge is remarkable not only for the daringness and ingenuity of its design, but also for the gracefulness of its appearance.

Brunel's Maidenhead bridge design shows his supreme self-confidence, a feature that he shared with Blumlein, Edison, Steinmetz and Tesla. All inspired confidence in others. Steinmetz[11] at General Electric became known as 'the Doctor' and 'the Supreme Court'. He was a 'wonderful teacher. His understanding was lucid to an extraordinary degree. He had a wonderful command of language, he could draw diagrams and make calculations which made everything clear in the simplest way, and he took a great interest in giving his help to anyone who wanted it'. Possibly his greatest achievement was his effect on other people: One writer has opined that 'his greatest work was to start the General Electric engineers upon the use of proper methods of calculation'.

Likewise, when Blumlein worked at EMI he had complete freedom to engage in the resolution of the many problems on which the research and development staff were engaged. One of his colleagues wrote of his seminal effect on all the work and of how 'being so full of ideas himself, he went out of his way to give credit to others. This, in turn, promoted complete trust so that [the staff] would discuss with him even half-baked ideas'[12].

Another colleague, E A Nind[13], remembered, as previously stated, Blumlein as 'a delightful man', who was 'very good at explaining anything. He did not get exasperated if you did not understand, and would go through a point again and again'. He had an 'inexhaustible patience' and a 'great facility for converting quite complicated mathematics into very simple circuit elements'.

Brunel also had an ability to describe the principles of his designs in elementary terms. His son, in his biography of the great engineer, tells an anecdote about the building of the Maidenhead Bridge[14]:

'The contractor being alarmed at learning that the arch was the flattest known in brick, Brunel pointed out to him that the weight which he feared would crush the bricks, would be less than in the wall which he, the contractor, had recently built, and he convinced him by geometry, made easy by diagrams, that the bridge must stand.'

Blumlein's, Brunel's, Tesla's and Edison's capacity for hard work was immense. For them, normal working hours meant nothing. One of Brunel's assistants once recalled:

'I never met his equal for sustained power of work. After a hard day spent in preparing and delivering evidence, and a hasty dinner, he would attend consultations till a late hour; and then, secure against interruption, sit down to his papers, and draw specifications, write letters or reports, or make calculations all through the night. If at all pressed for time he slept in his armchair for two or three hours, and at early dawn he was ready for the work of the day.'

Tesla always claimed that he never slept more than two hours a night[15]. His daily routine was usually to work until 6.00 p.m., or sometimes later, then change into evening clothes for dinner, at exactly 8.00 p.m., at the Waldorf hotel. On finishing his meal, at precisely 10.00 p.m., he would leave the hotel and return either to his rooms or his laboratories to work through the night.

Edison, too, would work very long hours. His biographer says that 'he called his men together at any hour of the day or night. There were neither hours nor days in the week, and he seemed to be at his best at night. The best time to reach him, even to go over his mail with him, was at midnight'.

Several anecdotes have already been related about Blumlein's insensitivity to normal working hours, namely his request to Hardwick (when he was playing tennis one Sunday) to return with him to the laboratory in order to resolve a certain problem; his all-night consideration, with a few colleagues, of the specification of the 405-line television waveform; his collapse through exhaustion during the construction of the Alexandra Palace London television station; his solution, at 10.00 p.m., in one of EMI's laboratories, of a problem which had been troubling Lubszynski and Turk for several days; and his wartime practice of sometimes sleeping in the labs. when a particularly urgent problem arose. On occasions, during the bombing attacks on London, he would work long hours at Hayes and then, late at night, proceed to an AA gun-battery at Gunnersbury Park to observe during the long air raids the VIE sound locators in operation. At other times he would undertake night-time fire-watching duties. Often during the war he would arrive home (Figure 17.1) at 3.00 a.m. His wife never knew at what time he would be back home; time meant nothing to him. But despite the arduous schedule of work which he set himself, he never lost his sparkle and his ability to work with people.

Figure 17.1   Blumlein's house at 31 The Ridings, Ealing, London with the GLC 'blue plaque'

Source: the author

A remarkable memory was a characteristic of Blumlein, Steinmetz and Tesla. Of these three creative engineers of genius, Tesla was the most idiosyncratic. He would summon one of his machinists and draw a sketch of what he required, in the middle of a large sheet of paper. Irrespective of the complexity of the task to be undertaken, the sketch was always less than one inch in its largest dimension. After details had been given to the workman Tesla would destroy the drawing. He required the machinists to work from memory. As O'Neill has written[15]:

'Tesla depended entirely on his memory for all details, he never reduced his mentally completed plans to paper for guidance in construction, and he believed others could achieve this ability if they would make sufficient effort. . . .

'All those who worked with Tesla greatly admired him for his remarkable ability to keep track of a vast number of finest details concerning every phase of the many projects he had underway simultaneously.'

It seems that no employee was ever given any more information than was necessary for him to complete the given task. Tesla was secretive about his objectives, since he claimed that Edison received more ideas from his associates than he contributed. Since Tesla did not wish to be accused of taking ideas from others, he bent over backwards to avoid this position. Interestingly, he refused to be the co-recipient with Edison of the 1912 Nobel Prize in Physics, possibly because he regarded himself as a creative engineer and Edison as a mere inventor.

A vignette of Blumlein's ability to recall facts and events has been given by a former junior EMI engineer[16]:

'What I remember most [in 1991] was his fantastic memory for detail and his ability to understand and analyse by quick observation and minimum use of words, followed by an almost abrupt appraisal of what one was doing and suggestions of what to try next to solve the problem. In particular, I recall him looking over one of my experimental rigs.

"Does it work?"

"Yes, but not as well as it could."

"Let me fiddle with it."

A few minutes later:

"It's not right — try this mod[ification]. If that doesn't fix it try this. I'll be thinking about it."

A week or so later, he looked over my shoulder again, needing no reminder of the details of the system, which was working reasonably well with his second suggested mod.

"Does it do the job?"

"Yes, but I think we could get some improvement if I carried on a little longer."

"Don't waste your time — if it works well enough leave it alone. Time isn't free."

Another EMI engineer, using computer jargon, remembered Blumlein had 'an extraordinary store in his head and a very rapid address system'.

Blumlein, Brunel and Steinmetz shared a great love of smoking. Tobacco was a luxury which they indulged in to excess. Brunel, like the great British Prime Minister W S Churchill, and Steinmetz, smoked cigars: 'At all times, even in bed, a cigar was in his mouth, and whenever he was engaged, there, near at hand, was the enormous leather cigar case so well known to his friends, and out of which he was quite as ready to supply their wants as his own.' Blumlein's preference, which was shared by another British Prime Minister, H Wilson, was to smoke a pipe — that 'symbol of tranquillity in times of turmoil'. He could not work without it, and if, due to an accident, his pipe broke, as on one occasion at Alexandra Palace, immediate steps had to be taken to remedy the situation. During the war, when tobacco was not in plentiful supply, Mrs Blumlein, who was evacuated with Simon and David to Penzance, Cornwall, had to make every attempt to buy as much as she could obtain. Blumlein's first question on meeting his wife and children at Paddington station was usually, 'How much?', meaning how much tobacco had she purchased.

Pipe smoking conjures up an image of avuncular affability, of good humour, of reasonableness and of calmness. The pipe can be seen as an accessory of the thinker. During intellectual argument, the straight stem of a 'billiard' pipe can be used to emphasise a point, and a smoker's ruminative puffing, which might appear to connote profound cogitation, can be useful in delaying an answer to an awkward question.

Smoking is a hazardous vice, though the nicotine in tobacco smoke may have a beneficial effect since it is rapidly delivered to the brain and can enhance concentration. Anecdotal evidence from smokers has existed for quite some time that smoking aids concentration. Recent research has shown that nicotine mimics acetycholine, one of the brain's natural signalling molecules.

Whatever the benefits or dangers of smoking may be, Blumlein, Brunel, Steinmetz, Churchill, Wilson and others exemplify the remark of Emile Zola[17]: 'I do not believe that intelligence and creative thinking are injured by smoking'.

From the preceding discussion it seems that the primary factors which can influence discovery and invention are:

1 commitment and perseverance—the progression of a task even when adversity arises, as exemplified by Baird's work on television;
2 curiosity—the disposition to be inquisitive, which characterised much of Faraday's work;
3 imagination—the prime determinant of creativity in science, engineering and the arts;
4 originality and faith—the investigation of new fields and a belief that success will follow, as illustrated by Blumlein's work on stereophony and slot antennas;
5 insight and intuition—the finding of solutions based on experience, comprehension and perception, as represented by Brunel's various bridge and ship designs;
6 appreciation of the problem or the need—the recognition of which may lead to inspiration and thence to realisation, for example, Tesla's work on polyphase machines.

Secondary factors include:

1 serendipity, luck and good fortune—the finding of a discovery or useful fact or thing not sought for, as typified by Edison's invention of the phonograph; and
2 the favourable accident—as highlighted by Goodyear's work on the vulcanisation of rubber. 'I was encouraged in my efforts by the reflection that what is hidden and unknown and cannot be discovered by scientific research, will most likely be discovered by accident, if at all, by the man who applies himself most perseveringly to the subject and is most observing of everything related thereto.'

Of all these factors the greatest is imagination. It is the basis of all creative work. The originality which marks the efforts of great engineers and scientists is to be ascribed to a tireless activity of their imaginations followed by a profound intellectual analysis of the possibilities of their ideas. Imagination distinguishes the creative geniuses of civilisation from their commonplace fellows. As Chesterton said: 'It isn't that the latter can't see the solutions, it is that they cannot see the problems'.

When Newton was once asked how he made his discoveries, he replied: 'By always thinking unto them. I keep the subject constantly before me and wait till the first dawnings open little by little into the full light'. Similarly, single-mindedness of thought and purpose typified the efforts of Blumlein, Brunel, Edison, Tesla and others.

Problems are part of our existence—their solution leads to truth or enrichment. Mankind advances by the labours of those for whom imagination, faith, and intuition are ever vital components of their human nature. The immortal Newton put the matter eloquently when he wrote:

'I know not what I may appear to the world, but to myself I seem to have been only like a boy playing on the sea-shore, and diverting myself in now and then finding a smoother pebble or a prettier shell than ordinary, whilst the great ocean of truth lay all undiscovered before me.'

Blumlein's life and work epitomises that of the pioneer engineer who has a brilliant intuitive imagination for new ideas—ideas founded on curiosity rather than ideas evolved by deduction, an analytical intellect and a sound appreciation of his art which can select and advance those ideas which have a realistic chance of achieving finality, and a commitment and perseverance which overwhelms difficulties to enable success to be accomplished. Blumlein's achievements in both peace and war were impressive. As Shoenberg said: 'There was not a single subject to which he turned his mind that he did not enrich extensively.'

Newspaper accounts of Blumlein's death referred to the 'national loss':

Figure 17.2   Alan Dower Blumlein (1903–42), probably the greatest British electronics engineer of the twentieth century

Source: Mr S J L Blumlein

'It can be truly said that Alan Blumlein's loss is a national one, and leaves a gap in the world of scientific research which it will indeed be hard to fill. The value of his work was accepted by all those in government circles who were familiar with it. He had a burning passion for the survival of justice and humanity in the world. He drove himself to the utmost limits of endurance and his modesty was such that few people realised how much he was doing in the struggle. It is a tragic and cruel waste of such a fine intellect just as he had reached meridian in length of days, at the age of 38, leaving a young widow and two sons, Simon and David, as yet too young to understand the loss of a brilliant father, who gave his life with no thought of self, in the sacred cause against oppression, having by his genius been the means of saving thousands of lives.' (Figure 17.2)

Therefore:

> Let us now praise famous men,
> Men of little showing,
> For their work continueth,
> And their work continueth
> Broad and deep, continueth
> Longer than their knowing.
> (Rudyard Kipling)

References

1 Quoted in 'A dictionary of scientific quotations', by A L Mackay (Institute of Physics Publishing, Bristol, 1991)
2 Article on 'Genius', 'Encyclopaedia Britannica', 1972, **10**, p. 181
3 WOLPERT, L.: 'The unnatural nature of science' (Faber and Faber, London, 1992), chapter 4 on 'Creativity'
4 TREVELYAN MILLER, F.: 'Thomas A. Edison' (Stanley Paul, London, 1932)
5 Article on 'Edison', 'The New Encyclopaedia Britannica', 1991, 15th edition
6 JOSEPHSON, M.: 'Edison: a biography' (Eyre and Spottiswoode, New York, 1961)
7 CONOT, R.: 'Thomas A. Edison: a stroke of luck' (Da Capo, New York, 1979)
8 Ref. 6
9 BRUNEL, I,: 'The life of Isambard Kingdom Brunel, Civil Engineer (1870)' (David Charles Reprints, Newton Abbot, 1971)
10 NEWMAN, E. A.: taped interview with the author, personal collection
11 ALGER, P. L.: 'The human side of engineering: tales of General Electric engineers over 80 years' (Mohawk Development Service, 1972)

12 LUBSZYNSKI, H. G.: 'Some early developments of television camera tubes at EMI research laboratories', *IEE Conf. Pub.*, 271, 'The history of television', 1986, p. 62
13 NIND, E. A.: taped interview with the author, personal collection
14 Ref. 9
15 O'NEILL, J.: 'Prodigal genius. The life of Nikola Tesla' (Neville Spearman, London, 1968), p. 208
16 CALLICK, E. B.: letter to the author, 25th May 1991, personal collection
17 LANGAN, S.: 'Puffs instead of huffs', *Daily Telegraph*, 15th June 1991

Appendix 1

A D Blumlein and stereo sound recording (J A Lodge)

Introduction

For approximately ten years until 1995 the late Mr J A Lodge, formerly of Thorn–EMI's Central Research Laboratories, had charge of all CRL's pre-war archives. Prior to his death in 1995 he had carried out an exhaustive examination of the extant files relating to Blumlein's work on sterophony and had written a comprehensive account of the experiments which Blumlein and his colleagues had undertaken from 1933 to about 1935. Ill-health prevented Mr Lodge from publishing his work. However, knowing of the author's intention to write a biography of Blumlein, Mr Lodge suggested that his findings might be included as an appendix. The author was pleased to accept this generous offer. The following seven sections of the paper are given verbatim. (To avoid unnecessary duplication the sections headed 'Introduction' and 'Early theoretical studies' have been omitted.)

Building the experimental apparatus

On the 8th January 1933 there is the first description in the file of experimental work. The heading is 'Polar curves of two WE [Western Electric] CTs for binaural work'. CTs are, of course, condenser transmitters—i.e. microphones. Many have expressed surprise that Blumlein did not use his own moving coil microphones, but EMI was in the middle of re-equipping their studios with the moving coil system, and almost certainly there would have been a shortage of moving coil instruments, and a surplus of Western equipment—not only microphones but amplifiers and wax cutters. And Blumlein knew that the Western equipment was adequate. It was being replaced mainly for commercial reasons.

Then on 13th February, Blumlein designed a 'shuffler' with the main phase-shifting element now including a bridging inductance, and designed the low phase-shift transformers for it the following day. Then starting 20th February, there are various tests of microphones, circuits and the shuffler.

The object of the experimental programme would have been to verify Blumlein's theory. This could have been done simply by a loudspeaker/microphone arrangement, but the results would have been ephemeral and it would have been difficult to compare experimental results. Blumlein therefore decided that as an interim measure, he would use the disc recording technique included in his patent 394 325 for his main first set of experiments.

The recording section files are not the only documents surviving; the weekly progress reports of the Recording Section starting 23rd September 1933 have also come to light, written by Turnbull, Holman and H A M Clark. The first entry refers to a new design of a binaural pick-up being made, which was completed and tested by the 2nd December. It had 20 dB separation up to 4 kHz after the damping and tracking had been improved. How the measurements were made is not mentioned; there were no complex cut test discs at this time, so separate lateral and vertical discs must be assumed.

Two binaural gramophone pick-ups have survived, although four are mentioned in the weekly reports. The earlier survivor is of the moving armature type, with a surprisingly massive armature. This is probably the PU3A of late 1933. Some recent notes suggest that it was tested in June 1933, but it is more likely that this test was of the second pick-up, PU2. A further improved pick-up, using two moving coils, is mentioned in the progress report, starting January 1934. It was probably designed by Holman, and is the other survivor. It seems to have been designated PU4A. Nothing is known of PU1—if it ever existed other than as a paper design.

The stereo wax cutter survives as well. It was made from two Western Electric moving-armature units coupled to a single stylus by a lightweight lever system, so that one unit moved the stylus vertically and the other horizontally. The first calibration of the recorder is believed to have been on 12th July 1933. Bandwidth is reported to have been about 4 kHz.

There are no references in the progress reports, before January 1934, to the construction of 'shuffling' amplifiers, except for the 'shuffling' circuit, when improved ones were started, so that it is reasonable to assume that the first experiments used redundant Western Electric amplifiers; in fact there would be two complete Western recording channels plus the shuffler. As mentioned above, Blumlein's detailed design of the shuffler is dated to 12th February 1933.

Experiments start

Drawn up sometime before August 1933, (the item is undated) there is a diagram of a complete microphones–shuffler–amplifiers–loudspeakers system, with the microphones in the large Auditorium which was at the rear of the Laboratory; the amplifiers on the ground floor of the Laboratory; and the loudspeakers in the listening room on the first floor. The loudspeakers appear to have been 12 inch moving coil paper cone units mounted on large balsa wood baffle boards. The difference channel could be cancelled remotely from the listening room to compare stereo and mono; the switch is marked 'ADB'. The diagram shows clearly that Western Electric condenser microphones were used. For some reason, although earlier tests had been done on microphones nos. 7 and 8, now nos. 6 and 8 were used. But then the early condenser microphones were very apt to develop crackles, and recording engineers have many stories of repairing them in the most unlikely places.

The Auditorium was built in 1928 as part of the new Research Laboratory, when the Gramophone Company started its sound-on-film development. It was 100 ft by 50 ft by 30 ft, and had a projection room, a proscenium arch and stage, and a polished maple floor. The decor was that of a cinema but the floor had no rake and there was no fixed seating.

There are no results from these back-to-back trials in the files, but as the work continued, presumably there was at least a modicum of success. So far, Blumlein had restricted his binaural work to below 750 Hz, arguing that above this frequency there would be ambiguity of phase difference, and this set the spacing of his microphones. However, on 23rd October, he filed a further patent, BP 429 022, in which he proposed the use of a second pair of microphones with much less spacing to deal with the binaural effect at higher frequencies. This may have been as a result of the back-to-back trials, but for the time being for experiments, the equipment was unchanged.

The next circuit in the file, dated 30th August 1933 is a very much expanded version of that described above, allowing switching to the recorder or to a pick-up. Furthermore, an amplifier with 26 dB gain and a variable loss pad in its output, originally designed for a strip (ribbon) microphone, has been added in the difference channel before feeding into the 'shuffler'; obviously the signal in the difference channel was less than expected. The diagram still indicates Western Electric condenser microphones. The wax recorder was probably in the ground floor equipment room.

On or before the 9th September, the wax cutter was tried out for the first time. Two commercial recordings were played, one into each channel, one vertical and the other lateral. Good separation was achieved, and it was judged that binaural recording was practicable. When the Prince of Wales visited the Laboratory the following year, this record was demonstrated to him.

The first recordings

It was with this set-up that the well-known 'walking' and 'talking' records, the first complex-cut stereo records ever, were made some time before 16th December 1933. The Auditorium was the studio and there is no indication of special acoustic treatment. Nor is there any record of the positions of microphones, which were fastened to the two sides of a 7 inch block of wood. Anecdotes have stated that Holman and Blumlein's HB moving-coil microphones were used, but the files quite definitely show WE microphones.

The signals feeding the two cutters were sum, for the lateral cutter, and difference for the vertical. Although attempts were made to make the two cutters identical, the degree to which this was achieved is rather doubtful, so that when playing these records with a modern stereo pick-up, it is desirable to combine the two signals to give the original lateral and vertical signals, and to equalise them, before recombining to feed loudspeakers.

The progress report for 16th December reports making these discs, and records a 'definite binaural effect'. The taped list of records is dated 19th December 1933. There were two matrix numbers, 5767 and 5768, with six subs of the former and four of the latter, i.e. ten 10 inch sides in all. The tests were used by Blumlein to try to optimise the settings of his shuffling, then to see whether binaural provided any of the benefits for which he hoped. Single speakers—often Blumlein himself—walked across the 'sound stage', talking all the time, and then multiple simultaneous conversations were recorded, to see whether the 'cocktail party' effect came into play. Several times Blumlein starts by speaking from the centre to facilitate balancing on playback, and then walks across the stage whilst asking for changes in the difference channel gain and in the shuffler. Surprisingly, no records appear to have been made simply recording the output of the microphones for shuffling experiments to be done later at leisure.

The main demonstration to Shoenberg and his colleagues of these records is recorded in the 30th December Progress Report, with widely spaced loudspeakers and alternatively with one each side of a radiogram cabinet. At least some of the demonstrations are likely to have been in the Auditorium. Quite good results appear to have been obtained, and it was decided to carry out trials recording music at the Abbey Road Studios. We know now that the binaural gramophone pick-up used for these demonstrations had only fair performance, and this explains the rather equivocal comments, for example, by H A M Clark, 'A definite binaural effect'.

By 11th January 1934, the equipment—in effect two Western recording channels plus monitoring facilities—had been transferred to the Abbey Road Studios by Turnbull, Clark and his assistants, and further 'walking and talking' records were made, probably in the big No. 1 studio.

The following day, the first stereo music discs were made, again in No. 1 studio. They were of a dance band, possibly Ray Noble's, with the microphone pair approximately 45 ft back. Heavy shuffling was used, and 6 dB brightening at 5000 Hz, i.e. a narrow band boost. Two cuts were made, on masters TT1557-1 and -2, the second at 6 dB lower level.

Then six more discs were cut, again in No. 1 studio, this time of a combination of three pianos; again the artistes are unknown. For comparison purposes, there was a standard HB microphone equidistant from each piano, and the pianos were arranged to converge on the HB. There was a 60° angle between the outer pianos, and the binaural microphones were about 1½ ft from the tip of each piano. Six waxes were cut, WT.5769 1 to 6, all with heavy shuffling and at varying levels. For one only, sub. 4, the binaural microphone was moved 6 ft forward. Subs. 1, 2 and 3 are of a Brahms Hungarian Rhapsody; 4 and 5 of the Ride of the Valkyries, and 6 is of rehearsal snippets.

A week later, on the 19th January, further binaural recordings were made, this time of a rehearsal of a performance of Mozart's Jupiter Symphony, played by the London Philharmonic Orchestra conducted by Sir Thomas Beecham. Nine sides were cut, of four excerpts, all on 10 inch waxes, on WT. 5571 subs. 1 to 9, in all cases with heavy shuffling and 6 dB sharp boost at 5000 Hz. For the first two, the microphones were [placed] 13 ft high and 12 ft back from the foremost first violin; for the third and fourth approximately 1 ft higher and 6 ft further back; and for the remaining five, approximately 25 ft, from the nearest violin. For four, five and eight, the levels, both sum and difference, were increased by 2 dB, for sides six and nine by a further 2 dB, and for seven 2 dB more.

In their progress report of 13th January, Turnbull, or perhaps Clark, reporting the 'dance band' and 'three piano' records, wrote that 'Binaural gives opening out on music, but the effect is more marked for speech'. Similarly on 20th January, they reported of the Beecham recordings that there is 'solidity, but less effective than on speech'. When H A M Clark wrote in 1946 a further report on the future of stereo, he repeated his earlier comments, presumably made from playback using the PU4A.

In December 1974, P B Vanderlyn, who as a junior had taken part in the original experiments, and who was joint author of the 1957 paper on the 'Stereosonic' system, transferred some of the 1933–4 pressings to tape, allowing for poling and balancing errors which had crept into the 1933–4 experiments. He used a Shure M7SB pick-up with a 0.0025 inch stylus. On the evidence of these transfers, there is no doubt that at least some of the discs made on 12th January, both of dance band and of three pianos, are good stereo, much better than the 1934 comments would suggest, but there does not appear to be even one of the 19th January records — that is, the Beecham discs — of equal standard.

Construction of sound movie equipment

When the music trials were over, in January 1934, work started on designing a new binaural recording chain for use in making experimental films.

By this time, in other parts of the Laboratory, the work on television was building up, and would have occupied a great deal of Blumlein's time; in early 1934, the first of EMI's electronic camera tubes was working; that vital component of the television system, the standard waveform, was being developed; techniques for maintaining a black level were being invented; and a series of experiments on the propagation of television signals at 44 MHz had started. Also, the extension of the moving-coil wax cutter to mobile operation was under way, and thus it was inevitable that although Blumlein was a man of great energy and industry, by this stage he could not devote a great deal of effort to the binaural project.

And by this time, the project had grown. It was to be all-new; it would use HB1B moving-coil microphones up to 750 Hz, and in accordance with Blumlein's October 1933 patent, a second pair of microphones, separated by about an inch, would be added to extend the binaural effect to about 7 kHz. The amplifiers would be designed to use a.c. heated valves and there would be new low phase-shift transformers. Each pair would have a separate shuffler.

It would not be possible to use moving-coil microphones for the high frequency units because of the small spatial separation needed. The approach adopted was to make a ribbon velocity microphone with two ribbons with the high frequency spacing, using a single field. To convert the microphone from velocity to pressure operation, acoustic absorption lines would be placed behind the ribbons. These consisted of rectangular copper funnels feeding into tubes packed with cotton wool. These lines were of manageable length because the microphones were not intended to operate below 750 Hz. The system had many problems, and was eventually abandoned.

A sound-on-film recorder with two tracks would also be needed, and this was probably arranged to be adapted from the sound camera developed by C O Browne in 1929, for the HMV sound film system.

It seems likely that the trials carried out the previous year had shown that stereo in a large auditorium, and in fact any good stereo, depended on the positioning of microphones and loudspeakers, and a series of experiments on this subject was also planned, and for these the microphone system was needed, making it the critical item. The main aim was still the cinema.

Microphones and amplifiers for binaural

Throughout the work, with a small exception when crystal microphones were the subject of experiments, moving coil microphones, type HB1B, were used for the spectrum up to 750 Hz, with their own shuffler. However, for the higher frequencies, five microphones were made, numbered BM1A to BM5A according to Blumlein's numbering method. (B meant binaural; M meant microphone; 1 etc. was the design number, and A was the version.) There was also a BM6 which never got beyond the design stage.

Examination of the polar diagrams of EMI's ribbon velocity microphone of that era shows that although it had the theoretical cosine characteristic at low frequencies, at a few kHz, it became more directional, and this may be the reason that Blumlein decided to make his higher frequency microphones pressure operated, although he had already described a system based on cosine microphones, and should he have wished to do so, he could have made binaural recordings going up to 3500 Hz with them. It is notable that there is no record of experiments to determine the relative merits of various frequency bands in binaural listening, nor any tests involving simple tones to guide the system design.

For preliminary experiments on the HF microphones, an absorber tube was added to one of the normal ribbon microphones designed in the

Laboratories, and in use at the Abbey Road studios, and frequency response measurements were made. The results were not very satisfactory, with the characteristic almost a sine wave, with peaks every 500 Hz. No contemporary comments survive, beyond the redesign of the packing of the absorbers, after measuring their acoustic impedance to the flow of air from an aspirator. This had little effect on the frequency response.

For BM1A, the first binaural microphone, a permanent magnet was designed, and the casting received for grinding in February; the ribbons and mounts had already been designed, with experimental absorber lines to make the velocity microphones into pressure instruments The expected output from both the LF and HF microphone amplifiers was calculated; the calculations for the strip microphone seem to indicate that the ribbons would be 9.6 mm square, although this does not seem very likely and the HF microphone and amplifier and the BA1A amplifier using triodes (B = binaural; A = preamplifier; 1 = first design; A = first version) were calculated to have 5 dB more output than that LF unit based on HB1Bs. A BA2A was designed by using screen grid valves with a higher gain. Blumlein's binaural system changed phase differences into amplitude differences, so that the phase responses were important. These were calculated with care, and in some cases, transformers were redesigned. They took similar care over the selection and matching of the two HB1B microphones for the lower frequencies, and after selection they were both carefully tuned to 500 Hz bass resonance, the signal circuits were padded to the same resistance, and the fields were connected in series with a shunt resistor on one to equalise the fields. One field was reversed, as was the speech coil, to minimise stray signals in the weaker difference channels. First microphone tests were done at the end of March.

The strip binaural microphone ran into many difficulties. The pole pieces and ribbon mounts had a much greater effect on the polar diagrams than had been expected. Again there was a standing wave in the absorbers, the sensitivity was low, partly due to low field from the permanent magnet, and the two channels were very much out of balance. The aluminium foil used was the thinnest obtainable, and trials with varying tensions and corrugations were ineffective but recalculation of the absorbers improved matters somewhat. In desperation, gold leaf was tried, as being the thinnest material available, and calculations showed that the Johnson noise from the resistance of the gold would be acceptable. This gave almost satisfactory results after the magnet and pole pieces were redesigned, but the strip was so frail that consistent results could not be achieved; this made it virtually impossible to balance the two channels; and it was always difficult to balance the two absorbers.

In June, H A M Clark calculated the theoretical ratio of difference signal to the sum for strip spacings of 8.1 and 1.3 cm for various frequencies and for a sound source displaced 10° from the axis. He obtained results of 40 dB at low and medium frequencies, down to 16 dB at 7 kHz, indicating the importance of low cross-talk and unbalanced distortion.

In September, a second permanent magnet was obtained (BM2), which improved sensitivity and polar response somewhat, but the other problems remained. By now the strips were of a more orthodox aspect ratio.

In October, H A M Clark carried out independent calculations on the extent to which velocity microphones could be used. Polar response tests on BM2A showed promise, and experiments started. Then the BM3A with an energised field, giving much more flux and hence sensitivity, and aluminium strips with three corrugations at each end, was designed and in October it was under test. The two strips were at right angles, but when tested the two polar diagrams were at 135°. This was reduced by 30° by redesigning the pole pieces but the process could not be carried farther. H A M Clark made many calculations regarding velocity microphones for binaural, including techniques for correcting for skewed polar diagrams, and in particular the effect of shuffling. Unfortunately he did not describe his conclusions explicitly, but the BM3A, operating as a pair of shuffled velocity microphones, seems to have been the most successful set-up.

Quite early on in the project, piezo crystal microphones had been considered, as they were quite compact, particularly useful for the high frequency microphone, but were deemed to have insufficiently good sound quality for professional standard recording. Then, in January 1935, two microphone sets using crystal microphone units were made. The first, BM4A, used the crystals only for the high frequency units, but the other, BM5A, used them for the low frequency units as well. This is the measure of how difficult the development of the strip microphones was proving to be. Variability of characteristics with time and temperature, however, turned this approach into a sideline, and although these units were used quite extensively in listening tests to compare the position of the perceived sound source with its real position, by Spring 1935, the BM3A with energised strips for high frequencies and HBs for the low was the usual arrangement.

A further microphone, the BM6A, was designed, but never carried beyond the field magnet test stage. It was all-ribbon, and had a large energised magnet, with four gaps in series. The two outer ones were for low frequencies, with strips spaced four or five inches, and the two inner ones about an inch for high frequencies.

Binaural was a new field, so that there was no background of knowledge on which to base experiments. As a result, there was a great deal of calculation during the microphone development—a thick file, in fact, and the above summary of the microphone work can give only a rough idea of the total labour involved. Blumlein designed the shuffler transformers and the first circuit, but as time went on, the design was somewhat varied by various people. The monitor loudspeaker amplifiers had a single driving triode and a PX4 5W power triode, whilst the main power amplifiers had a PX25. There was, as well, a full monitoring and control panel. All this equipment appears to have been ready around the end of 1934.

Experiments on loudspeaker systems

Early in 1934, experiments started to establish how to lay out loudspeakers for Blumlein's 2-channel binaural system. Earlier tests had shown that a 4-loudspeaker system with an outer and an inner pair helped to avoid a phenomenon called 'flickover', and which nowadays is well-known as 'hole-in-the-middle'. The arrangement used then consisted of two main 12 inch speakers as the outers and two smaller speakers fitted in the sides on an old radiogram cabinet, facing outwards, for the inner pair, driven mainly by higher frequencies.

This had been a temporary expedient, and now a permanent arrangement was to be made. The loudspeakers were all 12 inch diameter, and the pair on each side were to be driven by a single transformer. By careful design, at above 140 or 250 Hz the outers had about 3 dB loss, and the inners an equal amount of boost, or the boost could be switched out. The measurements made on the loudspeakers, which had paper and aluminium cones, showed that up to 2 kHz they had quite a smooth response, but above this, there were peaks and troughs up to 4 or 5 dB— perhaps more, for the only characteristic to survive is the mean of measurements on three [loudspeakers]. An alternative transformer design was to be fed with sum and difference signals, the combining taking place in the transformers. The fact that this worked well demonstrates the care with which the phase responses of the two channels had been balanced.

When the binaural microphone BM2A became available in September 1934, to use in experimental work, a series of trials was planned. Two rooms on the first floor of the Research building were to be used, one as a studio and the other as a listening room, both probably about 18 ft square, but there are no records of serious experiments.

In March 1935, the main listening programme started. Listening was transferred to the Auditorium, which was about 40 ft wide, with acoustics much closer to those of a cinema. Most of the listening was done 32 ft back from the loudspeakers. Almost all trials used recorded speech as the test signal; there were two old gramophone records, one of Winifred Hughes speaking on Walpole, and the other of Sir Lawrence Bragg, and latterly live speech was used. Also there were test records carrying 500 and 2000 Hz tones. A gramophone playing one of these records would be placed in several positions, with the binaural microphone fixed, and feeding the loud-speakers via a shuffler and gain controls. The loudspeakers were of the energised type, as was usual in those days, before the development of good magnetic materials, and the sensitivities of the individual instruments were varied by varying the field current, with the voltage drop across the field as the parameter recorded. They were fitted to large flat baffles.

The first task was to find an arrangement of the four loudspeakers which gave an accurate sound image to a centrally placed listener as the source was displaced laterally. The variations were of relative sensitivity and crossover frequency between the inner and outer speakers, the addition of the reflecting and absorbing material around and between them, and [alterations of] their angles. There are many graphs of actual vs. perceived position, with two young engineers, Westlake and Trott, the listeners most of the time, although Clark and Turnbull both did runs early in the work, and Blumlein did run no. 26. His run has noticeably more scatter than the others.

After 47 runs, over a period of a week, the plots of real vs. perceived position were pretty linear, with little scatter, and switchable pairs of 1000 Hz filters, one high pass the other low pass, were added, then the listeners moved their position 6.5 ft to one side. Immediately scatter in their graphs became much worse, with the mean curves decidedly non-linear and lopsided and divergence between the results from the two observers. About this time, a third observer, Vanderlyn, joined the team.

On 28th March, they summarised the week's results, showing that for an off-axis listener, the inner speakers running at about two-thirds field more or less eliminated flickover, but reduced the width of the image. They also found that variation of the crossover frequency between inner and outer speakers, at least within the range available, had little or no effect, but that sharpness of location was better with high frequencies, i.e. above 1000 Hz, and for the 2000 test disc. They also tested the hearing threshold of the three observers at 100, 1000, and 4000 Hz. Trott's threshold at 4000 Hz was significantly higher than that of the other two, but this appears to have had no noticeable effect on his location test results.

The tests continued, but the experimenters were unable to find a loud-speaker layout which would give appreciably better off-axis results. They do not comment on this, but it must have been disappointing.

The next series of tests may have used only the inner loudspeakers, pointing about 30° outward; the records are ambiguous, the layout sketches appearing to show two speakers, whilst field voltages are recorded for all four. The binaural image was restricted somewhat, and was not quite as linear as before, but was fairly good for a central listener. Certainly later tests used all four speakers. Then a light curtain was suspended across the Auditorium, and direct tests were made to determine how good aural location was. The graphs are good straight lines, with little scatter, even when the listener was off-centre.

All the system trials so far seem to have been made with the BM2A microphone, i.e. the low frequency unit was two HB moving coil pressure microphones, and the higher frequency unit a twin strip microphone made pressure sensitive by added absorbers. In April, experiments started using the BM3A, i.e. the microphone with plain velocity sensitivity, and an energised magnet After a false start, perceived location was about as accurate as it had been with the BM2A; the 'speaker' was placed various distances from the microphone

A few days later, a new series of tests started, with a group of people arranged in a 5 ft circle, talking, as the test source, and this time, the BM5A microphone, with all-crystal elements, was used, with the shuffling circuits part of the coupling between the power amplifiers and the loud-speakers. The graphs are for a group of listeners from the centre to about 5 ft to one side, and again are surprisingly linear. The group of listeners was about 32 ft back from the loudspeakers.

Using the same loudspeaker and microphone arrangements, the ability to perceive depth in the sound image was tested. The test positions were in four rows, 5, 7.5, 10, and 12.5 ft back from the microphone, and for each row there were three test positions, centre, five feet left and five feet right. Again there were four listening points, centre and up to 5 ft to one side, and this time as well as recording the results for individual observers, they were also averaged — how is not described. The results were quite fair, and when the BM5A was replaced with the BM3A, similar results were obtained. Blumlein is believed to have commented that they were at least as good as Bell Labs. achieved with their three-channel system.

One trial was also carried out in which a single signal was fed to both loudspeakers through complementary attenuators, thus simulating the two binaural signals, which, according to Blumlein's theory, should differ in amplitude, not phase. As far as the writer is aware, this must be the first trial of 'panpotting', and the graph of perceived versus intended position is a good straight line. However, the idea was not pursued further.

This concluded the loudspeaker system trials.

The sound-on-film system

Blumlein's patent describes the system used for adding the extra sound-track to the film, with the soundtrack in the normal position carrying the 'sum' signal and an additional track the 'difference'. This ensured that the stereo film could be played in an unmodified cinema. The task of making the recording equipment was given to C O Browne, one of the founding members of the Gramophone Company's research team, whose first task, in 1928, had been to develop a sound-on-film system, before talkies became universal. Technically, the HMV sound system was very successful, but commercial considerations seem to have ended the project, and by 1933, Browne was heavily involved in the development of television.

In April 1934, however, he was asked to make the binaural recorder. As far as can be determined, the film transport was that developed in 1928, but to get the two recorded tracks sufficiently close together, new galvanometers were needed. Variable area recording was to be used, without squeeze tracks. For playback, a double cathode photocell was needed, and this task was passed to the other leader of the 1928 project, W F Tedham who had developed photocells for the earlier equipment, and who was now also heavily involved in television, developing cathode ray tubes.

By April 1935, the twin recorder was being tested. It had been designed with an upper resonance at about 9000 Hz, but there were subsidiary resonances at lower frequencies which were to be damped by immersing the movement in oil. At first this was not very effective because sufficient power was dissipated to heat the oil and reduce its viscosity. Castor oil gave sufficient damping, and the galvanometer was flat to 1 kHz with a gentle rise of 4 dB to 9 kHz. However, the castor oil was insufficiently transparent to allow enough light through to expose the film fully, but by the end of May, with a better optical system and new mirrors the results were satisfactory. At the same time, the reproducer sound head was checked and found to be 12 dB down at 8 kHz; of this, 6 dB could be accounted for in the slit and pre-amplifier.

In June, the first tests were run on the binaural film sound system, with the results that although there was some frequency modulation, arising from vibration, there was excess noise because a 'squeeze' track had not yet been incorporated; and the channels were unbalanced, giving rise to some overloading. The equipment was capable of performing as desired. So a movie camera, probably also a relic of the HMV sound-on-film project, was borrowed from C O Browne and set up in the Auditorium. The intention was to determine how far the use of a binaural system would override the reverberation problems. At this stage, both the crystal and ribbon microphones were being used for experiments and pictures were being projected on a butter muslin screen.

But then in June, logistic problems started. First, the amplifiers were borrowed for several days to contribute to a major demonstration of the Company's television system, and then staff had to be found to design the sound system for the Alexandra Palace television station contract, with priority, and as the BBC's equipment was to have novel features, such as a.c. heating for valve cathodes, and running valves at the BBC's very conservative ratings, experimental work was needed.

In the middle of the month, the first binaural sound recordings on to film were made in the Auditorium, both with the crystal and velocity microphones, and then with a 14 ft wide set, 16 kW of lighting, and using the velocity microphone, which gave better results in film conditions, a few test films were made. Sound was regarded as fairly good, but picture quality was poor. It is possible that the 'walking and talking' film including Blumlein was made at this time. A clapperboard system to synchronise sound and film was produced, and the microphone was improved so that it could be mounted close to the camera and record from the whole stage.

At the end of July, acoustic damping had been put in the Auditorium and the 'Move the orchestra' film was made on a Friday evening, to be processed at Humphries Laboratory over the weekend. It lasted four minutes, and used 30 kW of lighting on a 20 ft wide set. There were two takes, the second with the sound recorded at higher level. It was a simple playlet, performed by the Company's amateur theatrical group, done as a single take and with no movement of the camera. The 'Orchestra moving' effect was achieved by two stalwart juniors carrying a gramophone from one side of the set to the other, behind the backdrop.

In the words of the weekly report, the binaural sound was 'fairly satisfactory technically', but with the benefit of hindsight, it is easy to see the major development which would be needed to apply the system to an 'A' movie.

In August, a simple 'squeeze' track and noise reduction circuit were added, and after one of the galvanometer ribbons broke from the large bias current needed to position them correctly new ones with mechanical adjustment which had already been made were installed. Also, a purpose-made cine screen was installed, to give a 50% increase in picture brightness.

Making test films continued, too, with the 'fire engine' and 'train films', and there were the investigations into the 'haziness' of position in the centre of the sound field. One scheme was to vary the field currents of the loudspeakers, but this was effective only on tone tests. Improvements to both the crystal and the shuffled velocity microphones were made, too, but in mid-September, the work was laid aside, almost certainly because of the overwhelming need for skilled engineers on the television project, now less than a year away from the first broadcast from Alexandra Palace.

In 1946, H A M Clark wrote a report on the future of stereo, and repeated the opinion that it had little to offer in music recording. However, the Blumlein stereo equipment was taken over by G F Dutton, who carried out a series of rather inconclusive experiments on music, or so it must be assumed from indirect evidence.

The equipment then passed back to Clark, and a new stereo microphone was made from a pair of the latest twin diaphragm instruments with cosine characteristics. A great deal of analysis and experiment followed, with specially arranged music sessions and a great deal of experimentation on microphone location, some of it described in the 1957 paper by Clark, Dutton, and Vanderlyn, and culminating in the issue of 'Stereosonic' music tape records and, later, discs.

Appendix 2

A D Blumlein's patents

(Author's classification based on patents listed on pages 220 and 221 of *Electronics and Power*, June 1967)

On telephony

291 511, March 1927 (with J. P. John). Phantom loading coils for reduction of crosstalk

337 134, June 1929. Matching coaxial submarine cable to land telephone line (circuits used having high attenuation in common mode and low attenuation in push–pull)

334 652, June 1929. Telephone loading and phantoming coil (using the closely coupled inductor ratio-arm bridge technique)

335 935, July 1929. Submarine telephone cable carrying speech, with telegraph signals going down the unused speech return path

357 229, June 1930. Double shielding to reduce interference at ends of submarine telephone cable

402 483, June 1932. Shielded loading coil for telephones

470 495, November 1935. Time-division multiplex of telephone signals (using techniques based on the television waveform)

On monophonic and stereophonic recording

350 954, March 1930 (with H. E. Holman). Cutting head for gramophone recording (mechanical arrangement of moving parts)

350 998, March 1930 (with H. E. Holman). Cutting head for gramophone recording (electromagnetic damping arrangement)

361 468, September 1930 (with P. W. Willans). Gramophone pick-up (mechanical arrangement of parts)

363 627, September 1930. Cutting head for gramophone recording (electromagnetic damping of main resonance combined with mechanical damping of minor resonances

368 336, December 1930 (with H. E. Holman). Gramophone pick-up (mechanical arrangement of parts)

369 063, May 1931 (with H. E. Holman). Moving-coil microphone (with electromagnetic damping of main resonance)

394 325, December 1931. Stereophonic recording and reproduction (the system description and details of techniques)

417 718, March 1933. Gramophone pick-up or cutting head for recording (mechanical arrangement of parts)

429 022, October 1933. Stereophonic sound (use of differently spaced microphones for different frequency bands)

429 054, February 1934. Stereophonic sound (two channels obtained from sum and difference of outputs of pressure and velocity microphone)

456 444, February 1935. Arrays of microphones with outputs mixed to give various polar diagrams (these ideas derived from the techniques used in stereophonic sound)

On measurements

323 037, September 1928. Closely coupled inductor ratio-arm bridge: the basic design

338 588, August 1929. Circuit using variable resistors to give a fixed mutual inductor the appearance of a variable mutual inductor

362 472, July 1930. Constant-impedance variable-attenuation network

461 324, August 1935. Arrangements to tap loads into a constant-impedance line along its length, with proper matching (a version of the closely coupled inductance ratio-arm bridge)

477 392, May 1936. Diode–triode valve voltmeter (negative feedback to give high input impedance)

507 665, November 1937 (with C. S. Bull). Valve used as potential divider for high voltages (grid earthed and input to anode: a reduced output appears at the cathode)

515 044, March 1938 (with J. Hardwick and C. O. Browne). Measuring waveforms on a cathode ray oscilloscope by moving the trace past a datum line, using a variable bias

581 164, January 1940. Improvements to closely coupled inductor ratio-arm bridge (using two different voltages, tapped from an autotransformer, across the impedances to be compared)

587 878, June 1940. Improvement to closely coupled inductor ratio-arm bridge

541 942 June 1940. Bridge for comparing low resistance with separate current and voltage terminals (a version of the closely coupled inductor ratio-arm bridge)

On antennas and cables

432 978, February 1934 (with J. L. Pawey). Aerial array

445 968, August 1934 (with E. C. Cork and E. L. C. White). Combined aerial system for two widely separated frequencies (small aerial for higher frequency has a coaxial feeder which is arranged to serve as the aerial for the lower frequency)

452 713, December 1934 (with E. C. Cork). A cable for VHF (the central wire is kinked and is thus self-supporting in a tube so that the dielectric is mostly air)

452 772, February 1935 (with E. C. Cork). A low-impedance cable for VHF (similar to that described in 452 713)

452 791, February 1935. VHF aerial consisting of concentric coaxial half-wave elements end-to-end (the outer of each element acts as a radiator, and the inner and outer together serve as a feed for the elements beyond)

455 492, March 1935 (with J. Hardwick). Loading a transmission line, for use at the lower radio frequencies, with series and shunt elements to give frequency-independent characteristics

463 111, September 1935 (with J. L. Pawsey). Frequency-selective coaxial cables (coil of fine insulated wire, of suitable length to constitute a resonant rejector, wound round centre conductor)

464 443, October 1935 (with J. L. Pawsey). Network to match a wideband aerial to a transmission line

466 418, November 1935. Long interference-free (coaxial line with auxiliary conductor outside: at low frequencies, the two conductors act as balanced line rejecting inductive interference: at high frequencies, screening gives protection)

470 408, February 1936. Reduction of inductive interference on audio-frequency coaxial lines (driving them from a high-impedance source, but feeding them into a load of characteristic impedance)

491 490, December 1936. Improvement of coupling between a cable and its termination, to pass wide band of signals but not DC

505 079, October 1937. Vertically and horizontally polarised transmitting stations in adjacent areas, to reduce co-channel interference

503 765, October 1937. High-impedance twin feeder, using very thin wires (the wires are kinked and are supported in tubes: when stretched they will concertina rather than break)

515 684, March 1938. Tube with longitudinal slot, used as a slot aerial or as a feeder

581 167, May 1941 (with C. O. Browne). Feeder to provide a balanced output at VHF from a single-sided input (using transforming properties of lines)

On power supplies

421 546, June 1933. Compensation of power supply having reactive elements, making it appear, to load, as pure resistance

461 004, July 1935. Decoupling AC loads from the common impedance of the power supply (the loads are taken from the ends of an autotransformer, the supply is tapped between them: a power-engineering version of the closely coupled inductor ratio/arm bridge)

462 583, July 1935. Network whereby resistor required to dissipate high power and have low capacitance to earth is replaced by two resistors, one dissipating high power and one having low capacitance to earth

462 584, July 1935. Network whereby battery with high capacitance to earth is used to supply a DC bias between points where this capacitance cannot be tolerated

462 530, July 1935. Network to feed power into points that cannot tolerate high capacitance to earth, through closely coupled inductors

462 823, August 1935. Power supplies which appear to load as constant resistance: an improvement to a previous patent (421 546), using resonant circuits

473 276, March 1936. Power-rectifier unit (compensation for leakage reactance of transformer by tuned circuit in AC supply path)

475 729, March 1936. Voltage-regulating system based on principle of previous patent (461 004). The 'generator' is replaced by a synchronous alternator; the true load and the mains, considered as a negative load, are the two 'loads': the true load is thus decoupled from the variations of the mains

474 607, April 1936 (with E. L. C. White). Stabilised HT supply (using series triode)

577 817, October 1939 (with E. A. Nind). Generator with DC output (for valve amplifiers) and AC output (to provide EHT for cathode ray tubes via voltage doublers)

On cathode ray tubes

432 485, December 1933. Multigun CRT for television receivers, giving increased brightness (spots follow one another across the screen; modulation is tapped off a delay line)

446 661, August 1934 (with J. D. McGee). An important patent for the Emitron television camera (improves on iconoscope by keeping mosaic at cathode potential and interposing a high-potential grid, through which the electron beam scans the mosaic)

447 754, October 1934 (with H. E. Holman). Drawing down a glass tube round a wire to insulate it (for weaving insulated grids: see 447 824)

447 824, October 1934 (with H. E. Holman). Double-sided signal plate for TV camera (a grid woven of insulated wire, with holes in the mesh plated through)

449 533, October 1934 (with M. Bowman-Manifold). Deflection-coil yoke for cathode ray tube

482 195, September 1936. CRT deflection coils (arranged as a continuous winding, so that tapping across one diameter gives line deflection and, across another, frame deflection)

483 650, October 1936. CRT deflection coils (arranged so that whole circumference of neck is occupied by windings)

495 724, June 1937. Noncircular spot for cathode ray tube, to eliminate line structure and other unpleasant visual effects

520 646, October 1938. Television-camera tube, in which mosaic is scanned by a light spot instead of an electron beam

On cathode ray tube circuits

400 976, April 1932. Scanning circuit (suitable for hard valves, though a thyratron version is also described)

436 734, April 1934 (with C. O. Browne). Method of coupling photocell to amplifier, giving wideband output down to DC

479 113, April 1936. Drive amplifier for CRT deflection coils (negative feedback taken from a separate network rather than from a resistor in series with coils)

480 355, July 1936 (with R. E. Spencer). EHT supply for two-anode
cathode ray tube (resistive network to ensure constant ratio between the
voltages on the anodes)

482 370, August 1936 (with E. L. C. White). The economy diode (makes
use of pulses produced when current through CRT deflector coils is
sharply cut off)

514 825, March 1938 (with E. L. C. White). Protection against surges
caused by final anode of cathode ray tube short-circuiting to earth

534 465, July 1939 (with F. Blythen). Compensating circuit to vary
focusing current of cathode ray tube in accordance with anode-voltage
variations

On DC restoration

422 914, July 1933 (with C. O. Browne and J. Hardwick). DC restoration
in television in presence of a subcarrier that might affect a simple diode
clamp

434 876, February 1934. Improvement to method of getting black-level
signal from television camera by cutting off electron beam during
playback (this patent solves the problem of transients in general illumination level)

449 242, September 1934 (with C. O. Brown and F. Blythen). Black-level
clamping, for DC restoration of television signals

507 239, November 1937 (with E. L. C. White). Black-level-clamp circuit
for television (synchronising pulses, slightly delayed, are used to open a
valve that clamps the black level)

508 377, December 1937. Method of combining output from a television
camera, giving the AC level of the signal, with output from a single
photocell giving DC level

515 360, May 1938. Improved method of establishing a black-level datum
in signals from television cameras

515 364, May 1938. Improvements to DC restoration circuits for television

On modulation

455 858, April 1935. Television synchronising signal (combined line and
frame pulses on one channel)

456 135, April 1935 (with E. A. Nind). Modulations of carrier, first by synchronising signal then again by composite television signal to ensure
thorough cutoff in the blacker-than-black region

On automatic gain control circuits

458 585, March 1935. Automatic gain control of television signal (by inserting black level into waveform)

476 935, May 1936. Television waveform to allow automatic gain control (contains both blacker-than-black and whiter-than-white levels)

515 348, March 1938. Variable-gain amplifier (automatic gain control keeps the level of a pilot signal constant at the output; then the amplification varies inversely as the amplitude of the pilot signal at the input)

515 361, May 1938. Improvements to patent 458 585 for AGC circuit

515 362, May 1938 (with J. Hardwick and F. Blythen). Improvement to patent 458 585 for AGC circuit

529 590, May 1939 (with A. H. Cooper). DC amplifier using valve, where anode is fed with raw AC (for amplified AGC)

On miscellaneous television circuits

446 663, September 1934. Inverting tips of whiter-than-white signal peaks, whether caused by interference or due to whiter-than-white synchronism signals, and amplifying them to produce black marks on screen rather than white

490 205, November 1936 (with C. O. Browne) Elimination of 'tilt', an unwanted sawtooth waveform of line frequency, in the output of a television camera (a grating is arranged to put vertical black bars on the camera screen, and the signal is DC restored at these bars)

490 150, March 1937. Television-receiver circuits, to separate line and frame signals

497 060, June 1937. Gamma-correction amplifier for television signal, using non-linear elements (diodes or triodes) in the negative-feedback path

501 966, June 1937 (with R. E. Spencer). Improvements to 'overlay' technique, where an object viewed by one television camera is superimposed on a scene viewed by another

503 555, October 1937. High-definition television system compatible with a low-definition system (a fine spot is used, and carefully synchronised spot wobble allows extra detail to be received on special receivers)

507 417, December 1937. Anti-ghosting device for television receiver (main signal delayed and fed back as required, so as to neutralise ghosts)

536 089, July 1939. Valve circuits to correct for high gamma, when films are being televised

On electronic circuits

425 553, September 1933 (with H. A. M. Clark). Negative-feedback power amplifier (with both current and voltage feedback)

448 421, September 1934. Cathode follower for increasing input impedance of valve

482 740, July 1936. The long-tailed pair (push–pull circuit with unde-coupled cathode: negative feedback discriminates against common-mode signals)

479 599, July 1936 (with W. H. Connell). Variable-selectivity circuit (using negative feedback)

489 950, February 1937. Circuit for coupling between valves, using tapped coil (input impedance and output voltage remain constant; output impedance falls at higher frequencies)

496 883, June 1937. 'Ultralinear' amplifier circuit (screen tapped in on output transformer, to improve linearity)

505 480, November 1937. Pentode cathode follower, with decoupled suppressor grid

512 109, December 1937 (with E. L. C. White). Bidirectional switch (utilising three diodes with strapped anodes)

514 065, April 1938. Using the long-tailed pair for mixing pulse trains from different sources

554 715, November 1939. FM modulator (coaxial line with diaphragm, driven to follow modulation waveform, varying its length)

563 464, June 1940. Method of amplitude control for a valve oscillator (negative feedback at cathode is made operative, by means of a diode, when amplitude reaches a certain value)

On miscellaneous topics

419 284, March 1933. Improvements to powdered-iron-core coils (dielectric loss in powdered iron removed by electrostatic screening)

496 139, May 1937. Synchronous induction motor with series-wound DC exciter, to improve characteristics when running up to speed and synchronising

On delay lines

517 516, June 1938 (with H. E. Kallmann and W. S. Percival). Method of designing filters to have specified amplitude and phase characteristics (using a tapped delay line)

528 310, February 1939. Production of pulses using reflection from the open-circuit or short-circuit end of a line

567 227, January 1940 (with C. O. Browne). Multistud switches for tapping into impedances or delay lines (by an arrangement of tapping, a fine-step switch is made to interpolate accurately into the steps of a coarse-step switch)

577 942, October 1941. Improvement to delay-line circuit for power pulses or modulation (using sections of different characteristic impedance along the length)

589 127, October 1941. The delay-line circuit for supplying power pulses to magnetrons

574 133, June 1942. Improvement to previous patent (517 516) for designing filters of desired characteristics using delay lines

On radar and circuits associated with radar

543 602, June 1939. System for indicating when two versions of the same signal have zero relative time delay (using dynamometer as multiplying device)

581 920, July 1939 (with E. L. C. White). Direction finding on modulated signals, or reflections of modulated signals, by comparing time of arrival of modulation at separated aerials

585 906, November 1939. Receiver for pulses whose expected time of arrival is known (valves are driven to high instantaneous condition at this time and switched off at other times)

585 907, December 1939 (with E. L. C. White). Improvement of reception of a carrier modulated by a signal of known repetition rate (by sampling at the repetition rate and integrating)

585 908, December 1939. Reproducing a repetitive waveform at a lower frequency (by sampling different parts in successive cycles)

589 228, December 1939 (with E. L. C. White). Adjustment of timing of pulses with an accuracy that is high relative to the repetition rate (by generating a train of pulses at an integral multiple of the repetition rate, adjusting their timing with moderate accuracy and gating the two trains together)

579 725, January 1940. Generation of power-supply pulses for an intermittently operated valve (current in an inductor allowed to build up and then cut off)

581 161, January 1940. Low-level altimeter based on earth capacitance (local capacitances at aircraft being dealt with by a self-balancing version of the closely coupled inductor ratio-arm bridge)

589 229, January 1940. Homodyne receiver insensitive to phase of receiver carrier (by using two local signals in phase quadrature and combining the resultant signals after separate detection)

579 154, March 1940. Radar with continuous transmission switched between two frequencies (the returning echo of frequency A is hetero-dyned by frequency B at the receiver over a period which depends on the delay of the echo)

581 561, June 1940. Damping for a pulsed oscillator, to remove the tail of the oscillation (damping is switched in after a delay)

574 708, September 1940. Improvement to klystron (resonator lies partly inside and partly outside vacuum: it is so shaped that, although part of the valve envelope lies inside it, the envelope is in the position of minimum field, to reduce losses)

587 562, March 1941. Receiver for detecting pulses in presence of CW interference (a self-adjusting clipper of long time constant is not respon-sive to pulses, but clips the CW down to almost zero)

585 789, November 1941 (with H. G. Holloway and W. H. Connell). Differential capacitor, having suitable law for self-balancing bridge as used in the low-level altimeter (see 581 161)

580 527, June, 1942. The Miller integrator (circuit to linearise a sawtooth waveform by means of a valve with a capacitor from anode to grid)

582 503, October 1943 (with E. L. C. White and F. C. Williams). Radar system for searching (in range) and locking onto any echo found

595 509, May 1945 (with E. L. C. White). Monostable circuit

Index

Printed in the USA
CPSIA information can be obtained
at www.ICGtesting.com
JSHW011507221024
72173JS00005B/1232